Lecture Notes in Mathematics

2259

More information about this series at http://www.springer.com/series/304

Chaire Jean-Morlet

The CIRM Jean-Morlet Series is a collection of scientific publications centering on the themes developed by successive holders of the Jean Morlet Chair.

This chair has been hosted by the *Centre International de Rencontres Mathématiques* (CIRM, Luminy, France) since its creation in 2013. The Chair is named in honour of Jean Morlet (1931–2007). He was an engineer at the French oil company Elf (now Total) and, together with the physicist Alex Grossman, conducted pioneering work in wavelet analysis. This theory has since become a building block of modern mathematics. It was at CIRM that they met on several occasions, and the center then played host to some of the key conferences in this field.

Appointments to the *Jean-Morlet* Chair are made to world-class researchers based outside France and who work in collaboration with local project leaders in order to conduct original and ambitious scientific programs. The Chair is supported financially by CIRM, Aix-Marseille Université and the City of Marseille.

A key feature of the Chair is that it does not focus solely on the research themes developed by Jean Morlet. The idea is to support the freedom of pioneers in mathematical sciences and to nurture the enthusiasm that comes from opening new avenues of research.

CIRM: a beacon for international cooperation

Situated at the heart of the *Parc des Calanques*, an area of outstanding natural beauty, CIRM is one of the largest conference centers dedicated to mathematical and related sciences in the world, with close to 3500 visitors per year. Jointly supervised by SMF (the French Mathematical Society) and CNRS (French National Center for Scientific Research), CIRM has been a hub for international research in mathematics since 1981. CIRM's *raison d'être* is to be a venue that fosters exchanges, pioneering research in mathematics in interaction with other sciences and the dissemination of knowledge to the younger scientific community.

www.chairejeanmorlet.com
www.cirm-math.fr

Kerrie L. Mengersen • Pierre Pudlo •
Christian P. Robert

Editors

Case Studies in Applied Bayesian Data Science

CIRM Jean-Morlet Chair, Fall 2018

 Springer

Editors
Kerrie L. Mengersen
Mathematical Sciences
Queensland University of Technology
Brisbane, QLD, Australia

Pierre Pudlo
I2M, CNRS, Centrale Marseille
Aix-Marseille University
Marseille, France

Christian P. Robert
CEREMADE
Université Paris Dauphine
Paris, France

ISSN 0075-8434 ISSN 1617-9692 (electronic)
Lecture Notes in Mathematics
ISBN 978-3-030-42552-4 ISBN 978-3-030-42553-1 (eBook)
https://doi.org/10.1007/978-3-030-42553-1

Mathematics Subject Classification (2020): 62R07, 62F15, 60GXX, 62H30, 62P10, 62M40, 62G05, 60J10

Jointly published with Société Mathématique de France (SMF); sold and distributed to its members by the SMF, http://smf.emath.fr; ISBN SMF: 978-2-85629-914-2

This Springer imprint is published by the registered company Springer Nature Switzerland AG.
The registered company address is: Gewerbestrasse 11, 6330 Cham, Switzerland

Contents

Part I
Surveys

Chapter 1
Introduction

Kerrie L. Mengersen, Pierre Pudlo, and Christian P. Robert

Abstract This chapter is an introduction to this Lecture Note. We briefly describe the contents of this book. Both parts are introduced, namely part A which deals with Bayesian modeling and part B which presents real-world case studies. The last part of the chapter details the organization of the various events related to the Jean-Morlet Chair. It ends with the issues and research directions identified by the participants of the Conference on Bayesian Statistics in the Big Data Era.

Keywords Bayesian inference · Big data · Computational statistics · Conferences · Workshop · Statistical models · Bayesian modeling · Bayesian computation · Case studies

1.1 Overview

The field of Bayesian statistics has exploded over the past 30 years and is now an established field of research in mathematical statistics and computer science, a key component of data science, and an underpinning methodology in many domains of science, business and social science. Moreover, while remaining naturally entwined, the three arms of Bayesian statistics, namely modelling, computation and inference, have grown into independent research fields. Examples of Bayesian models that have matured during this timeframe include hierarchical models, latent variable models, spatial and temporal models, network and systems models, and models

K. L. Mengersen (✉)
Queensland University of Technology, Brisbane, QLD, Australia
e-mail: k.mengersen@qut.edu.au

P. Pudlo
I2M, CNRS, Centrale Marseille, Aix-Marseille University, Marseille, France

C. P. Robert
Université Paris-Dauphine, Paris, France

K. L. Mengersen et al. (eds.), *Case Studies in Applied Bayesian Data Science*, Lecture Notes in Mathematics 2259, https://doi.org/10.1007/978-3-030-42553-1_1

for dimension reduction. Bayesian computational statistics is now an established discipline in its own right, with a wealth of extensions to the original Markov chain Monte Carlo algorithms, likelihood-free approaches such as Approximate Bayesian computation, and optimization methods such as Variational Bayes and Hamiltonian Monte Carlo. In the domain of Bayesian inference, progress continues to be made on many fronts, including the role and influence of priors, model choice and model robustness, hypothesis testing and so on.

While the research arms of Bayesian statistics continue to grow in many directions, they are harnessed when attention turns to solving substantive applied problems. Each such problem set has its own challenges and hence draws from the suite of research a bespoke solution. It is often useful for both theoretical and applied statisticians, as well as practitioners, to inspect these solutions in the context of the problems, in order to draw further understanding, awareness and inspiration.

The aim of this book is to contribute to the field by presenting a range of such problems and their Bayesian solutions. The book arises from a research program at CIRM in France in the second semester of 2018, which supported Kerrie L. Mengersen (Queensland University of Technology, Australia) as a visiting Jean-Morlet Chair and Pierre Pudlo (Aux-Marseille University) as the local Research Professor. Mengersen was also supported by the Australian Research Council (ARC) through a Laureate Fellowship. Various events were held during the course of this semester, including a Masterclass on Bayesian Statistics, a conference on Bayesian Methods in the Big Data Era, a workshop on Bayes and Big Data for Social Good, and a number of Research in Pairs activities. Summaries of the masterclass, conference and workshop are presented later in this chapter.

1.2   Outline of Book

This book comprises two main parts. In Part A, the state of the art of modern Bayesian statistics is reflected through a set of surveys on topics of current interest in the field.

The first chapters of Part A focus on Bayesian modelling, including a general literature survey and evaluation of Bayesian statistical models in the context of big data by Jahan et al. and a more in-depth discussion by Goan of a popular modelling approach, namely Bayesian neural networks. The survey by Jahan et al. explores the various approaches to Bayesian modelling, in particular those that are motivated by the advent of so-called 'big data'. The authors conclude their survey by considering the question of whether focusing only on improving computational algorithms and infrastructure will be sufficient to face the challenges of this 'big data era'. The chapter of Goan complements this general overview. Unlike their frequentist counterparts, Bayesian neural networks can naturally and formally allow for uncertainty in their predictions, which can lead to richer inferences for detection, classification and regression. Goan introduces these models, discusses the common algorithms used to implement them, compares various approximate inference schemes and highlights opportunities for future research.

The third chapter of Part A focuses on Bayesian computation, with a survey of Markov chain Monte Carlo (MCMC) algorithms for Bayesian computation by Wu and Robert. The authors provide a brief, general overview of Monte Carlo computational methods, followed by a more detailed description of common and leading edge MCMC approaches. These include Metropolis-Hastings and Hamilton Monte Carlo algorithms, as well as scalable versions of these, and continuous time MCMC samplers based on piecewise deterministic Markov processes (PDMP), the Zig-Zag process Sampler and the Bouncy Particle Sampler. Wu and Robert then introduce a generalization of the latter algorithm in terms of its transition dynamics. Their new Generalised Bouncy Particle Sampler is perceived as a bridge between bouncy particle and zig-zag processes that avoids some of the tuning requirements.

The final survey chapters of Part A focus on two illustrative challenge in Bayesian data science that merge the fields of modelling and computation, namely variable selection and model choice in high dimensional regression, and posterior inference for intractable likelihoods. Sutton addresses the first challenge by describing three common priors that are used for sparse variable selection, namely model space priors, spike and slab priors and shrinkage priors, with corresponding computational approaches and software solutions. The second challenge is surveyed by Moores et al., who defines an intractable likelihood as one for which the likelihood function is unavailable in closed form or which is infeasible to evaluate. The approaches covered by Moores et al. include pseudo-marginal methods, approximate Bayesian computation (ABC), the exchange algorithm, thermodynamic integration and composite likelihood, with particular attention paid to advancements in scalability for large datasets.

Part B of the book consists of a set of real-world case studies that aim to illustrate the wide variety of ways in which Bayesian modelling and analysis can enhance understanding and inference in practice. Three fields have been chosen for exposition, namely health, environmental health and ecology. The value of Bayesian data science in modelling the brain is described in the first pair of chapters by Cespedes et al., who focus on a joint model of cortical thickness and network connections, and White et al., who aim to cluster action potential spikes. The second pair of chapters address public health issues of vector-borne diseases (Aswi et al.), cancer (Cramb et al.). The next chapter explores the link between environmental exposures and the neurodegenerative Parkinson's disease (Thomas et al.), followed by two chapters that address challenges in workplace health (Harden et al., Tierney et al.).

In the last four chapters on ecological applications, the authors showcase the use of diverse data sources to address challenges in conservation and biosecurity. Davis et al. use data elicited from experts and citizens to explore factors involved in conservation of cheetahs in Southern Africa and jaguars in South America, while Sequeira et al. and Vercelloni et al. employ observational data to gain insights into marine conservation. In contrast, Ullah et al. employ satellite imagery to model the risk of fire-ant incursion.

Each of these case studies has a complication that motivates a rich range of Bayesian solutions. The models considered by the authors include hierarchical models (Sequeira *et al.*, Tierney *et al.*, Vercelloni *et al.*), parametric and nonparametric mixture models (White *et al.*), spatial models (Aswi *et al.*, Cespedes *et al.*, Cramb *et al.*, Ullah *et al.*) and Bayesian network approaches (Davis *et al.*, Harden *et al.*, Thomas *et al.*).

Cespedes *et al.* propose a new Bayesian generative model for analysis of MRI data that allows for more complete insight into the morphological processes and organization of the human brain. Unlike current models that typically perform independent analyses, their proposed model uses a form of wombling to perform joint statistical inference on biomarkers and connectivity covariance networks. The new model provides posterior probabilities for the connectivity matrix, accounting for the uncertainty of each connection, and enables estimation of the spatial covariance among regions as well as global cortical thickness. These features are critical in the assessment of the pathology of neuro-degenerative diseases such as Alzheimers. White *et al.* also consider a case study in neuroscience research, this time focusing on the analysis of action potentials or 'spike sorting', which aims to characterize neural activities in subjects exposed to different stimuli or other experimental conditions. This problem is cast as an unsupervised clustering problem to which two types of mixture models are applied. The complications in these models include the choice of the number of identified clusters and classification uncertainty.

In a quite different spatial setup, Aswi *et al.* compare the performance of six Bayesian spatio-temporal models in their investigation of dengue incidence in Makassar, Indonesia, taking into account the challenges that are typically faced in practice but not typically catered for by the models, namely a small number of areas and limited number of time periods. These types of geographic spatial models for small area estimation of disease are also considered by Cramb *et al.*, with a focus on the choice of model under different scenarios of rare and common cancers over different types of spatial surfaces. The authors reveal the dramatic impact of model choice on posterior estimates and recommend comparing several different models, especially when analyzing very sparse data.

Returning to neurodegenerative diseases, but from a different perspective, Thomas *et al.* focus on the challenge of understanding the association between environmental exposure to organochloride pesticide (OCP) and age at onset of Parkinson's disease. The authors explore this complicated association via an ensemble model comprised of a meta-analysis and a Bayesian network, whereby odds ratios and other information extracted from the literature are merged with clinical data to probabilistically quantify the network model. The authors acknowledge the limitations of this approach but suggest its merit for future investigation as a mechanism for integrating disparate, sparse data sources to address environmental health questions.

The utility of Bayesian approaches in modelling different aspects of a problem is highlighted in the two chapters by Harden *et al.* and Tierney *et al.*, who both focus on workplace health. Harden *et al.* use an approach similar to that of Thomas *et al.*, in that they employ published information to characterize, quantify and compare features of workplace health and workplace wellness programs. In contrast, Tierney *et al.* utilise records of routine medical examinations, which are characterized by substantive missing data, to facilitate early detection of disease amongst workplace employees. Whereas Harden *et al.* adopt a Bayesian network approach to combine and quantify their workplace health and wellness systems, Tierney *et al.* use a Bayesian hierarchical regression model to create a workplace health surveillance program.

Turning to the case studies in ecology, one of the major constraints in conservation research and practice is the sparsity of data and the complex interaction of contributing factors. An example is given by Davis *et al.*, who tackle the sensitive issue of human-wildlife conflict and illustrate the utility of a Bayesian network for these types of problems. Two iconic wildlife species are considered and the implications for different conservation management strategies are discussed.

In a quite different ecological setup, Sequeira *et al.* employ a series of uninformed and informed priors to investigate the issue of space-time misalignment of responses and predictors in hierarchical Bayesian regression models. The particular models of interest are predictive biodiversity distribution models, which are used to understand the structure and functioning of ecological communities and to facilitate their management in the face of anthropogenic disturbances. The focal challenge is to predict fish species richness on Australia's Great Barrier Reef. Vercelloni *et al.* also take the Great Barrier Reef as their study area, with the aim of estimating long-term trajectories of habitat forming coral cover as a function of three different spatial scales and environmental disturbances. A hierarchical Bayesian model was also adopted by these authors, but in a semi-parametric framework.

The final chapter in this part, by Ullah *et al.*, focuses on the problem of classification of features of interest in large images. The approach proposed by these authors is to fit a Bayesian non-parametric mixture model to multiple stratified random samples of the image data, followed by the formation of consensus posterior distribution which is used for inference. The method is applied to the challenge of employing remote sensing for plant and animal biosecurity surveillance, with a particular focus on using satellite data to identify high risk areas for fire ants in the Brisbane region of Australia.

Together, these chapters provide a rich tapestry of activity in applied Bayesian data science, motivated by a wide range of real-world problems. It is hoped that these case studies will inspire and expand both research and practice in Bayesian data science.

1.3   Jean-Morlet Research Semester Activities

As described above, the Jean-Morlet semester at CIRM in the second half of 2018 included the organization of a masterclass, conference and workshop at the CIRM Research Centre in Marseille, France. A brief summary of each of these three activities is given in this section, with a focus on highlighting research directions and illustrating applications of Bayesian statistical modelling and analysis.

1.3.1   Masterclass on Modern Bayesian Statistics

A clear indicator of the establishment of Bayesian statistics is the increased number of graduate courses on the topic. Such courses are more common in statistical science, computer science and data science, but they are also now appearing in a wide range of other fields of science, social science and business. Intensive short courses on Bayesian statistics are also popular mechanisms for training for graduates as well as academics, other researchers and practitioners. These courses are presented either in-person or online.

One such course was the Masterclass in Bayesian Statistics, presented at CIRM on 22–26 October, 2018. Videos and slides of the presentations given in this Masterclass are publicly available on the CIRM website: https://www.chairejeanmorlet. com/2018-2-mengersen-pudlo-1854.html.

The topics presented in this Masterclass can be broadly categorised into Bayesian modelling and Bayesian computation. As an example of the former, Chris Holmes addressed the problem of Bayesian learning at scale. His argument was that Bayesian learning from data is predicated on the likelihood being true, whereas in reality all models are false. If the data are simple and small, and the models are sufficiently rich, then the consequences of model misspecification may not be severe. However, since data are increasingly being captured at scale, Bayesian theory as well as computational methods are required that accommodate and respect the approximate nature of scalable models. A proposed approach is to include the uncertainty of the model in the analysis, via a principled nonparametric representation. Other approaches to model assessment were discussed by Aki Vehtari, who covered cross-validation and projection predictive approaches for model assessment, inference after model selection, and Pseudo-BMA and Bayesian stacking for model averaging. This discussion was complemented by R notebooks using rstanarm, bayesplot, loo, and projpred packages.

Problem-specific Bayesian models were also presented. For example, the presentation by Adeline Samson focused on various types of stochastic models in biology, including point processes, discrete time processes, continuous time processes and models with latent variables, and elaborated on some of the statistical challenges associated with their application.

The many directions of current research in Bayesian computational statistics were highlighted in the presentation by Christian P. Robert, who discussed more efficient simulation via accelerating MCMC algorithms, to approximation of the posterior or prior distributions via partly deterministic Markov processes (PDMP) like the bouncy particle and zigzag samplers. The focus of this presentation was on the evaluation of the normalising constants and ratios of normalising constants in such methods.

Two algorithms of strong current interest are Sequential Monte Carlo (SMC) and Variational Inference (VI) or Variational Bayes (VB). VI algorithms were addressed by Simon Barthelmé, who suggested methods for correcting variational approximations to improve accuracy, including importance sampling and perturbation series. SMC was introduced by Nicolas Chopin, who motivated the approach by state-space (hidden Markov) models and their sequential analysis, and touched on the analysis of non-sequential problems. The presentation also included a description of the formal underpinnings of SMC, building on concepts of Markov kernels and Feynman-Kac distributions, and a discussion of Monte Carlo ingredients including importance sampling and resampling. Standard bootstrap, guided and auxiliary particle filters were then described, followed by estimation methods via PMCMC and SMC^2. SMC was also discussed by Marie-Pierre Etienne in the context of partially observed stochastic differential equations applied to ecology, and by Adam Johansen in the context of defining a genealogy of SMC algorithms. Similarly, Sebastian Reich proposed a unifying mathematical framework and algorithmic approaches for state and parameter estimation of particular types of partially observed diffusion processes.

Scalable algorithms for Bayesian inference are also of great interest. This was reflected in the presentation by Giacomo Zanella, who focused on scalable importance tempering and Bayesian variable selection.

Another indication of the mainstream status of Bayesian statistics is the proliferation of dedicated R packages and analogies in Python and other software, as well as an increase in the number of stand-alone statistical software packages. While many of the Masterclass presentations referred to specific packages, some presentations focused on the stand-alone software. For example, Harvard Rue presented a tutorial on Bayesian computing with INLA, with a focus on estimation of the distribution of unobserved nodes in large random graphs from the observation of very few edges and a derivation of the first non-asymptotic risk bounds for maximum likelihood estimators of the unknown distribution of the nodes for this sparse graphical model. This tutorial was complemented by a presentation on the same topic by Sylvain le Corff. A tutorial on JASP was presented by Eric-Jan Wagenmakers and the software package STAN was used by Bruno Nicenboim to implement a cognitive model of memory processes in sentence comprehension.

Finally, as in all areas of computational statistics, good practice in dealing with data and coding Bayesian algorithms is essential. Julien Stoehr and Guillaume Kon Kam King presented a tutorial on this topic, with reference to writing R code, R packages and R Markdown and knitr documents.

1.3.2 Conference on Bayesian Statistics in the Big Data Era

This conference aimed to bring together an international and interdisciplinary group of researchers and practitioners to share insights, research, challenges and opportunities in developing and using Bayesian statistics in the Big Data era.

As expected, a major focus of the conference was on scalable methods, i.e. models and algorithms that cope with or adapt to increasing large datasets. As illustration, scalable nonparametric clustering and classification were proposed by Peter Muller. Two strategies were discussed: one based on a consensus Monte Carlo approach that splits the data into shards and then combines subset posteriors to recover joint inference, and another that exploits predictive recursion to build up posterior inference for the complete data. Ming-Ngoc Tran canvassed a range of topics such as intractable likelihood and its connection with Big Data problems, subsampling-based MCMC, HMC and SMC for models with tall data, and Variational Bayes estimation methods for extremely high-dimensional models. A quite different compositional approach to scalable Bayesian computation and probabilistic programming was described by Darren Wilkinson.

Sub-sampling, approximations and related methods for dealing with large datasets was discussed by a range of authors. Pierre Alquier proposed techniques for sub-sampling MCMC and associated approximate Bayesian inference for large datasets, while Tamara Broderick proposed a different approach to automated scalable Bayesian inference via data summarisation. David Dunson also contributed to this discussion, describing new classes of scalable MCMC algorithms based on biased subsampling and multiscale representations that, instead of converging to an exact posterior distribution, employ approximations to speed up computation and achieve more robust inference in big data settings. Stéphane Robin also used deterministic approximations to accelerate Sequential Monte Carlo (SMC) for posterior sampling via a so-called shortened bridge sampler. Approximate Bayesian Computation (ABC) was discussed by Pierre Pudlo in the context of model choice, and Jean-Michel Marin described a method of improving ABC through the use of random forests.

With respect to modelling, nonparametric approaches were a popular topic of discussion. In addition to Muller's presentation described above, Amy Herring described centred partition processes for sparse data. Alternative approaches to defining nonparametric priors were also proposed, for example by Antonio Lijoi in the context of covariate-dependent data, and Igor Prünster through the use of hierarchies of discrete random probabilities.

Other models of great international interest included high-dimensional spatial and spatio-temporal models, discussed by Sudipto Banerjee and Noel Cressie, optimal transport described by Marco Cuturi, and high dimensional inference for graphical models presented by Reza Mohammadi. High dimensional regression was addressed by Akihiko Nishimura, who described computational approaches for "large n and large p" sparse Bayesian regression in the context of binary and survival outcomes, and Benoit Liquet, who focused on Bayesian variable selection and regression of multivariate responses for group data. Related design questions were also a priority issue, since efficient sampling, survey and experimental designs can dramatically reduce the number of observations and variables required for inference and the associated computational cost of analysis. To this end, Jia Liu proposed a Bayesian model-based spatiotemporal survey design for log-Guassian Cox processes.

Approaches for high dimensional time series data, motivated by applications in economics, marketing and finance, were promoted by Sylvia Frühwirth-Schnatter, Gregor Kastner and Gary Koop. Fruwirth-Schnatter focused on Markov chain mixture models to describe time series with discrete states, and showed that these models are able to capture both persistence in the individual time series as well as cross-sectional unobserved heterogeneity. Koop described a different approach, focusing on composite likelihood methods for Bayesian vector autoregressive (VAR) models with stochastic volatility, presented by Gary Koop. Kastner also considered VAR models with time-varying contemporaneous correlations that are reportedly capable of handling vast dimensional information sets.

A wide range of applied problems were tackled in the conference, with attendant novel methodology. For example, challenges in public health ranged from nonparametric approaches to modelling sparse health data, by Amy Herring, to methods for including residential history in mapping long-latency diseases such as mesothelioma, by Christel Faes. Graphical models for brain connectivity were also discussed by Reza Mohammadi, as mentioned above. In the genetics field, the problem of high-throughput sequencing data in genomics was addressed using Bayesian multiscale Poisson models by Heejung Shim, while Zitong Li proposed non-parametric regression using Gaussian Processes for analysing time course quantitative genetic data, in particular quantitative trait loci (QTL) mapping of longitudinal traits. Time-course data prediction for repeatedly measured gene expression was also discussed by Atanu Bhattacharjee. Business-related applications included Bayesian preference learning, described by Marta Crispino, Bayesian generalised games in choice form as a new definition of a stochastic game in the spirit of the competitive economy, by Monica Patriche, and econometric models by Fruwirth-Schnatter. As mentioned above, an environmental problem, namely estimating the extent of arctic sea-ice, was addressed by Noel Cressie using a hierarchical spatiotemporal generalised linear model, where data dependencies are introduced through a latent, dynamic spatiotemporal mixed-effects model using a fixed number of spatial basis functions. The model was implemented via a combination of EM and MCMC.

Other issues that were addressed at the conference included data privacy and security (presented by Louis Aslett) and causality in modern machine learning (Logan Graham). In the latter presentation, Graham argued that while much current attention has focused on using machine learning to improve causal inference, there is opportunity for the inverse, namely to use tools from causal inference to improve the learning, efficiency, and generalisation of machine learning approaches to machine learning problems.

1.3.3 Workshop on Bayes, Big Data and Social Good

There is increasing international interest and engagement in the concept of 'data and statistics for social good', with volunteers and organisations working on issues such as human rights, migration, social justice and so on. This interest is generating a growing number of workshops on the topic.

One such workshop on "Young Bayesians and Big Data for Social Good" was held at CIRM on 23–26 November 2018. The workshop showcased some of the organisations that are dedicated to social good and are employing data science in general, and Bayesian statistics for this purpose. It also provided opportunity for Bayesian statisticians to discuss methods and applications that are aligned to social good. As indicated by the title of the workshop, the participants were primarily, but not exclusively, early career researchers.

Dedicated social good organisations that were represented at the workshop included Peace at Work (peace-work.org, represented by David Corliss) and Element AI (elementai.com, represented by Julien Cornebise). For example, David Corliss, a spokesperson for Peace at Work, provided an overview of the state of Data for Good, Bayesian methodology as an important area of new technological development, and experiences and opportunities for students to get involved in making a difference by applying their developing analytic skills in projects for the greater good.

The workshop exposed a wide range of social good problems and associated Bayesian statistical solutions. For example, Jacinta Holloway focused on the utility of satellite imagery to inform the United Nations and World Bank Sustainable Development Goals related to quality of human life and environment by 2030. In a similar vein, Matthew Rushworth described the use of underwater imagery to inform statistical models of the health of the Great Barrier Reef, a UNESCO World Heritage site under threat in Australia. From a computational perspective, Tamara Broderick related her research into the development of simple, general and fast local robustness measures for variational Bayes in order to measure the sensitivity of posterior estimates to variation in choices of choices of priors and likelihoods, to the issue of analysing microcredit data which impacts on small business success in developing countries.

An important issue of trust in data was raised by Ethan Goan in the context of deep learning. Although these models are able to learn combinations of abstract and low level patterns from increasingly larger datasets, the inherent nature of these models remain unknown. Goan proposed that a Bayesian framework can be employed to gain insight into deep learning systems, in particular their attendant uncertainty, and how this information can be used to deliver systems that society can trust.

A number of presentations focused on entity resolution (record linkage and de-duplication of records in one or more datasets) in order in order to accurately estimate population size, with application to estimating the number of victims killed in recent conflicts. Different statistical approaches to address this problem were presented by Andrea Tencredi and Brunero Liseo, Rebecca Steorts, Bihan Zhuang and David Corliss. For example, Bayesian capture-recapture methods were proposed by David Corliss to estimate numbers of human trafficking victims and estimate the size of hate groups in the analysis of hate speech in social media. Tencredi and Liseo took another approach, by framing the linkage problem as a clustering task, where similar records are clustered to true latent individuals. The statistical model incorporated both the linking process and the inferential process, including the features of the record as well as the variables needed for inference. Paramount to their approach is the key observation that the prior over the space of linkages can be written as a random partition model. In particular, the Pitman-Yor process was used as the prior distribution regarding the cluster assignment of records. The method is able to account for the matching uncertainty in the inferential procedures based on linked data, and can also generate a feedback mechanism of the information provided by the working statistical model on the record linkage process, thereby eliminating potential biases that can jeopardize the resulting post-linkage inference.

The use of Bayesian statistics and big data for health was also a common theme in the workshop. For example, Akihiko Nishimura proposed new sparse regression methods for analyzing binary and survival data; Gajendra Vishwakarma described the use of Bayesian state-space models for gene expression data analysis with application to biomarker prediction; and Antonietta Mira detailed a Bayesian spatio-temporal model to predict cardiac risk, creating a corresponding risk map for a city, and using this to optimize the position of defibrillators.

A different problem tackled by Cody Ross was the resolution of apparent paradoxes in analyses of racial disparities in police use-of-force against unarmed individuals. For example, although anti-black racial disparities in U.S. police shootings have been consistently documented at the population level, new work has suggested that racial disparities in encounter-conditional use of lethal force by police are reversed relative to expectations, with police being more likely to shoot white relative to black individuals, and use non-lethal as opposed to lethal force on black relative to white individuals. Ross used a generative stochastic model of encounters and use-of-force conditional on encounter to demonstrate that if even a small subset of police more frequently encounter and use non-lethal force against black individuals than white individuals, then analyses of pooled encounter-conditional data can fail to correctly detect racial disparities in the use of lethal force.

Finally, as noted above, good practice in statistical computation can provide substantial benefits for both researchers and practitioners. To this end, Charles Gray described the use of github and the R package 'tidyverse' for improved collaborative workflow, with reference to an application in maternal child health research.

1.4   The Future of Bayesian Statistics

Given that Bayesian statistics is now an established field of research, computation and application, it is of interest to consider the future of the profession, particularly in the era of 'big data'. This was the question posed to the participants of the Conference on Bayesian Statistics in the Big Data Era held at CIRM, Marseille, France on 26–30th November 2018.

The participants collectively identified major issues and directions for Bayesian statistics. These were collated into four key themes: data, computation, modelling; and training. An overall statement on each theme is presented below.

1.4.1 Data

1. Policies like GDPR will need mathematical and statistical formalizations and implementation. This will become an increasingly important issue also beyond the regions where GDPR formally applies.
2. Addressing grand challenges will increasingly require the use of multiple data sources from diverse locations, and need approaches to deal with the resulting heterogeneous.
3. Issues of quality assurance and persistence will increase, as official statistics tend to be replaced by commercial services.
4. While traditional questions of experimental design are becoming less relevant, other experimental design questions will arise, related to subsampling big data.
5. Recognition of the provenance of the data is becoming important, including in particular social media data and derived data from climate models etc.

1.4.2 Computation

1. Despite the exponential growth in computational Bayesian statistics, new algorithms are still required that are targeted to big data.
2. There will be continuing interest in approximations and subsampling strategies, as well as methods for taking advantage of sparse data.
3. Bayesian software will become faster and more intuitive for users to use. This will benefit from active online communities.
4. Current software, such as Tensorflow, R and C++, differ with respect to ease and computational speed, and need to be able to talk to each other better.
5. However, software alone cannot help. Bayesian statisticians will also need to understand more about hardware and decentralised data in order to fine tune algorithms for specific problems.

1.4.3 Models

1. Bayesian models will continue to evolve in the 'big data era'. Three major directions of evolution are in priors, model setup and model choice.
2. With respect to priors, on the one hand, informative priors such as those that induce shrinkage will play an increasingly important role, but on the other hand,

priors on high dimensional data tend to become very influential so development of objective priors for high dimensional data will continue to be of great interest. Overall, we need better ways of choosing priors.

3. With respect to model setup, overall the future will see the development of better Bayesian frameworks, which ignore unnecessary information from data before modelling, determine relevant information for modelling, automatically determine the required complexity of the model, and include generalized methods for Bayesian model selection and diagnostics. There is little doubt that models need to evolve to cope with new kinds and quantities of data. On the other hand, perhaps progress could come from being able to ignore certain aspects of the data. After all, having to fully specify every aspect of a data generating process for a complex dataset can be tedious at best, impossible or harmful at worst.

4. With respect to model choice, we need to learn to handle model misspecification in better ways and develop robust Bayesian modelling approaches. There will be co-existence of parametric and nonparametric models in the future, where the application and utility will depend on specific domains of application. Model-free methodologies will also become more important. On one hand, Bayesian models will become more sophisticated, more flexible (taking advantage of Bayesian nonparametrics), bigger and better, as enabled by data and computational advances.

1.4.4 Training

1. The future will see the development of Bayesian tools for non-expert modellers, with plug and play type models for easy application. When compared to 10 years ago, a huge amount of students now study statistics and machine learning. A few of them are indeed really interested in mathematics, modelling and computer science, but others are more in quest of user-friendly software to use easily in their jobs. If we want to attract these non-specialist students, we need to provide more user-friendly tools for Bayesian learning: Bayesian equivalents of TensorFlow for neural networks. On the other hand, we should not sacrifice the statistical part of the training: modelling, theory, understanding the methods, interpretation of the results. Indeed, there is and should continue to be a role for statisticians and data scientists.

2. We should also talk about artificial intelligence (AI), but also about the world in which AI resides and alternatives to AI. AI will lead to personalized medicine, but this cannot be done without a sound knowledge of biostatistics. Similarly, environmental and economic problems cannot be solved without statistics. We live in a complex world. We should warn people that it will become impossible to understand these topics without a strong statistical background and show them how the Bayesian approach is flexible enough to tackle these problems.

Chapter 2
A Survey of Bayesian Statistical Approaches for Big Data

Farzana Jahan, Insha Ullah, and Kerrie L. Mengersen

Abstract The modern era is characterised as an era of information or Big Data. This has motivated a huge literature on new methods for extracting information and insights from these data. A natural question is how these approaches differ from those that were available prior to the advent of Big Data. We present a survey of published studies that present Bayesian statistical approaches specifically for Big Data and discuss the reported and perceived benefits of these approaches. We conclude by addressing the question of whether focusing only on improving computational algorithms and infrastructure will be enough to face the challenges of Big Data.

Keywords Bayesian statistics · Bayesian modelling · Bayesian computation · Scalable algorithms

2.1 Introduction

Although there are many variations on the definition of Big Data [51, 52, 91, 184], it is clear that it encompasses large and often diverse quantitative data obtained from increasing numerous sources at different individual, spatial and temporal scales, and with different levels of quality. Examples of Big Data include data generated from social media [22]; data collected in biomedical and healthcare informatics research such as DNA sequences and electronic health records [114]; geospatial data generated by remote sensing, laser scanning, mobile mapping, geo-located sensors, geo-tagged web contents, volunteered geographic information (VGI), global navigation satellite system (GNSS) tracking and so on [103]. The volume and complexity of Big Data often exceeds the capability of the

F. Jahan (✉) · I. Ullah · K. L. Mengersen
School of Mathematical Sciences, ARC Centre of Mathematical and Statistical Frontiers, Science and Engineering Faculty, Queensland University of Technology, Brisbane, QLD, Australia
e-mail: f.jahan@hdr.qut.edu.au

K. L. Mengersen et al. (eds.), *Case Studies in Applied Bayesian Data Science*, Lecture Notes in Mathematics 2259, https://doi.org/10.1007/978-3-030-42553-1_2

standard analytics tools (software, hardware, methods and algorithms) [70, 92]. The concomitant challenges of managing, modelling, analysing and interpreting these data have motivated a large literature on potential solutions from a range of domains including statistics, machine learning and computer science. This literature can be grouped into four broad categories of articles. The first includes general articles about the concept of Big Data, including the features and challenges, and their application and importance in specific fields. The second includes literature concentrating on infrastructure and management, including parallel computing and specialised software. The third focuses on statistical and machine learning models and algorithms for Big Data. The final category includes articles on the application of these new techniques to complex real-world problems.

In this chapter, we classify the literature published on Big Data into finer classes than the four broad categories mentioned earlier and briefly reviewed the contents covered by those different categories. But the main focus of the chapter is around the third category, in particular on statistical contributions to Big Data. We examine the nature of these innovations and attempt to catalogue them as modelling, algorithmic or other contributions. We then drill further into this set and examine the more specific literature on Bayesian approaches. Although there is an increasing interest in this paradigm from a wide range of perspectives including statistics, machine learning, information science, computer science and the various application areas, to our knowledge there has not yet been a survey of Bayesian statistical approaches for Big Data. This is the primary contribution of this chapter.

This chapter provides a survey of the published studies that present Bayesian statistical models specifically for Big Data and discusses the reported and perceived benefits of these approaches. We conclude by addressing the question of whether focusing only on improving computational algorithms and infrastructure will be enough to face the challenges of Big Data.

The chapter proceeds as follows. In the next section, literature search and inclusion criteria for this chapter is outlined. A classification of Big Data literature along with brief survey of relevant literature in each class is presented in Sect. 2.3. Section 2.4 consists of a brief survey of articles discussing Big Data problems from statistical perspectives, followed by a survey of Bayesian approaches applied to Big Data. The final section includes a discussion of this survey with a view to answering the research question posed above.

2.2   Literature Search and Inclusion Criteria

The literature search for this survey paper was undertaken using different methods. The search methods implemented to find the relevant literature and the criteria for the inclusion of the literature in this chapter are briefly discussed in this section.

2.2.1   Inclusion Criteria

Acknowledging the fact that there has been a wide range of literature on Big Data, the specific focus in this chapter was on recent developments published in the last 5 years, 2013–2019.

For quality assurance reasons, of the literature only peer reviewed published articles, book chapters and conference proceedings were included in the chapter. Some articles were also included from arXiv and pre-print versions for those to be soon published and from well known researchers working in that particular area of interest.

2.2.2   Search Methods

Database Search   The database "Scopus" was used to initiate the literature search. To identify the availability of literature and broadly learn about the broad areas of concentration, the following keywords were used: Big Data, Big Data Analysis, Big Data Analytics, Statistics and Big Data.

The huge range of literature obtained by this initial search was complemented by a search of "Google Scholar" using more specific key words as follows: Features and Challenges of Big Data, Big Data Infrastructure, Big Data and Machine Learning, Big Data and Cloud Computing, Statistical approaches/methods/models in Big Data, Bayesian Approaches/Methods/Models in Big Data, Big Data analysis using Bayesian Statistics, Bayesian Big Data, Bayesian Statistics and Big Data.

Expert Knowledge   In addition to the literature found by the above Database search, we used expert knowledge and opinions in the field and reviewed the works of well known researchers in the field of Bayesian Statistics for their research works related to Bayesian approaches to Big Data and included the relevant publications for survey in this chapter.

Scanning References of Selected Literature   Further studies and literature were found by searching the references of selected literature.

Searching with Specific Keywords   Since the focus of this chapter is to survey the Bayesian approaches to Big Data, more literature was sourced by using specific Bayesian methods or approaches found to be applied to Big Data: Approximate Bayesian Computation and Big Data, Bayesian Networks in Big Data, Classification and regression trees/Bayesian Additive regression trees in Big Data, Naive Bayes Classifiers and Big Data, Sequential Monte Carlo and Big Data, Hamiltonian Monte Carlo and Big Data, Variational Bayes and Big Data, Bayesian Empirical Likelihood and Big Data, Bayesian Spatial modelling and Big Data, Non parametric Bayes and Big Data.

This last step was conducted in order to ensure that this chapter covers the important and emerging areas of Bayesian Statistics and their application to Big Data. These searches were conducted in "Google Scholar" and up to 30 pages of results were considered in order to find relevant literature.

2.3   Classification of Big Data Literature

The published articles on Big Data can be divided into finer classes than the four main categories described above. Of course, there are many ways to make these delineations. Table 2.1 shows one such delineation, with representative references from the last 5 years of published literature. The aim of this table is to indicate the wide ranging literature on Big Data and provide relevant references in different categories for interested readers.

The links between these classes of literature can be visualised as in Fig 2.1 and a brief description of each of the classes and the contents covered by the relevant references listed are provided in Table 2.2. The brief surveys presented in Table 2.2 can be helpful for interested readers to develop a broad idea about each of the classes mentioned in Table 2.1. However, Table 2.2 does not include brief surveys of the last two classes, namely, Statistical Methods and Bayesian Methods, since these classes are discussed in detail in Sects. 2.4 and 2.5. We would like to acknowledge the fact that Bayesian methods are essentially part of statistical methods, but in this chapter, the distinct classes are made intentionally to be able to identify and discuss the specific developments in Bayesian approaches.

Table 2.1 Classes of big data literature

Topic	Representative references
Features and challenges	[51, 52, 63, 65, 70, 140, 160, 168, 184, 200]
Infrastructure	[10, 50, 96, 108, 117, 132, 137, 142, 165, 183, 193, 206, 207]
Cloud computing	[11, 32, 38, 54, 113, 120, 130, 139, 148, 175, 201, 205]
Applications (3 examples)	Social science: [5, 22, 37, 39, 121, 164]
	Health/medicine/medical science: [8, 9, 16, 19, 21, 28, 34, 46, 82, 118, 153, 158, 159, 182, 202]
	Business: [2, 31, 36, 60, 66, 122, 154, 172]
Machine learning methods	[3, 4, 26, 27, 55, 64, 89, 97, 138, 173]
Statistical methods	[44, 45, 58, 61, 67, 84, 86, 87, 111, 136, 143, 162, 174, 188, 191, 192, 194, 198, 204]
Bayesian methods	[7, 80, 81, 100, 102, 105, 109, 110, 115, 128, 129, 151, 163, 170, 180, 199, 210, 211]

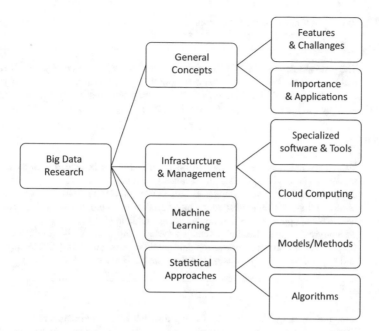

Fig. 2.1 Classification of big data literature

2.4   Statistical Approaches to Big Data

The importance of modelling and theoretical considerations for analysing Big Data
are well stated in the literature [86, 198]. These authors pointed out that blind
trust in algorithms without proper theoretical considerations will not result in valid
outputs. The emerging challenges of Big Data are beyond the issues of processing,
storing and management. The choice of suitable statistical methods is crucial in
order to make the most of the Big Data [67, 87]. Dunson [61] highlighted the role of
statistical methods for interpretability, uncertainty quantification, reducing selection
bias in analysing Big Data.

In this section we present a brief survey of some of the published research on
statistical perspectives, methods, models and algorithms that are targeted to Big
Data. As above, the survey is confined to the last 5 years, commencing with the
most recent contributions. Bayesian approaches are reserved for the next section.

Among the brief surveys of the relevant literature in Table 2.3, we include
detailed surveys of three papers which are more generic in explaining the role of
statistics and statistical methods in Big Data along with recent developments in this
area.

Wang et al. [191] summarised the published literature on recent methodological
developments for Big Data in three broad groups: subsampling, which calculates a
statistic in many subsamples taken from the data and then combining the results
[144]; divide and conquer, the principle of which is to break a dataset into

Table 2.2 Brief survey of relevant literature under identified classes

Features and challenges

- The general features of Big Data are volume, variety, velocity, veracity, value [52, 160] and some salient features include massive sample sizes and high dimensionality [160].
- Many challenges of Big Data regarding storage, processing, analysis and privacy are identified in the literature [52, 63, 65, 140, 160].

Infrastructure

- To manage and analyse Big Data, infrastructural support is needed such as sufficient storage technologies and data management systems. These are being continuously developed and improved. MangoDB, Terrastore and RethhinkDb are some examples of the storage technologies; more on evolution technologies with their strengths, weaknesses, opportunities and threats are available in [165].
- To analyse Big Data, parallel processing systems and scalable algorithms are needed. MapReduce is one of the pioneering data processing systems [206]. Some other useful and popular tools to handle Big Data are Apache, Hadoop, Spark [10].

Cloud computing

- Cloud computing, the practice of using a network of remote servers hosted on the Internet rather than a local server or a personal computer, plays a key role in Big Data analysis by providing required infrastructure needed to store, analyse, visualise and model Big Data using scalable and adaptive systems [11].
- Opportunities and challenges of cloud computing technologies, future trends and application areas are widely discussed in the literature [32, 175, 201] and new developments on cloud computing are proposed to overcome known challenges, such as collaborative anomaly detection [130], hybrid approach for scalable sub-tree anonymisation using MapReduce on cloud [205] etc.

Applications (3 examples)

- Big Data has made it possible to analyse social behaviour and an individual's interactions with social systems based on social media usage [5, 37, 164]. Discussions on challenges and future of social science research using Big Data have been made in the literature [39, 164].
- Research involving Big Data in medicine, public health, biomedical and health informatics has increased exponentially over the last decade [19, 28, 46, 114, 153, 158]. Some examples include infectious disease research [16, 82], developing personalised medicine and health care [9, 182] and improving cardiovascular care [159].
- Analysis of Big Data is used to solve many real world problems in business, in particular, using Big Data analytics for innovations in leading organisations [122], predictive analytics in retail [31], analysis of business risks and benefits [154], development of market strategies [60] and so on. The opportunities and challenges of Big Data in e-commerce and Big Data integration in business processes can be found in the survey articles by Akter and Wamba [2] and Wamba et al. [184].

(continued)

Table 2.2 (continued)

Machine learning methods

- Machine learning is an interdisciplinary field of research primarily focusing on theory, performance, properties of learning systems and algorithms [149]. Traditional machine learning is evolving to tackle the additional challenges of Big Data [4, 149].
- Some examples of developments in machine learning theories and algorithms for Big Data include high performance machine learning toolbox [3], scalable machine learning online services for Big Data real time analysis [14].
- There is a large and increasing research on specific applications of machine learning tools for Big Data in different disciplines. For example, [138] discussed the future of Big Data and machine learning in clinical medicine; [13] discussed a classifier specifically for medical Big Data and [26] reviewed the state of art and future prospects of machine learning and Big Data in radiation oncology.

smaller subsets to analyse these in parallel and combine the results at the end [169]; and online updating of streaming data [162], based on online recursive analytical processing. He summarised the following methods in the first two groups: subsampling based methods (bag of little bootstraps, leveraging, mean log likelihood, subsample based MCMC), divide and conquer (aggregated estimating equations, majority voting, screening with ultra high dimension, parallel MCMC). The authors, after reviewing existing online updating methods and algorithms, extended the online updating of stream data method by including criterion based variable selection with online updating. The authors also discussed the available software packages (open source R as well as commercial software) developed to handle computational complexity involving Big Data. For breaking the memory barrier using R, the authors cited and discussed several data management packages (sqldf, DBI, RSQLite, filehash, bigmemory, ff) and packages for numerical calculation (speedglm, biglm, biganalytics, ffbase, bigtabulate, bigalgebra, bigpca, bigrf, biglars, PopGenome). The R packages for breaking computing power were cited and discussed in two groups: packages for speeding up (compiler, inline, Rcpp, RcpEigen, RcppArmadilo, RInside, microbenchmark, proftools, aprof, lineprof, GUIprofiler) and packages for scaling up (Rmpi, snow, snowFT, snowfall, multicore, parallel, foreach, Rdsm, bigmemory, pdpMPI, pbdSLAP, pbdBASE, pbdMAT, pbdDEMO, Rhipe, segue, rhbase, rhdfs, rmr, plymr, ravroSparkR, pnmath, pnmath0, rsprng, rlecuyer, doRNG, gputools, bigvis). The authors also discussed the developments in Hadoop, Spark, OpenMP, API and using FORTRAN and C++ from R in order to create flexible programs for handling Big Data. The article also presented a brief summary about the commercial statistical software, e.g., SAS, SPSS, MATLAB. The study included a case study of fitting a logistic model to a massive data set on airline on-time performance data from the 2009 ASA Data Expo mentioning the use of some R packages discussed earlier to handle the problem with memory and computational capacity. Overall, this study provided a comprehensive

Table 2.3 Brief survey and classification of literature on statistical approaches to Big Data

---

Topic: Discussion article

Author: Dunson [61]

- Discussed the background of Big Data from the perspectives of the machine learning and statistics communities.
- Listed the differences in the methods and inferences as replicability, uncertainty quantification, sampling, selection bias and measurement error drawn from statistical perspectives to those of machine learning.
- Identified the statistical challenges for high dimensional complex data (Big Data) in quantifying uncertainty, scaling up sampling methods and selection of priors in Bayesian methods.

Topic: Survey

Author: Nongxa [136]

- Identified challenges of Big Data as: high dimensionality, heterogeneity and incompleteness, scale, timeliness, security and privacy.
- Pointed out that mathematical and statistical challenges of Big Data require updating the core knowledge areas (i.e., linear algebra, multivariable calculus, elementary probability and statistics, coding or programming) to more advanced topics (i.e., randomised numerical linear algebra, topological data analysis, matrix and tensor decompositions, random graphs; random matrices and complex networks) in mathematical and statistical education.

Author: Franke et al. [67]

- Reviewed different strategies of analysis as: data wrangling, visualisation, dimension reduction, sparsity regularisation, optimisation, measuring distance, representation learning, sequential learning and provided detailed examples of applications.

Author: Chen et al. [45]

- Emphasised the importance of statistical knowledge and skills in Big Data Analytics using several examples.
- Discussed some statistical methods that are useful in the context of Big Data as: confirmatory and exploratory data analysis tools, data mining methods including supervised learning (classification, regression/prediction) and unsupervised learning (cluster analysis, anomaly detection, association rule learning), visualisation techniques etc.
- Elaborated on the computational skills needed for statisticians in data acquisition, data processing, data management and data analysis.

Author: Hoerl et al. [87]

- Provided a background of Big Data reviewing relevant articles.
- Discussed the importance of statistical thinking in Big Data problems reviewing some misleading results produced by sophisticated analysis of Big Data without involving statistical principles.
- Elaborated on the roles of statistical thinking for data quality, domain knowledge, analysis strategies in order to solve complex unstructured problems involving Big Data.

(continued)

Table 2.3 (continued)

Topic: Survey of methods & extension

Author: Wang et al. [191]

- Reviewed statistical methods and software packages in R and recently developed tools to handle Big Data, focusing on three groups: sub-sampling, divide and conquer and online processing.
- Extended the online updating approach by employing variable selection criteria.

Topic: Methods survey, new methods

Author: Genuer et al. [72]

- Reviewed proposals dealing with scaling random forests to Big Data problems.
- Discussed subsampling, parallel implementations, online processing of random forests in detail.
- Proposed five variants of Random Forests for Big Data.

Author: Wang and Xu [185]

- Reviewed different clustering methods applicable to Big Data situations.
- Proposed a clustering procedure with adaptive density peak detection applying multivariate kernel estimation and demonstrated the performance through simulation studies and analysis of a few benchmark gene expression data sets.
- Developed a R-package "ADPclust" to implement the proposed methods.

Author: Wang et al. [192]

- Proposed a method and algorithm for online updating implementing bias corrections with extensions for application in a generalised linear model (GLM) setting.
- Evaluated the proposed strategies in comparison with previous algorithms [162].

Topic: New methods and algorithms

Author: Liu et al. [111]

- Proposed a novel sparse GLM with L0 approximation for feature selection and prediction in big omics data scenarios.
- Provided novel algorithm and software in MATLAB (L0ADRIDGE) for performing L0 penalised GLM in ultra high dimensional Big Data.
- Comparison of performance with other methods (SCAD, MC+) using simulation and real data analysis (mRNA, microRNA, methylation data from TGCA ovarian cancer).

Author: Schifano et al. [162]

- Developed new statistical methods and iterative algorithms for analysing streaming data.
- Proposed methods to enable update of the estimations and models with the arrival of new data.

(continued)

Table 2.3 (continued)

Author: Allen et al. [6]

- Proposed generalisations to Principal Components Analysis (PCA) to take into account structural relationships in Big Data settings.
- Developed fast computational algorithms using the proposed methods (GPCA, sparse GPCA and functional GPCA) for massive data sets.

Topic: New algorithms

Author: Wang and Samworth [188]

- Proposed a new algorithm "inspect" (informative sparse projection for estimation of change points) to estimate the number and location of change points in high dimensional time series.
- The algorithm, starting from a simple time series model, was extended to detect multiple change points and was also extended to have spatial or temporal dependence, assessed using simulation studies and real data application.

Author: Yu and Lin [203]

- Extended the alternating direction method of multipliers (ADMM) to solve penalised quantile regression problems involving massive data sets having faster computation and no loss of estimation accuracy.

Author: Zhang and Yang [204]

- Proposed new algorithms using ridge regression to make it efficient for handling Big Data.

Author: Doornik and Hendry [58]

- Discussed the statistical model selection algorithm "autometrics" for econometric data [57] with its application to fat Big Data (having larger number of variables than the number of observations).
- Extended algorithms for tackling computational issues of fat Big Data applying block searches and re-selection by lasso for correlated regressors.

Author: Sysoev et al. [174]

- Presented efficient algorithms to estimate bootstrap or jackknife type confidence intervals for fitted Big Data sets by Multivariate Monotonic Regression.
- Evaluated the performance of the proposed algorithms using a case study on death in coronary heart disease for a large population.

Author: Pehlivanlı [143]

- Proposed a novel approach for feature selection from high dimensional data.
- Tested the efficiency of the proposed method using sensitivity, specificity, accuracy and ROC curve.
- Demonstrated the approach on micro-array data.

survey and discussion of state-of-the-art statistical methodologies and software development for handling Big Data.

Chen et al. [45] presented their views on the challenges and importance of Big Data and explained the role of statistics in Big Data Analytics based on a survey of relevant literature. This study emphasised the importance of statistical knowledge and skills in Big Data Analytics using several examples. As detailed in Table 2.3, the authors broadly discussed a range of statistical methods which can be really helpful in better analysis of Big Data, such as, the use of exploratory data analysis principle in Statistics to investigate correlations among the variables in the data or establish causal relationships between response and explanatory variables in the Big Data. The authors specifically mentioned hypothesis testing, predictive analysis using statistical models, statistical inference using uncertainty estimation to be some key tools to use in Big Data analysis. The authors also explained that the combination of statistical knowledge can be combined with the Data mining methods such as unsupervised learning (cluster analysis, Association rule learning, anomaly detection) and supervised learning (regression and classification) can be beneficial for Big Data analysis. The challenges for the statisticians in coping with Big Data were also described in this article, with particular emphasis on computational skills in data acquisition (knowledge of programming languages, knowledge of web and core communication protocols), data processing (skills to transform voice or image data to numeric data using appropriate software or programming), data management (knowledge about database management tools and technologies, such as NoSQL) and scalable computation (knowledge about parallel computing, which can be implemented using MapReduce, SQL etc.).

As indicated above, many of the papers provide a summary of the published literature which is not replicated here. Some of these surveys are based on large thematic programs that have been held on this topic. For example, the paper by Franke et al. [67] is based on presentations and discussions held as part of the program on Statistical Inference, Learning and Models for Big Data which was held in Canada in 2015. The authors discussed the four V's (volume, variety, veracity and velocity) of Big Data and mentioned some more challenges in Big Data analysis which are beyond the complexities associated with the four V's. The additional "V" mentioned in this article is veracity. Veracity refers to biases and noise in the data which may be the result of the heterogeneous structure of the data sources, which may make the sample non representative of the population. Veracity in Big Data is often referred to as the biggest challenge compared with the other V's. The paper reviewed the common strategies for Big Data analysis starting from data wrangling which consists of data manipulation techniques for making the data eligible for analysis; visualisation which is often an important tool to understand the underlying patterns in the data and is the first formal step in data analysis; reducing the dimension of data using different algorithms such as Principal Component Analysis (PCA) to make Big Data models tractable and interpretable; making models more robust by enforcing sparsity in the model by the use of regularisation techniques such as variable selection and model fitting criteria; using optimisation methods based on different distance measures proposed for high dimensional data and by using different learning algorithms such as representation learning and sequential learning. Different applications of Big Data were shown in public health, health policy, law and order, education, mobile application security, image recognition and labelling, digital humanities and materials science.

There are few other research articles focused on statistical methods tailored to specific problems, which are not included in Table 2.3. For example, Castruccio and Genton [40] proposed a statistics-based algorithm using a stochastic space-time model with more than 1 billion data points to reproduce some features of a climate model. Similarly, [123] used various statistical methods to obtain associations between drug-outcome pairs in a very big longitudinal medical experimental database (with information on millions of patients) with a detailed discussion on the big results problem by providing a comparison of statistical and machine learning approaches. Finally, Hensman et al. [84] proposed stochastic variational inference for Gaussian processes which makes the application of Gaussian process to huge data sets (having millions of data points).

From the survey of some relevant literature related to statistical perspectives for analysing Big Data, it can be seen that along with scaling up existing algorithms, new methodological developments are also in progress in order to face the challenges associated with Big Data.

2.5   Bayesian Approaches in Big Data

As described in the Introduction, the intention of this survey is to commence with a broad scope of the literature on Big Data, then focus on statistical methods for Big Data, and finally to focus in particular on Bayesian approaches for modelling and analysis of Big Data. This section consists of a survey of published literature on the last of these.

There are two defining features of Bayesian analysis: (1) the construction of the model and associated parameters and expectations of interest, and (2) the development of an algorithm to obtain posterior estimates of these quantities. In the context of Big Data, the resultant models can become complex and suffer from issues such as unavailability of a likelihood, hierarchical instability, parameter explosion and identifiability. Similarly, the algorithms can suffer from too much or too little data given the model structure, as well as problems of scalability and cost. These issues have motivated the development of new model structures, new methods that avoid the need for models, new Markov chain Monte Carlo (MCMC) sampling methods, and alternative algorithms and approximations that avoid these simulation-based approaches. We discuss some of the concomitant literature under two broad headings, namely computation and models realising that there is often overlap in cited papers.

2.5.1   Bayesian Computation

In Bayesian framework a main-stream computational tool has been the Markov chain Monte Carlo (MCMC). The traditional MCMC methods do not scale well because they need to iterate through the full data set at each iteration to evaluate the likelihood [199]. Recently several attempts have been made to scale MCMC methods up to massive data. A widely used strategy to overcome the computational cost is to distribute the computational burden across a number of machines. The strategy is generally referred to as divide-and-conquer sampling. This approach breaks a massive data set into a number of easier to handle subsets, obtains posterior samples based on each subset in parallel using multiple machines and finally combines the subset posterior inferences to obtain the full-posterior estimates [169]. The core challenge is the recombination of sub-posterior samples to obtain true posterior samples. A number of attempts have been made to address this challenge.

Neiswanger et al. [134] and White et al. [195] approximated the sub-posteriors using kernel density estimation and then aggregated the sub-posteriors by taking their product. Both algorithms provided consistent estimates of the posterior. Neiswanger et al. [134] provided faster MCMC processing since it allowed the machine to process the parallel MCMC chains independently. However, one limitation of the asymptotically embarrassing parallel MCMC algorithm [134] is that it only works for real and unconstrained posterior values, so there is still scope of works to make the algorithm work under more general settings.

Wang and Dunson [187] adopted a similar approach of parallel MCMC but used a Weierstrass transform to approximate the sub-posterior densities instead of a kernel density estimate. This provided better approximation accuracy, chain mixing rate and potentially faster speed for large scale Bayesian analysis.

Scott et al. [163] partitioned the data at random and performed MCMC independently on each subset to draw samples from posterior given the data subset. To obtain consensus posteriors they proposed to average samples across subsets and showed the exactness of the algorithm under a Gaussian assumption. This algorithm is scalable to a very large number of machines and works in cluster, single multi core or multiprocessor computers or any arbitrary collection of computers linked by a high speed network. The key weakness of consensus MCMC is it does not apply to non Gaussian posterior.

Minsker et al. [128] proposed dividing a large set of independent data into a number of non-overlapping subsets, making inferences on the subsets in parallel and then combining the inferences using the median of the subset posteriors. The median posterior (M-posterior) is constructed from the subset posteriors using Weiszfeld's algorithm, which provides a scalable algorithm for robust estimation.

Guhaniyogi and Banerjee [77] extended this notion to spatially dependent data, provided a scalable divide and conquer algorithm to analyse big spatial data sets named spatial meta kriging. The multivariate extension of spatial meta kriging has been addressed by Guhaniyogi and Banerjee [78]. These approaches of meta kriging are practical developments for Bayesian spatial inference for Big Data, specifically with "big-N" problems [98].

Wu and Robert [199] proposed a new and flexible divide and conquer framework by using re-scaled sub-posteriors to approximate the overall posterior. Unlike other parallel approaches of MCMC, this method creates artificial data for each subset, and applies the overall priors on the artificial data sets to get the subset posteriors. The sub-posteriors are then re-centred to their common mean and then averaged to approximate the overall posterior. The authors claimed this method to have statistical justification as well as mathematical validity along with sharing same computational cost with other classical parallel MCMC approaches such as consensus Monte Carlo, Weierstrass sampler. Bouchard-Côté et al. [30] proposed a non-reversible rejection-free MCMC method, which reportedly outperforms state-of-the-art methods such as: HMC, Firefly by having faster mixing rate and lower variances for the estimators for high dimensional models and large data sets. However, the automation of this method is still a challenge.

Another strategy for scalable Bayesian inference is the sub-sampling based approach. In this approach, a smaller subset of data is queried in the MCMC algorithm to evaluate the likelihood at every iteration. Maclaurin and Adams [116] proposed to use an auxiliary variable MCMC algorithm that evaluates the likelihood based on a small subset of the data at each iteration yet simulates from the exact posterior distribution. To improve the mixing speed, Korattikara et al. [95] used an approximate Metropolis Hastings (MH) test based on a subset of data. A similar approach is used in [17], where the accept/reject step of MH evaluates the likelihood of a random subset of the data. Bardenet et al. [18] extended this approach by replacing a number of likelihood evaluations by a Taylor expansion centred at the maximum of the likelihood and concluded that their method outperforms the previous algorithms [95].

The scalable MCMC approach was also improved by Quiroz et al. [150] using a difference estimator to estimate the log of the likelihood accurately using only a small fraction of the data. Quiroz et al. [151] introduced an unbiased estimator of the log likelihood based on weighted sub-sample which is used in the MH acceptance step in speeding up based on a weighted MCMC efficiently. Another scalable adaptation of MH algorithm was proposed by Maire et al. [119] to speed up Bayesian inference in Big Data namely informed subsampling MCMC which involves drawing of subsets according to a similarity measure (i.e., squared L2 distance between full data and maximum likelihood estimators of subsample) instead of using uniform distribution. The algorithm showed excellent performance in the case of a limited computational budget by approximating the posterior for a tall dataset.

Another variation of MCMC in Big Data has been made by Strathmann et al. [170]. These authors approximated the posterior expectation by a novel Bayesian inference framework for approximating the posterior expectation from a different perspective suitable for Big Data problems, which involves paths of partial posteriors. This is a parallelisable method which can easily be implemented using existing MCMC techniques. It does not require the simulation from full posterior, thus bypassing the complex convergence issues of kernel approximation. However, there is still scope for future work to look at computation-variance trade off and finite time bias produced by MCMC.

Hamiltonian Monte Carlo (HMC) sampling methods provide powerful and efficient algorithms for MCMC using high acceptance probabilities for distant proposals [44]. A conceptual introduction to HMC is presented by Betancourt [25]. Chen et al. [44] proposed a stochastic gradient HMC using second-order Langevin dynamics. Stochastic Gradient Langevin Dynamics (SGLD) have been proposed as a useful method for applying MCMC to Big Data where the accept-reject step is skipped and decreasing step size sequences are used [1]. For more detailed and rigorous mathematical framework, algorithms and recommendations, interested readers are referred to [178].

A popular method of scaling Bayesian inference, particularly in the case of analytically intractable distributions, is Sequential Monte Carlo (SMC) or particle filters [24, 48, 80]. SMC algorithms have recently become popular as a method to approximate integrals. The reasons behind their popularity include their easy implementation and parallelisation ability, much needed characteristics in Big Data implementations [100]. SMC can approximate a sequence of probability distributions on a sequence of spaces with an increasing dimension by applying resampling, propagation and weighting starting with the prior and eventually reaching to the posterior of interest of the cloud of particles. Gunawan et al. [80] proposed a sub-sampling SMC which is suitable for parallel computation in Big Data analysis, comprising two steps. First, the speed of the SMC is increased by using an unbiased and efficient estimator of the likelihood, followed by a Metropolis within Gibbs kernel. The kernel is updated by a HMC method for model parameters and a block-pseudo marginal proposal for the auxiliary variables [80]. Some novel approaches of SMC include: divide-and-conquer SMC [105], multilevel SMC [24], online SMC [75] and one pass SMC [104], among others.

Stochastic variational inference (VI, also called Variational Bayes, VB) is a faster alternative to MCMC [88]. It approximates probability densities using a deterministic optimisation method [109] and has seen widespread use to approximate posterior densities for Bayesian models in large-scale problems. The interested reader is referred to [29] for a detailed introduction to variational inference designed for statisticians, with applications. VI has been implemented in scaling up algorithms for Big Data. For example, a novel re-parameterisation of VI has been implemented for scaling latent variable models and sparse GP regression to Big Data [69].

There have been studies which combined the VI and SMC in order to take advantage from both strategies in finding the true posterior [56, 133, 152]. Naesseth et al. [133] employed a SMC approach to get an improved variational approximation, Rabinovich et al. [152] by splitting the data into block, applied SMC to compute partial posterior for each block and used a variational argument to get a proxy for the true posterior by the product of the partial posteriors. The combination of these two techniques in a Big Data context was made by Donnet and Robin [56]. Donnet and Robin [56] proposed a new sampling scheme called Shortened Bridge Sampler, which combines the strength of deterministic approximations of the posterior that is variational Bayes with those of SMC. This sampler resulted in reduced computational time for Big Data with huge numbers of parameters, such as data from genomics or network.

Guhaniyogi et al. [79] proposed a novel algorithm for Bayesian inference in the context of massive online streaming data, extending the Gibbs sampling mechanism for drawing samples from conditional distributions conditioned on sequential point estimates of other parameters. The authors compared the performance of this conditional density filtering algorithm in approximating the true posterior with SMC and VB, and reported good performance and strong convergence of the proposed algorithm.

Approximate Bayesian computation (ABC) is gaining popularity for statistical inference with high dimensional data and computationally intensive models where the likelihood is intractable [125]. A detailed overview of ABC can be found in [167] and asymptotic properties of ABC are explored in [68]. ABC is a likelihood free method that approximates the posterior distribution utilising imperfect matching of summary statistics [167]. Improvements on existing ABC methods for efficient estimation of posterior density with Big Data (complex and high dimensional data with costly simulations) have been proposed by Izbicki et al. [90]. The choice of summary statistics from high dimensional data is a topic of active discussion; see, for example, [90, 166]. Pudlo et al. [147] provided a reliable and robust method of model selection in ABC employing random forests which was shown to have a gain in computational efficiency.

There is another aspect of ABC recently in terms of approximating the likelihood using Bayesian Synthetic likelihood or empirical likelihood [59]. Bayesian synthetic likelihood arguably provides computationally efficient approximations of the likelihood with high dimensional summary statistics [126, 196]. Empirical likelihood, on the other hand is a non-parametric technique of approximating the likelihood empirically from the data considering the moment constraints; this has been suggested in the context of ABC [127], but has not been widely adopted. For further reading on empirical likelihood, see [141].

Classification and regression trees are also very useful tools in data mining and Big Data analysis [33]. There are Bayesian versions of regression trees such as Bayesian Additive Regression Trees (BART) [7, 47, 93]. The BART algorithm has also been applied to the Big Data context and sparse variable selection by Rocková and van der Pas [157], van der Pas and Rockova [181], and Linero [106].

Some other recommendations to speed up computations are to use graphics processing units (see, e.g., [101, 171]) and parallel programming approaches (see, e.g., [42, 71, 76, 197]).

2.5.2 Bayesian Modelling

The extensive development of Bayesian computational solutions has opened the door to further developments in Bayesian modelling. Many of these new methods are set in the context of application areas. For example, there have been applications of ABC for Big Data in many different fields [62, 102]. For example, Dutta et al. [62] developed a high performance computing ABC approach for estimation of parameters in platelets deposition, while Lee et al. [102] proposed ABC methods for inference in high dimensional multivariate spatial data from a large number of locations with a particular focus on model selection for application to spatial extremes analysis. Bayesian mixtures are a popular modelling tool. VB and ABC techniques have been used for fitting Bayesian mixture models to Big Data [29, 88, 124, 129, 177].

Variable selection in Big Data (wide in particular, having massive number of variables) is a demanding problem. Liquet et al. [107] proposed multivariate extensions of the Bayesian group lasso for variable selection in high dimensional data using Bayesian hierarchical models utilising spike and slab priors with application to gene expression data. The variable selection problem can also be solved employing ABC type algorithms. Liu et al. [112] proposed a sampling technique, ABC Bayesian forests, based on splitting the data, useful for high dimensional wide data, which turns out to be a robust method in identifying variables with larger marginal inclusion probability.

Bayesian non-parametrics [131] have unbounded capacity to adjust unseen data through activating additional parameters that were inactive before the emergence of new data. In other words, the new data are allowed to speak for themselves in non-parametric models rather than imposing an arguably restricted model (that was learned on an available data) to accommodate new data. The inherent flexibility of these models to adjust with new data by adapting in complexity makes them more suitable for Big Data as compared to their parametric counterparts. For a brief introduction to Bayesian non-parametric models and a nontechnical overview of some of the main tools in the area, the interested reader is referred to Ghahramani [73].

The popular tools in Bayesian non-parametrics include Gaussian processes (GP) [156], Dirichlet processes (DP) [155], Indian buffet process (IBP) [74] and infinite hidden Markov models (iHMM) [20]. GP have been used for a variety of applications [35, 41, 49] and attempts have been made to scale it to Big Data [53, 84, 85, 179]. DP have seen successes in clustering and faster computational algorithms are being adopted to scale them to Big Data [71, 104, 115, 186, 189]. IBP are used for latent feature modeling, where the number of features are determined in a data-driven fashion and have been scaled to Big Data through variational inference algorithms [211]. Being an alternative to classical HMM, one of the distinctive properties of iHMM is that it infers the number of hidden states in the system from the available data and has been scaled to Big Data using particle filtering algorithms [180].

Gaussian Processes are also employed in the analysis of high dimensional spatially dependent data [15]. Banerjee [15] provided model-based solutions employing low rank GP and nearest neighbour GP (NNGP) as scalable priors in a hierarchical framework to render full Bayesian inference for big spatial or spatio temporal data sets. Zhang et al. [208] extended the applicability of NNGP for inference of latent spatially dependent processes by developing a conjugate latent NNGP model as a practical alternative to onerous Bayesian computations. Use of variational optimisation with structured Bayesian GP latent variable model to analyse spatially dependent data is made in in Atkinson and Zabaras [12]. For a survey of methods of analysis of massive spatially dependent data including the Bayesian approaches, see Heaton et al. [83].

Another Bayesian modelling approach that has been used for big and complex data is Bayesian Networks (BN). This methodology has generated a substantial literature examining theoretical, methodological and computational approaches, as well as applications [176]. BN belong to the family of probabilistic graphical models and based on direct acyclic graphs which are very useful representation of causal relationship among variables [23]. BN are used as efficient learning tool in Big Data analysis integrated with scalable algorithms [190, 209]. For a more detailed understanding of BN learning from Big Data, please see Tang et al. [176].

Classification is also an important tool for extracting information from Big Data and Bayesian classifiers, including Naive Bayes classifier (NBC) are used in Big Data classification problems [94, 110]. Parallel implementation of NBC has been proposed by Katkar and Kulkarni [94]. Moreover, Liu et al. [110] evaluated the scalability of NBC in Big Data with application to sentiment classification of millions of movie survey and found NBC to have improved accuracy in Big Data. Ni et al. [135] proposed a scalable multi step clustering and classification algorithm using Bayesian nonparametrics for Big Data with large n and small p which can also run in parallel.

The past 15 years has also seen an increase in interest in Empirical Likelihood (EL) for Bayesian modelling. The idea of replacing the likelihood with an empirical analogue in a Bayesian framework was first explored in detail by Lazar [99]. The author demonstrated that this Bayesian Empirical Likelihood (BEL) approach increases the flexibility of EL approach by examining the length and coverage of BEL intervals. The paper tested the methods using simulated data sets. Later, Schennach [161] provided probabilistic interpretations of BEL exploring moment condition models with EL and provided a non parametric version of BEL, namely Bayesian Exponentially Tilted Empirical Likelihood (BETEL). The BEL methods have been applied in spatial data analysis in Chaudhuri and Ghosh [43] and Porter et al. [145, 146] for small area estimation.

We acknowledge that there are many more studies on the application of Bayesian approaches in different fields of interest which are not included in this survey. There are also other survey papers on overlapping and closely related topics. For example, Zhu et al. [210] describes Bayesian methods of machine learning and includes some of the Bayesian inference techniques reviewed in the present study. However, the scope and focus of this survey is different from that of Zhu et al. [210], which was focused around the methods applicable to machine learning.

2.6   Conclusions

We are living in the era of Big Data and continuous research is in progress to make most use of the available information. The current chapter has attempted to survey the recent developments made in Bayesian statistical approaches for handling Big Data along with a general overview and classification of the Big Data literature with brief survey in last 5 years. This survey chapter provides relevant references in Big Data categorised in finer classes, a brief description of statistical contributions to the field and a more detailed discussion of the Bayesian approaches developed and applied in the context of Big Data.

On the basis of the surveys made above, it is clear that there has been a huge amount of work on issues related to cloud computing, analytics infrastructure and so on. However, the amount of research conducted from statistical perspectives is also notable. In the last 5 years, there has been an exponential increase in published studies focused on developing new statistical methods and algorithms, as well as scaling existing methods. These have been summarised in Sect. 2.4, with particular focus on Bayesian approaches in Sect. 2.5. In some instances citations are made outside of the specific period (see Sect. 2.2) to refer the origin of the methods which are currently being applied or extended in Big Data scenarios.

With the advent of computational infrastructure and advances in programming and software, Bayesian approaches are no longer considered as being very computationally expensive and onerous to execute for large volumes of data, that is Big Data. Traditional Bayesian methods are now becoming much more scalable due to the advent of parallelisation of MCMC algorithms, divide and conquer and/or subsampling methods in MCMC, and advances in approximations such as HMC, SMC, ABC, VB and so on. With the increasing volume of data, non-parametric Bayesian methods are also gaining in popularity.

This survey chapter aimed to survey a range of methodological and computational advancement made in Bayesian Statistics for handling the difficulties arose by the advent of Big Data. By not focusing to any particular application, this chapter provided the readers with a general overview of the developments of Bayesian methodologies and computational algorithms for handling these issues. The survey has revealed that most of the advancements in Bayesian Statistics for Big Data have been around computational time and scalability of particular algorithms, concentrating on estimating the posterior by adopting different techniques. However the developments of Bayesian methods and models for Big Data in the recent literature cannot be overlooked. There are still many open problems for further research in the context of Big Data and Bayesian approaches, as highlighted in this chapter.

Based on the above discussion and the accompanying survey presented in this chapter, it is apparent that to address the challenges of Big Data along with the strength of Bayesian statistics, research on both algorithms and models are essential.

Acknowledgments This research was supported by an ARC Australian Laureate Fellowship for project, Bayesian Learning for Decision Making in the Big Data Era under Grant no. FL150100150. The authors also acknowledge the support of the Australian Research Council (ARC) Centre of Excellence for Mathematical and Statistical Frontiers (ACEMS).

References

1. S. Ahn, B. Shahbaba, M. Welling, Distributed stochastic gradient MCMC, in *International Conference on Machine Learning* (2014), pp. 1044–1052
2. S Akter, S.F. Wamba, Big data analytics in e-commerce: a systematic review and agenda for future research. Electron. Mark. **26**(2), 173–194 (2016)
3. A. Akusok, K.M. Björk, Y. Miche, A. Lendasse, High-performance extreme learning machines: a complete toolbox for big data applications. IEEE Access **3**, 1011–1025 (2015)
4. O.Y. Al-Jarrah, P.D. Yoo, S. Muhaidat, G.K. Karagiannidis, K. Taha, Efficient machine learning for big data: a review. Big Data Res. **2**(3), 87–93 (2015)
5. K. Albury, J. Burgess, B. Light, K Race, R. Wilken, Data cultures of mobile dating and hook-up apps: emerging issues for critical social science research. Big Data Soc. **4**(2), 1–11 (2017)
6. G.I. Allen, L. Grosenick, J. Taylor, A generalized least-square matrix decomposition. J. Am. Stat. Assoc. **109**(505), 145–159 (2014)
7. G.M. Allenby, E.T. Bradlow, E.I. George, J. Liechty, R.E. McCulloch, Perspectives on Bayesian methods and big data. Cust. Needs Solut. **1**(3), 169–175 (2014)
8. S.G. Alonso, I. de la Torre Díez, J.J. Rodrigues, S. Hamrioui, M. López-Coronado, A systematic review of techniques and sources of big data in the healthcare sector. J. Med. Syst. **41**(11), 183 (2017)
9. A. Alyass, M. Turcotte, D. Meyre, From big data analysis to personalized medicine for all: challenges and opportunities. BMC Med. Genomics **8**(1), 33 (2015)
10. D. Apiletti, E. Baralis, T. Cerquitelli, P. Garza, F. Pulvirenti, L. Venturini, (2017) Frequent itemsets mining for big data: a comparative analysis. Big Data Res. **9**, 67–83
11. M.D. Assunção, R.N. Calheiros, S. Bianchi, M.A. Netto, R. Buyya, Big data computing and clouds: trends and future directions. J. Parallel Distrib. Comput. **79**, 3–15 (2015)
12. S. Atkinson, N. Zabaras, Structured Bayesian Gaussian process latent variable model: applications to data-driven dimensionality reduction and high-dimensional inversion. J. Comput. Phys. **383**, 166–195 (2019)
13. A.T. Azar, A.E. Hassanien, Dimensionality reduction of medical big data using neural-fuzzy classifier. Soft Comput. **19**(4), 1115–1127 (2015)
14. A. Baldominos, E. Albacete, Y. Saez, P. Isasi, A scalable machine learning online service for big data real-time analysis, in *2014 IEEE Symposium on Computational Intelligence in Big Data (CIBD)* (IEEE, Piscataway, 2014), pp. 1–8
15. S. Banerjee, High-dimensional Bayesian geostatistics. Bayesian Anal. **12**(2), 583 (2017)
16. S. Bansal, G. Chowell, L. Simonsen, A. Vespignani, C. Viboud, Big data for infectious disease surveillance and modeling. J. Infect. Dis. **214**(suppl_4), S375–S379 (2016)
17. R. Bardenet, A. Doucet, C. Holmes, Towards scaling up Markov chain Monte Carlo: an adaptive subsampling approach, in *International Conference on Machine Learning (ICML)* (2014), pp. 405–413
18. R. Bardenet, A. Doucet, C. Holmes, On Markov chain Monte Carlo methods for tall data. J. Mach. Learn. Res. **18**(1), 1515–1557 (2017)
19. D.W. Bates, S. Saria, L. Ohno-Machado, A. Shah, G. Escobar, Big data in health care: using analytics to identify and manage high-risk and high-cost patients. Health Aff. **33**(7), 1123–1131 (2014)
20. M.J. Beal, Z. Ghahramani, C.E. Rasmussen, The infinite hidden Markov model, in *Advances in Neural Information Processing Systems* (2002), pp. 577–584

21. A. Belle, R. Thiagarajan, S. Soroushmehr, F. Navidi, D.A. Beard, K. Najarian, Big data analytics in healthcare. BioMed. Res. Int. **2015**, 370194 (2015)
22. G. Bello-Orgaz, J.J. Jung, D. Camacho, Social big data: recent achievements and new challenges. Inf. Fusion **28**, 45–59 (2016)
23. I. Ben-Gal, Bayesian Networks. Encycl. Stat. Qual. Reliab. **1**, 1–6 (2008)
24. A. Beskos, A. Jasra, E.A. Muzaffer, A.M. Stuart, Sequential Monte Carlo methods for Bayesian elliptic inverse problems. Stat. Comput. **25**(4), 727–737 (2015)
25. M. Betancourt, A conceptual introduction to Hamiltonian Monte Carlo. Preprint, arXiv: 170102434 (2017)
26. J.E. Bibault, P. Giraud, A. Burgun, Big data and machine learning in radiation oncology: state of the art and future prospects. Cancer Lett. **382**(1), 110–117 (2016)
27. A. Bifet, Morales GDF Big data stream learning with Samoa, in *2014 IEEE International Conference on Data Mining Workshop (ICDMW)*, IEEE, pp. 1199–1202 (2014)
28. H. Binder, M. Blettner, Big data in medical science–a biostatistical view: Part 21 of a series on evaluation of scientific publications. Dtsch. Ärztebl Int. **112**(9), 137 (2015)
29. D.M. Blei, A. Kucukelbir, J.D. McAuliffe, Variational inference: a review for statisticians. J. Am. Stat. Assoc. **112**(518), 859–877 (2017)
30. A. Bouchard-Côté, S.J. Vollmer, A. Doucet, The bouncy particle sampler: a nonreversible rejection-free Markov chain Monte Carlo method. J. Am. Stat. Assoc. **113**, 1–13 (2018)
31. E.T. Bradlow, M. Gangwar, P. Kopalle, S. Voleti, The role of big data and predictive analytics in retail. J. Retail. **93**(1), 79–95 (2017)
32. R. Branch, H. Tjeerdsma, C. Wilson, R. Hurley, S. McConnell, Cloud computing and big data: a review of current service models and hardware perspectives. J. Softw. Eng. Appl. **7**(08), 686 (2014)
33. L. Breiman, *Classification and Regression Trees* (Routledge, Abingdon, 2017)
34. P.F. Brennan, S. Bakken, Nursing needs big data and big data needs nursing. J. Nurs. Scholarsh. **47**(5), 477–484 (2015)
35. F. Buettner, K.N. Natarajan, F.P. Casale, V. Proserpio, A. Scialdone, F.J. Theis, S.A. Teichmann, J.C. Marioni, O. Stegle, Computational analysis of cell-to-cell heterogeneity in single-cell RNA-sequencing data reveals hidden subpopulations of cells. Nat. Biotechnol. **33**(2), 155 (2015)
36. J. Bughin, Big data, big bang? J. Big Data **3**(1), 2 (2016)
37. R. Burrows, M. Savage, After the crisis? Big data and the methodological challenges of empirical sociology. Big Data Soc. **1**(1), 1–6 (2014)
38. H. Cai, B. Xu, L. Jiang, A.V. Vasilakos, Iot-based big data storage systems in cloud computing: perspectives and challenges. IEEE Internet Things J. **4**(1), 75–87 (2017)
39. J.N. Cappella, Vectors into the future of mass and interpersonal communication research: big data, social media, and computational social science. Hum. Commun. Res. **43**(4), 545–558 (2017)
40. S. Castruccio, M.G. Genton, Compressing an ensemble with statistical models: an algorithm for global 3d spatio-temporal temperature. Technometrics **58**(3), 319–328 (2016)
41. K. Chalupka, C.K. Williams, I. Murray, A framework for evaluating approximation methods for Gaussian process regression. J. Mach. Learn. Res. **14**(Feb), 333–350 (2013)
42. J. Chang, J.W. Fisher III, Parallel sampling of DP mixture models using sub-cluster splits, in *Advances in Neural Information Processing Systems* (2013), pp. 620–628
43. S. Chaudhuri, M. Ghosh, Empirical likelihood for small area estimation. Biometrika **98**, 473–480 (2011)
44. T. Chen, E. Fox, C. Guestrin, Stochastic gradient Hamiltonian Monte Carlo, in *Int. Conference on Machine Learning* (2014), pp. 1683–1691
45. J.J. Chen, E.E. Chen, W. Zhao, W. Zou, Statistics in big data. J. Chin. Stat. Assoc. **53**, 186–202 (2015)
46. A.S. Cheung, Moving beyond consent for citizen science in big data health and medical research. Northwest J. Technol. Intellect. Prop. **16**(1), 15 (2018)

47. H.A. Chipman, E.I. George, R.E. McCulloch et al., BART: Bayesian additive regression trees. Ann. Appl. Stat. **4**(1), 266–298 (2010)
48. N. Chopin, P.E. Jacob, O. Papaspiliopoulos, Smc2: an efficient algorithm for sequential analysis of state space models. J. R. Stat. Soc. Ser. B (Stat Methodol.) **75**(3), 397–426 (2013)
49. A. Damianou, N. Lawrence, Deep Gaussian processes, in *Artificial Intelligence and Statistics* (2013), pp. 207–215
50. T. Das, P.M. Kumar, Big data analytics: a framework for unstructured data analysis. Int. J. Eng. Sci. Technol. **5**(1), 153 (2013)
51. A. De Mauro, M. Greco, M. Grimaldi, What is big data? a consensual definition and a review of key research topics, in *AIP Conference Proceedings, AIP*, vol. 1644 (2015), pp. 97–104
52. A. De Mauro, M. Greco, M. Grimaldi A formal definition of big data based on its essential features. Libr. Rev. **65**(3), 122–135 (2016)
53. M.P. Deisenroth, J.W. Ng, Distributed Gaussian processes, in *Proceedings of the 32nd International Conference on International Conference on Machine Learning*, vol. 37, JMLR.org (2015), pp. 1481–1490
54. H. Demirkan, D. Delen Leveraging the capabilities of service-oriented decision support systems: putting analytics and big data in cloud. Decis. Support Syst. **55**(1), 412–421 (2013)
55. K.S. Divya, P. Bhargavi, S. Jyothi Machine learning algorithms in big data analytics. Int. J. Comput. Sci. Eng. **6**(1), 63–70 (2018)
56. S. Donnet, S. Robin Shortened bridge sampler: using deterministic approximations to accelerate SMC for posterior sampling. Preprint, arXiv 170707971 (2017)
57. J.A. Doornik, Autometrics, in *The Methodology and Practice of Econometrics, A Festschrift in Honour of David F. Hendry*, University Press, pp. 88–121 (2009)
58. J.A. Doornik, D.F. Hendry, Statistical model selection with "big data". Cogent Econ. Finan. **3**(1), 1045216 (2015)
59. C.C. Drovandi, C. Grazian, K. Mengersen, C. Robert, Approximating the likelihood in ABC, in *Handbook of Approximate Bayesian Computation*, ed. by S.A. Sisson, Y. Fan, M. Beaumont (Chapman and Hall/CRC, Boca Raton, 2018), pp. 321–368
60. P. Ducange, R. Pecori, P. Mezzina, A glimpse on big data analytics in the framework of marketing strategies. Soft Comput. **22**(1), 325–342 (2018)
61. D.B. Dunson, Statistics in the big data era: failures of the machine. Stat. Probab. Lett. **136**, 4–9 (2018)
62. R. Dutta, M. Schoengens, J.P. Onnela, A. Mira, Abcpy, in *Proceedings of the Platform for Advanced Scientific Computing Conference on - PASC* (2017)
63. C.K. Emani, N. Cullot, C. Nicolle, Understandable big data: a survey. Comput. Sci. Rev. **17**, 70–81 (2015)
64. A. Fahad, N. Alshatri, Z. Tari, A. Alamri, I. Khalil, A.Y. Zomaya, S. Foufou, A. Bouras, A survey of clustering algorithms for big data: taxonomy and empirical analysis. IEEE Trans. Emerg. Top. Comput. **2**(3), 267–279 (2014)
65. J. Fan, F. Han, H. Liu, Challenges of big data analysis. Natl. Sci. Rev. **1**(2), 293–314 (2014)
66. S. Fosso Wamba, D. Mishra, Big data integration with business processes: a literature review. Bus. Process Manag. J. **23**(3), 477–492 (2017)
67. B. Franke, J.F. Plante, R. Roscher, A. Lee, C. Smyth, A. Hatefi, F. Chen, E. Gil, A. Schwing, A. Selvitella et al., Statistical inference, learning and models in big data. Int. Stat. Rev. **84**(3), 371–389 (2016)
68. D.T. Frazier, G.M. Martin, C.P. Robert, J. Rousseau, Asymptotic properties of approximate Bayesian computation. Biometrika **105**(3), 593–607 (2018)
69. Y. Gal, M. Van Der Wilk, C.E. Rasmussen, Distributed variational inference in sparse Gaussian process regression and latent variable models, in *Advances in Neural Information Processing Systems* (2014), pp. 3257–3265
70. A. Gandomi, M. Haider, Beyond the hype: Big data concepts, methods, and analytics. Int. J. Inf. Manag. **35**(2), 137–144 (2015)
71. H. Ge, Y. Chen, M. Wan, Z. Ghahramani, Distributed inference for Dirichlet process mixture models, in *International Conference on Machine Learning* (2015), pp. 2276–2284

72. R. Genuer, J.M. Poggi, Tuleau-Malot C, N. Villa-Vialaneix, Random forests for big data. Big Data Res. **9**, 28–46 (2017)
73. Z. Ghahramani, Bayesian non-parametrics and the probabilistic approach to modelling. Phil. Trans. R. Soc. A. **371**(1984), 20110553 (2013)
74. Z. Ghahramani, T.L. Griffiths, Infinite latent feature models and the Indian buffet process, in *Advances in Neural Information Processing Systems* (2006), pp. 475–482
75. P. Gloaguen, M.P. Etienne, S. Le Corff Online sequential Monte Carlo smoother for partially observed diffusion processes. URASIP J. Adv. Signal Process. **2018**(1), 9 (2018)
76. S. Guha, R. Hafen, J. Rounds, J. Xia, J. Li, B. Xi, W.S. Cleveland, Large complex data: divide and recombine (D&R) with RHIPE. Stat **1**(1), 53–67 (2012)
77. R. Guhaniyogi, S. Banerjee, Meta-Kriging: scalable Bayesian modeling and inference for massive spatial datasets. Technometrics **60**(4), 430–444 (2018)
78. R. Guhaniyogi, S. Banerjee, Multivariate spatial meta kriging. Stat. Probab. Lett. **144**, 3–8 (2019)
79. R. Guhaniyogi, S. Qamar, D.B. Dunson, Bayesian conditional density filtering for big data. Stat **1050**, 15 (2014)
80. D. Gunawan, R. Kohn, M. Quiroz, K.D. Dang, M.N. Tran, Subsampling Sequential Monte Carlo for Static Bayesian Models. Preprint, arXiv:180503317 (2018)
81. H. Hassani, E.S. Silva, Forecasting with big data: a review. Ann. Data Sci. **2**(1), 5–19 (2015)
82. S.I. Hay, D.B. George, C.L. Moyes, J.S. Brownstein, Big data opportunities for global infectious disease surveillance. PLoS Med. **10**(4), e1001413 (2013)
83. M.J. Heaton, A. Datta, A. Finley, R. Furrer, R. Guhaniyogi, F. Gerber, R.B. Gramacy, D. Hammerling, M. Katzfuss, F. Lindgren et al., Methods for analyzing large spatial data: a review and comparison. Preprint, arXiv:171005013 (2017)
84. J. Hensman, N. Fusi, N.D. Lawrence, Gaussian processes for big data. Preprint, arXiv:13096835 (2013)
85. J. Hensman, A.G.d.G. Matthews, Z. Ghahramani, Scalable variational Gaussian process classification, in *18th International Conference on Artificial Intelligence and Statistics (AISTATS)* (2015), pp. 351–360
86. M. Hilbert, Big data for development: a review of promises and challenges. Dev. Policy Rev. **34**(1), 135–174 (2016)
87. R.W. Hoerl, R.D. Snee, R.D. De Veaux, Applying statistical thinking to "Big Data" problems. Wiley Interdiscip. Rev. Comput. Stat. **6**(4), 222–232 (2014)
88. M.D. Hoffman, D.M. Blei, C. Wang, J. Paisley, Stochastic variational inference. J. Mach. Learn. Res. **14**(1), 1303–1347 (2013)
89. H.H. Huang, H. Liu, Big data machine learning and graph analytics: Current state and future challenges, in *2014 IEEE International Conference on Big Data (Big Data)* (IEEE, Piscataway, 2014), pp. 16–17
90. R. Izbicki, A.B. Lee, T. Pospisil, ABC–CDE: toward approximate Bayesian computation with complex high-dimensional data and limited simulations. J. Comput. Graph. Stat. **28**, 1–20 (2019)
91. G. Jifa, Z. Lingling, Data, DIKW, big data and data science. Procedia Comput. Sci. **31**, 814–821 (2014)
92. S. Kaisler, F. Armour, J.A. Espinosa, W. Money, Big data: issues and challenges moving forward, in *2013 46th Hawaii International Conference on System Sciences* (IEEE, Piscataway, 2013), pp. 995–1004
93. A. Kapelner, J. Bleich bartMachine: machine learning with Bayesian additive regression trees. Preprint, arXiv:13122171 (2013)
94. V.D. Katkar, S.V. Kulkarni, A novel parallel implementation of Naive Bayesian classifier for big data, in *2013 International Conference on Green Computing, Communication and Conservation of Energy (ICGCE)* (IEEE, Piscataway, 2013), pp. 847–852
95. A. Korattikara, Y. Chen, M. Welling, Austerity in MCMC land: Cutting the Metropolis-Hastings budget, in *International Conference on Machine Learning* (2014), pp. 181–189

96. H. Kousar, B.P. Babu, Multi-Agent based MapReduce Model for Efficient Utilization of System Resources. Indones. J. Electr. Eng. Comput. Sci. **11**(2), 504–514 (2018)
97. S. Landset, T.M. Khoshgoftaar, A.N. Richter, T. Hasanin, A survey of open source tools for machine learning with big data in the hadoop ecosystem. J. Big Data **2**(1), 24 (2015)
98. G.J. Lasinio, G. Mastrantonio, A. Pollice, Discussing the "big n problem". Stat. Methods Appt. **22**(1), 97–112 (2013)
99. N.A. Lazar, Bayesian empirical likelihood. Biometrika **90**(2), 319–326 (2003)
100. A. Lee, N. Whiteley, Forest resampling for distributed sequential Monte Carlo. Stat. Anal. Data Min. **9**(4), 230–248 (2016)
101. A. Lee, C. Yau, M.B. Giles, A. Doucet, C.C. Holmes, On the utility of graphics cards to perform massively parallel simulation of advanced Monte Carlo methods. J. Comput. Graph. Stat. **19**(4), 769–789 (2010)
102. X.J. Lee, M. Hainy, McKeone JP, C.C. Drovandi, A.N. Pettitt, ABC model selection for spatial extremes models applied to South Australian maximum temperature data. Comput. Stat. Data Anal. **128**, 128–144 (2018)
103. S. Li, S. Dragicevic, F.A. Castro, M. Sester, S. Winter, A. Coltekin, C. Pettit, B. Jiang, J. Haworth, A. Stein et al., Geospatial big data handling theory and methods: a review and research challenges. ISPRS J. Photogramm. Remote Sens. **115**, 119–133 (2016)
104. D. Lin, Online learning of nonparametric mixture models via sequential variational approximation, in *Advances in Neural Information Processing Systems* (2013), pp. 395–403
105. F. Lindsten, A.M. Johansen, C.A. Naesseth, B. Kirkpatrick, T.B. Schön, J. Aston, A. Bouchard-Côté, Divide-and-conquer with sequential Monte Carlo. J. Comput. Graph. Stat. **26**(2), 445–458 (2017)
106. A.R. Linero, Bayesian regression trees for high-dimensional prediction and variable selection. J. Am. Stat. Assoc. **113**, 1–11 (2018)
107. B. Liquet, K. Mengersen, A. Pettitt, M. Sutton et al., Bayesian variable selection regression of multivariate responses for group data. Bayesian Anal. **12**(4), 1039–1067 (2017)
108. L. Liu, Computing infrastructure for big data processing. Front. Comput. Sci. **7**(2), 165–170 (2013)
109. Q. Liu, D. Wang, Stein variational gradient descent: a general purpose Bayesian inference algorithm, in *Advances In Neural Information Processing Systems* (2016), pp. 2378–2386
110. B. Liu, E. Blasch, Y. Chen, D. Shen, G. Chen, Scalable sentiment classification for big data analysis using Naive Bayes classifier, in *2013 IEEE International Conference on Big Data* (IEEE, Piscataway, 2013), pp. 99–104
111. Z. Liu, F. Sun, D.P. McGovern, Sparse generalized linear model with L0 approximation for feature selection and prediction with big omics data. BioData Min. **10**(1), 39 (2017)
112. Y. Liu, V. Ročková, Y. Wang, ABC variable selection with Bayesian forests. Preprint, arXiv:180602304 (2018)
113. C. Loebbecke, A. Picot, Reflections on societal and business model transformation arising from digitization and big data analytics: a research agenda. J. Strategic Inf. Syst. **24**(3), 149–157 (2015)
114. J. Luo, M. Wu, D. Gopukumar, Y. Zhao, Big data application in biomedical research and health care: a literature review. Biomed. Inform. Insights **8**, BII–S31559 (2016)
115. Z. Ma, P.K. Rana, J. Taghia, M. Flierl, A. Leijon, Bayesian estimation of Dirichlet mixture model with variational inference. Pattern Recognit. **47**(9), 3143–3157 (2014)
116. D. Maclaurin, R.P. Adams, Firefly Monte Carlo: exact MCMC with subsets of data, in *Twenty-Fourth International Joint Conference on Artificial Intelligence* (2014), pp. 543–552
117. T. Magdon-Ismail, C. Narasimhadevara, D. Jaffe, R. Nambiar, Tpcx-hs v2: transforming with technology changes, in *Technology Conference on Performance Evaluation and Benchmarking* (Springer, Berlin, 2017), pp. 120–130
118. L. Mählmann, M. Reumann, N. Evangelatos, A. Brand, Big data for public health policy-making: policy empowerment. Public Health Genomics **20**(6), 312–320 (2017)
119. F. Maire, N. Friel, P. Alquier, Informed sub-sampling MCMC: approximate Bayesian inference for large datasets. Stat. Comput. 1–34 (2017). https://doi.org/10.1007/s11222-018-9817-3

120. R. Manibharathi, R. Dinesh, Survey of challenges in encrypted data storage in cloud computing and big data. J. Netw. Commun. Emerg. Technol. **8**(2) (2018). ISSN:2395-5317
121. R.F. Mansour, Understanding how big data leads to social networking vulnerability. Comput. Hum. Behav. **57**, 348–351 (2016)
122. A. Marshall, S. Mueck, R. Shockley, How leading organizations use big data and analytics to innovate. Strateg. Leadersh. **43**(5), 32–39 (2015)
123. T.H. McCormick, R. Ferrell, A.F. Karr, P.B. Ryan, Big data, big results: knowledge discovery in output from large-scale analytics. Stat. Anal. Data Min. **7**(5), 404–412 (2014)
124. C.A. McGrory, D. Titterington, Variational approximations in Bayesian model selection for finite mixture distributions. Comput. Stat. Data Anal. **51**(11), 5352–5367 (2007)
125. T.J. McKinley, I. Vernon, I. Andrianakis, N. McCreesh, J.E. Oakley, R.N. Nsubuga, M. Goldstein, R.G. White et al., Approximate Bayesian computation and simulation-based inference for complex stochastic epidemic models. Stat. Sci. **33**(1), 4–18 (2018)
126. E. Meeds, M. Welling, GPS-ABC: Gaussian process surrogate approximate Bayesian computation. Preprint, arXiv:14012838 (2014)
127. K.L. Mengersen, P. Pudlo, C.P. Robert, Bayesian computation via empirical likelihood. Proc. Natl. Acad. Sci. **110**(4), 1321–1326 (2013)
128. S. Minsker, S. Srivastava, L. Lin, D.B. Dunson, Robust and scalable Bayes via a median of subset posterior measures. J. Mach. Learn. Res. **18**(1), 4488–4527 (2017)
129. M.T. Moores, C.C. Drovandi, K. Mengersen, C.P. Robert, Pre-processing for approximate Bayesian computation in image analysis. Stat. Comput. **25**(1), 23–33 (2015)
130. N. Moustafa, G. Creech, E. Sitnikova, M. Keshk, Collaborative anomaly detection framework for handling big data of cloud computing, in *Military Communications and Information Systems Conference (MilCIS), 2017* (IEEE, Piscataway, 2017), pp. 1–6
131. P. Müller, F.A. Quintana, A. Jara, T. Hanson, *Bayesian Nonparametric Data Analysis* (Springer, Berlin, 2015)
132. O. Müller, I. Junglas, J.v. Brocke, S. Debortoli, Utilizing big data analytics for information systems research: challenges, promises and guidelines. Eur. J. Inf. Syst. **25**(4), 289–302 (2016)
133. C.A. Naesseth, S.W. Linderman, R. Ranganath, D.M. Blei, Variational sequential Monte Carlo. Preprint, arXiv:170511140 (2017)
134. W. Neiswanger, C. Wang, E. Xing, Asymptotically exact, embarrassingly parallel MCMC. Preprint, arXiv:13114780 (2013)
135. Y. Ni, P. Müller, M. Diesendruck, S. Williamson, Y. Zhu, Y. Ji Scalable Bayesian nonparametric clustering and classification. J. Comput. Graph. Stat. 1–45 (2019). https://doi.org/10.1080/10618600.2019.1624366
136. L.G. Nongxa, Mathematical and statistical foundations and challenges of (big) data sciences. S. Afr. J. Sci. **113**(3–4), 1–4 (2017)
137. B. Oancea, R.M. Dragoescu et al., Integrating R and hadoop for big data analysis. Romanian Stat. Rev. **62**(2), 83–94 (2014)
138. Z. Obermeyer, E.J. Emanuel, Predicting the future—big data, machine learning, and clinical medicine. N. Engl. J. Med. **375**(13), 1216 (2016)
139. A. O'Driscoll, J. Daugelaite, R.D. Sleator, 'Big data', Hadoop and cloud computing in genomics. J. Biomed. Inform. **46**(5), 774–781 (2013)
140. D. Oprea, Big questions on big data. Rev. Cercet. Interv. Soc. **55**, 112 (2016)
141. A.B. Owen, *Empirical Likelihood* (Chapman and Hall/CRC, Boca Raton, 2001)
142. S. Pandey, V. Tokekar, Prominence of mapreduce in big data processing, in *2014 Fourth International Conference on Communication Systems and Network Technologies (CSNT)* (IEEE, Piscataway, 2014), pp. 555–560
143. A.Ç. Pehlivanlı, A novel feature selection scheme for high-dimensional data sets: four-staged feature selection. J. Appl. Stat. **43**(6), 1140–1154 (2015)
144. D.N. Politis, J.P. Romano, M. Wolf, *Subsampling* (Springer Science & Business Media, New York, 1999)

145. A.T. Porter, S.H. Holan, C.K. Wikle, Bayesian semiparametric hierarchical empirical likelihood spatial models. J. Stat. Plan. Inference **165**, 78–90 (2015)
146. A.T. Porter, S.H. Holan, C.K. Wikle, Multivariate spatial hierarchical Bayesian empirical likelihood methods for small area estimation. Stat **4**(1), 108–116 (2015)
147. P. Pudlo, J.M. Marin, A. Estoup, J.M. Cornuet, M. Gautier, C.P. Robert, Reliable ABC model choice via random forests. Bioinformatics **32**(6), 859–866 (2015)
148. F. Qi, F. Yang, Analysis of large data mining platform based on cloud computing, in *2018 4th World Conference on Control Electronics and Computer Engineering* (2018)
149. J. Qiu, Q. Wu, G. Ding, Y. Xu, S. Feng, A survey of machine learning for big data processing. EURASIP J. Adv. Signal Process. **2016**(1), 67 (2016)
150. M. Quiroz, M. Villani, R. Kohn, Scalable MCMC for large data problems using data subsampling and the difference estimator. SSRN Electron. J. (2015). arXiv:1507.02971
151. M. Quiroz, R. Kohn, M. Villani, M.N. Tran, Speeding up MCMC by efficient data subsampling. J. Am. Stat. Assoc. 1–13 (2018). https://doi.org/10.1080/01621459.2018.1448827
152. M. Rabinovich, E. Angelino, M.I. Jordan, Variational consensus Monte Carlo, in *Advances in Neural Information Processing Systems* (2015), pp. 1207–1215
153. W. Raghupathi, V. Raghupathi, Big data analytics in healthcare: promise and potential. Health Inf. Sci. Syst. **2**(1), 3 (2014)
154. E. Raguseo, Big data technologies: an empirical investigation on their adoption, benefits and risks for companies. Int. J. Inf. Manag. **38**(1), 187–195 (2018)
155. C.E. Rasmussen, The infinite Gaussian mixture model, in *Advances in Neural Information Processing Systems* (2000), pp. 554–560
156. C.E. Rasmussen, Gaussian processes in machine learning, in *Advanced Lectures on Machine Learning* (Springer, Berlin, 2004), pp. 63–71
157. V. Rocková, S. van der Pas, Posterior concentration for Bayesian regression trees and forests. Ann. Stat. (in revision) 1–40 (2017). arXiv:1708.08734
158. J. Roski, G.W. Bo-Linn, T.A. Andrews, Creating value in health care through big data: opportunities and policy implications. Health Aff. **33**(7), 1115–1122 (2014)
159. J.S. Rumsfeld, K.E. Joynt, T.M. Maddox, Big data analytics to improve cardiovascular care: promise and challenges. Nat. Rev. Cardiol. **13**(6), 350–359 (2016)
160. S. Sagiroglu, D. Sinanc, Big data: a review, in *2013 International Conference on Collaboration Technologies and Systems (CTS)* (IEEE, Piscataway, 2013), pp. 42–47
161. S.M. Schennach, Bayesian exponentially tilted empirical likelihood. Biometrika **92**(1), 31–46 (2005)
162. E.D. Schifano, J. Wu, C. Wang, J. Yan, M.H. Chen, Online updating of statistical inference in the big data setting. Technometrics **58**(3), 393–403 (2016)
163. S.L. Scott, A.W. Blocker, F.V. Bonassi, H.A. Chipman, E.I. George, R.E. McCulloch (2016) Bayes and big data: The consensus Monte Carlo algorithm. Int. J. Manag. Sci. Eng. Manag. **11**(2), 78–88
164. D.V. Shah, J.N. Cappella, W.R. Neuman, Big data, digital media, and computational social science: possibilities and perils. Ann. Am. Acad. Pol. Soc. Sci. **659**(1), 6–13 (2015)
165. A. Siddiqa, A. Karim, A. Gani, Big data storage technologies: a survey. Front. Inf. Technol. Electron. Eng. **18**(8), 1040–1070 (2017)
166. P. Singh, A. Hellander, Multi-statistic Approximate Bayesian Computation with multi-armed bandits. Preprint, arXiv:180508647 (2018)
167. S. Sisson, Y. Fan, M. Beaumont, Overview of ABC, in *Handbook of Approximate Bayesian Computation* (Chapman and Hall/CRC, New York, 2018), pp. 3–54
168. U. Sivarajah, M.M. Kamal, Z. Irani, V. Weerakkody, Critical analysis of big data challenges and analytical methods. J. Bus. Res. **70**, 263–286 (2017)
169. S. Srivastava, C. Li, D.B. Dunson, Scalable Bayes via barycenter in Wasserstein space. J. Mach. Learn. Res. **19**(1), 312–346 (2018)
170. H. Strathmann, D. Sejdinovic, M. Girolami, Unbiased Bayes for big data: paths of partial posteriors. Preprint, arXiv:150103326 (2015)

171. M.A. Suchard, Q. Wang, C. Chan, J. Frelinger, A. Cron, M. West, Understanding GPU programming for statistical computation: studies in massively parallel massive mixtures. J. Comput. Graph. Stat. **19**(2), 419–438 (2010)

172. Z. Sun, L. Sun, K. Strang, Big data analytics services for enhancing business intelligence. J. Comput. Inf. Syst. **58**(2), 162–169 (2018)

173. S. Suthaharan, Big data classification: problems and challenges in network intrusion prediction with machine learning. ACM SIGMETRICS Perform. Eval. Rev. **41**(4), 70–73 (2014)

174. O. Sysoev, A. Grimvall, O. Burdakov, Bootstrap confidence intervals for large-scale multivariate monotonic regression problems. Commun. Stat. Simul. Comput. **45**(3), 1025–1040 (2014)

175. D. Talia, Clouds for scalable big data analytics. Computer **46**(5), 98–101 (2013)

176. Y. Tang, Z. Xu, Y. Zhuang, Bayesian network structure learning from big data: a reservoir sampling based ensemble method, in *International Conference on Database Systems for Advanced Applications* (Springer, Berlin, 2016), pp. 209–222

177. A. Tank, N. Foti, E. Fox, Streaming variational inference for Bayesian nonparametric mixture models, in *Artificial Intelligence and Statistics* (2015), pp. 968–976

178. Y.W. Teh, A.H. Thiery, S.J. Vollmer, Consistency and fluctuations for stochastic gradient Langevin dynamics. J. Mach. Learn. Res. **17**(1), 193–225 (2016)

179. D. Tran, R. Ranganath, D.M. Blei, The variational Gaussian process. Preprint, arXiv:151106499 (2015)

180. N. Tripuraneni, S. Gu, H. Ge, Z. Ghahramani, Particle Gibbs for infinite hidden Markov models, in *Advances in Neural Information Processing Systems* (2015), pp. 2395–2403

181. S. van der Pas, V. Rockova, Bayesian dyadic trees and histograms for regression, in *Advances in Neural Information Processing Systems* (2017), pp. 2089–2099

182. M. Viceconti, P. Hunter, R. Hose, Big data, big knowledge: big data for personalized healthcare. IEEE J. Biomed. Health Inform. **19**(4), 1209–1215 (2015)

183. A. Vyas, S. Ram, Comparative study of MapReduce frameworks in big data analytics. Int. J. Mod. Comput. Sci. **5**(Special Issue), 5–13 (2017)

184. S.F. Wamba, S. Akter, A. Edwards, G. Chopin, D. Gnanzou, How "big data" can make big impact: findings from a systematic review and a longitudinal case study. Int. J. Prod. Econ. **165**, 234–246 (2015)

185. X.F. Wang, Fast clustering using adaptive density peak detection. Stat. Methods Med. Res. **26**(6), 2800–2811 (2015)

186. L. Wang, D.B. Dunson, Fast Bayesian inference in Dirichlet process mixture models. J. Comput. Graph. Stat. **20**(1), 196–216 (2011)

187. X. Wang, D.B. Dunson, Parallelizing MCMC via weierstrass sampler. Preprint, arXiv:13124605 (2013)

188. T. Wang, R.J. Samworth, High dimensional change point estimation via sparse projection. J. R. Stat. Soc. Ser. B (Stat Methodol.) **80**(1), 57–83 (2017)

189. C. Wang, J. Paisley, D. Blei, Online variational inference for the hierarchical Dirichlet process, in *Proceedings of the Fourteenth International Conference on Artificial Intelligence and Statistics* (2011), pp. 752–760

190. J. Wang, Y. Tang, M. Nguyen, I. Altintas, A scalable data science workflow approach for big data Bayesian network learning, in *2014 IEEE/ACM Int Symp. Big Data Comput.* (IEEE, Piscataway, 2014), pp. 16–25

191. C. Wang, M.H. Chen, E. Schifano, J. Wu, J. Yan, Statistical methods and computing for big data. Stat. Interface **9**(4), 399–414 (2016)

192. C. Wang, M.H. Chen, J. Wu, J. Yan, Y. Zhang, E. Schifano, Online updating method with new variables for big data streams. Can. J. Stat. **46**(1), 123–146 (2017)

193. H.J. Watson, Tutorial: big data analytics: concepts, technologies, and applications. Commun. Assoc. Inf. Syst. **34**, 65 (2014)

194. Y. Webb-Vargas, S. Chen, A. Fisher, A. Mejia, Y. Xu, C. Crainiceanu, B. Caffo, M.A. Lindquist, Big data and neuroimaging. Stat. Biosci. **9**(2), 543–558 (2017)

195. S. White, T. Kypraios, S.P. Preston, Piecewise Approximate Bayesian Computation: fast inference for discretely observed Markov models using a factorised posterior distribution. Stat. Comput. **25**(2), 289–301 (2015)
196. R. Wilkinson, Accelerating ABC methods using Gaussian processes, in *Artificial Intelligence and Statistics* (2014), pp. 1015–1023
197. S. Williamson, A. Dubey, E.P. Xing, Parallel Markov chain Monte Carlo for nonparametric mixture models, in *Proceedings of the 30th International Conference on Machine Learning (ICML-13)* (2013), pp. 98–106
198. A.F. Wise, D.W. Shaffer, Why theory matters more than ever in the age of big data. J. Learn. Anal. **2**(2), 5–13 (2015)
199. C. Wu, C.P. Robert, Average of recentered parallel MCMC for big data. Preprint, arXiv:170604780 (2017)
200. X.G. Xia, Small data, mid data, and big data versus algebra, analysis, and topology. IEEE Signal Process. Mag. **34**(1), 48–51 (2017)
201. C. Yang, Q. Huang, Z. Li, K. Liu, F. Hu, Big data and cloud computing: innovation opportunities and challenges. Int. J. Digit Earth **10**(1), 13–53 (2017)
202. C. Yoo, L. Ramirez, J. Liuzzi, Big data analysis using modern statistical and machine learning methods in medicine. Int. Neurourol. J. **18**(2), 50 (2014)
203. L. Yu, N. Lin, ADMM for penalized quantile regression in big data. Int. Stat. Rev. **85**(3), 494–518 (2017)
204. T. Zhang, B. Yang, An exact approach to ridge regression for big data. Comput. Stat. **32**, 1–20 (2017)
205. X. Zhang, C. Liu, S. Nepal, C. Yang, W. Dou, J. Chen, A hybrid approach for scalable sub-tree anonymization over big data using MapReduce on cloud. J. Comput. Syst. Sci. **80**(5), 1008–1020 (2014)
206. Y. Zhang, T. Cao, S. Li, X. Tian, L. Yuan, H. Jia, A.V. Vasilakos, Parallel processing systems for big data: a survey. Proc. IEEE **104**(11), 2114–2136 (2016)
207. Z. Zhang, K.K.R. Choo, B.B. Gupta, The convergence of new computing paradigms and big data analytics methodologies for online social networks. J. Comput. Sci. **26**, 453–455 (2018)
208. L. Zhang, A. Datta, S. Banerjee, Practical Bayesian modeling and inference for massive spatial data sets on modest computing environments. Stat. Anal. Data Min. **12**(3), 197–209 (2019)
209. L. Zhou, S. Pan, J. Wang, A.V. Vasilakos, Machine learning on big data: Opportunities and challenges. Neurocomputing **237**, 350–361 (2017)
210. J. Zhu, J. Chen, W. Hu, B. Zhang, Big learning with Bayesian methods. Natl. Sci. Rev. **4**(4), 627–651 (2017)
211. G. Zoubin, Scaling the Indian Buffet process via submodular maximization, in *International Conference on Machine Learning* (2013), pp. 1013–1021

Chapter 3
Bayesian Neural Networks:
An Introduction and Survey

Ethan Goan and Clinton Fookes

Abstract Neural Networks (NNs) have provided state-of-the-art results for many challenging machine learning tasks such as detection, regression and classification across the domains of computer vision, speech recognition and natural language processing. Despite their success, they are often implemented in a frequentist scheme, meaning they are unable to reason about uncertainty in their predictions. This article introduces Bayesian Neural Networks (BNNs) and the seminal research regarding their implementation. Different approximate inference methods are compared, and used to highlight where future research can improve on current methods.

3.1 Introduction

Biomimicry has long served as a basis for technological developments. Scientists and engineers have repeatedly used knowledge of the physical world to emulate nature's elegant solutions to complex problems which have evolved over billions of years. An important example of biomimicry in statistics and machine learning has been the development of the perceptron [1], which proposes a mathematical model based on the physiology of a neuron. The machine learning community has used this concept[1] to develop statistical models of highly interconnected arrays of neurons to create Neural Networks (NNs).

Though the concept of NNs has been known for many decades, it is only recently that applications of these network have seen such prominence. The lull in research and development for NNs was largely due to three key factors: lack of sufficient algorithms to train these networks, the large amount of data required to train

---

[1] While also relaxing many of the constraints imposed by a physical model of a natural neuron [2].

E. Goan (✉) · C. Fookes
Queensland University of Technology, Brisbane, QLD, Australia
e-mail: ej.goan@qut.edu.au

K. L. Mengersen et al. (eds.), *Case Studies in Applied Bayesian Data Science*,
Lecture Notes in Mathematics 2259, https://doi.org/10.1007/978-3-030-42553-1_3

complex networks and the large amount of computing resources required during the training process. In 1986, Rumelhart et al. [3] introduced the backpropagation algorithm to address the problem of efficient training for these networks. Though an efficient means of training was available, considerable compute resources was still required for the ever increasing size of new networks. This problem was addressed in [4–6] where it was shown that general purpose GPUs could be used to efficiently perform many of the operations required for training. As digital hardware continued to advance, the number of sensors able to capture and store real world data increased. With efficient training methods, improved computational resources and large data sets, training of complex NNs became truly feasible.

In the vast majority of cases, NNs are used within a frequentist perspective; using available data, a user defines a network architecture and cost function, which is then optimised to allow us to gain point estimate predictions. Problems arise from this interpretation of NNs. Increasing the number of parameters (often called weights in machine learning literature), or the depth of the model increases the capacity of the network, allowing it to represent functions with greater non-linearities. This increase in capacity allows for more complex tasks to be addressed with NNs, though when frequentist methodologies are applied, leaves them highly prone to overfitting to the training data. The use of large data sets and regularisation methods such as finding a MAP estimate can limit the complexity of functions learnt by the networks and aid in avoiding overfitting.

Neural Networks have provided state-of-the-art results for numerous machine learning and Artificial intelligence (AI) applications, such as image classification [6–8], object detection [9–11] and speech recognition [12–15]. Other networks such as the AlphaGo model developed by DeepMind [16] have emphasised the potential of NNs for developing AI systems, garnering a wide audience interested in the development of these networks. As the performance of NNs has continued to increase, the interest in their development and adoption by certain industries becomes more prominent. NNs are currently used in manufacturing [17], asset management [18] and human interaction technologies [19, 20].

Since the deployment of NNs in industry, there have been a number of incidents where failings in these systems has led to models acting unethically and unsafely. This includes models demonstrating considerable gender and racial bias against marginalised groups [21–23] or to more extreme cases resulting in loss of life [24, 25]. NNs are a statistical black-box models, meaning that the decision process is not based on a well-defined and intuitive protocol. Instead decisions are made in an uninterpretable manner, with hopes that the reasonable decisions will be made based on previous evidence provided in training data.[2] As such, the implementation of these systems in social and safety critical environments raises considerable ethical concerns. The European Union released a new regulation[3] which effectively states

---

[2]Due to this black-box nature, the performance of these models is justified entirely through empirical means.

[3]This regulation came into effect on the 25th of May, 2018 across the EU [26].

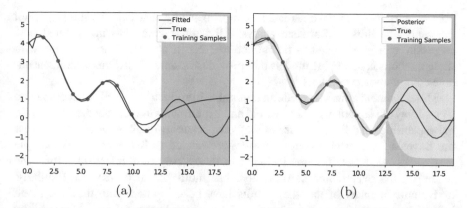

Fig. 3.1 Comparison of neural network to traditional probabilistic methods for a regression task, with no training data in the purple region. (**a**) Regression output using a neural network with 2 hidden layers; (**b**) Regression using a Gaussian Process framework, with grey bar representing ±2 std. from expected value

that users have a "right to an explanation" regarding decisions made by AI systems [26, 27]. Without clear understanding of their operation or principled methods for their design, experts from other domains remain apprehensive about the adoption of current technology [28–30]. These limitations have motivated research efforts into the field of Explainable AI [31].

Adequate engineering of NNs requires a sound understanding of their capabilities and limitations; to identify their shortcomings prior to deployment as apposed to the current practice of investigating these limitations in the wake of these tragedies. With NNs being a statistical black-box, interpretation and explanation of the decision making process eludes current theory. This lack of interpretation and over-confident estimates provided by the frequentist perspective of common NNs makes them unsuitable for high risk domains such as medical diagnostics and autonomous vehicles. Bayesian statistics offers natural way to reason about uncertainty in predictions, and can provide insight into how these decisions are made.

Figure 3.1 compares Bayesian methods for performing regression with that of a simple neural network, and illustrates the importance of measuring uncertainty. While both methods perform well within the bounds of the training data, where extrapolation is required, the probabilistic method provides a full distribution of the function output as opposed to the point estimates provided by the NN. The distribution over outputs provided by probabilistic methods allows for the development of trustworthy models, in that they can identify uncertainty in a prediction. Given that NNs are the most promising model for generating AI systems, it is important that we can similarly trust their predictions.

A Bayesian perspective allows us to address many of the challenges currently faced within NNs. To do this, a distribution is placed over the network parameters, and the resulting network is then termed a Bayesian Neural Network (BNN). The goal of a BNN is to have a model of high capacity that exhibits the important theoretical benefits of Bayesian analysis. Recent research has investigated how Bayesian approximations can be applied to NNs in practice. The challenge with these methods is deploying models that provide accurate predictions within reasonable computation constraints.[4]

This document aims to provide an accessible introduction to BNNs, accompanied by a survey of seminal works in the field and experiments to motivate discussion into the capabilities and limits of current methods. A survey of all research items across the Bayesian and machine learning literature related to BNNs could fill multiple text books. As a result, items included in this survey only intend to inform the reader on the overarching narrative that has motivated their research. Similarly, derivations of many of they key results have been omitted, with the final result being listed accompanied by reference to the original source. Readers inspired by this exciting research area are encouraged to consult prior surveys: [32] which surveys the early developments in BNNs, [33] which discusses the specifics of a full Bayesian treatment for NNs, and [34] which surveys applications of approximate Bayesian inference to modern network architectures.

This document should be suitable for all in the statistics field, though the primary audience of interest are those more familiar with machine learning concepts. Despite seminal references for new machine learning scholars almost equivalently being Bayesian texts [2, 35], in practice there has been a divergence between much of the modern machine learning and Bayesian statistics research. It is hoped that this survey will help highlight similarities between some modern research in BNNs and statistics, to emphasis the importance of a probabilistic perspective within machine learning and to encourage future collaboration/unison between the machine learning and statistics fields.

3.2 Literature Survey

3.2.1 Neural Networks

Before discussing a Bayesian perspective of NNs, it is important to briefly survey the fundamentals of neural computation and to define the notation to be used throughout the chapter. This survey will focus on the primary network structure of interest, the Multi-Layer Perceptron (MLP) network. The MLP serves as the basis for NNs, with modern architectures such as convolutional networks having an equivalent MLP representation. Figure 3.2 illustrates a simple MLP with a single hidden layer suitable for regression or classification. For this network with an input

---

[4]The term "reasonable" largely depends on the context. Many neural networks are currently trained using some of the largest computing facilities available, containing thousands of GPU devices.

Fig. 3.2 Example of a NN architecture with a single hidden layer for either binary classification or 1-D regression. Each node represents a neuron or a state where the summation and activation of input states is performed. Arrows are the parameters (weights) indicating the strength of connection between neurons

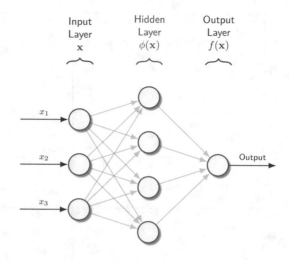

$\mathbf{x}$ of dimension N_1, the output of the $\mathbf{f}$ network can be modelled as,

$$\phi_j = \sum_{i=1}^{N_1} a(x_i w_{ij}^1), \tag{3.1}$$

$$f_k = \sum_{j=1}^{N_2} g(\phi_j w_{jk}^2). \tag{3.2}$$

The parameters w represent the weighted connection between neurons from subsequent layers, and the superscripts denoting the layer number. Equation (3.1) represents the output of the hidden layer, which will be of dimension N_2. The kth output of the network is then a summation over the N_2 outputs from the prior hidden layer. This modelling scheme can be expanded to include many hidden layers, with the input of each layer being the output of the layer immediately prior. A bias value is often added during each layer, though is omitted throughout this chapter in favour of simplicity.

Equation (3.1) refers to the state of each neuron (or node) in the hidden layer. This is expressed as an affine transform followed by a non-linear element wise transform $\phi(\cdot)$, which is often called an activation. For the original perceptron, activation function used was the sign($\cdot$) function, though the use of this function has ceased due to It's derivative being equal to zero.[5] More favourable activation functions such as the Sigmoid, Hyperbolic Tangent (TanH), Rectified Linear Unit (ReLU) and Leaky-ReLU have since replaced this the sign function [36, 37]. Figure 3.3 illustrates these functions along with their corresponding derivatives. When using

---

[5]When the derivative is defined, as is a piece-wise non-differentiable function at the origin.

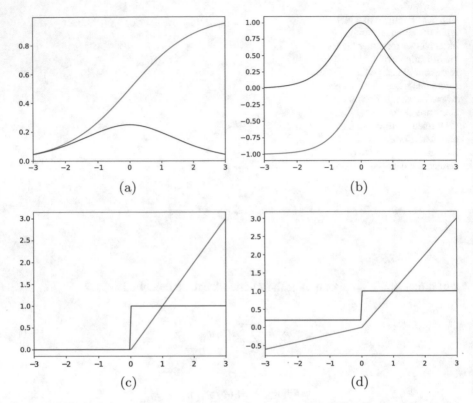

Fig. 3.3 Examples of commonly used activation functions in NNs. The output for each activation is shown in blue and the numerical derivative of each function is shown in red. These functions are (**a**) Sigmoid; (**b**) TanH; (**c**) ReLU; (**d**) Leaky-ReLU. Note the change in scale for the y-axis

the Sigmoid function, expression (3.1) is equivalent to logistic regression, meaning that the output of the network becomes the sum of multiple logistic regression models.

For a regression model, the function applied to the output $g(\cdot)$ will be the identity function,[6] and for binary classification will be a Sigmoid.

Equations (3.1) and (3.2) can be efficiently implemented using matrix representations, and is often represented as such in machine learning literature. This is achieved by stacking the input vector in our data set as a column in $\mathbf{X}$. Forward propagation can then be performed as,

$$\boldsymbol{\Phi} = a(\mathbf{X}^T\mathbf{W}^1), \tag{3.3}$$

$$\mathbf{F} = g(\boldsymbol{\Phi}\mathbf{W}^2). \tag{3.4}$$

---

[6]Meaning no activation is used on the output layer, $g(x) = x$.

Whilst this matrix notation is more concise, the choice to use the summation notation to describe the network here is deliberate. It is hoped that with the summation notation, relations to kernel and statistical theory discussed later in this chapter becomes clearer.

In the frequentist setting of NN learning, a MLE or MAP estimate is found through the minimisation of a non-convex cost function $J(x, y)$ w.r.t. network weights. Minimisation of this cost-function is performed through backpropagation, where the output of the model is computed for the current parameter settings, partial derivatives w.r.t parameters are found and then used to update each parameter,

$$w_{t+i} = w_t - \alpha \frac{\partial J(x, y)}{\partial w_t}. \tag{3.5}$$

Equation (3.5) illustrates how backpropagation updates model parameters, with α representing the learning rate and the subscripts indicate the iteration in the training procedure. Partial derivatives for individual parameters at different layers in the network is found through application of the chain rule. This leads to the preference of discontinuous non-linearities such as the ReLU for deep NNs, as the larger gradient of the ReLU assists in preventing vanishing gradients of early layers during training.

3.2.2 Bayesian Neural Networks

In the frequentist setting presented above, the model weights are not treated as random variables; weights are assumed to have a true value that is just unknown and the data we have seen is treated as a random variable. This may seem counterintuitive for what we want to achieve. We would like to learn what our unknown model weights are based of the information we have at hand. For statistical modelling the information available to us comes in the form of our acquired data. Since we do not know the value for our weights, it seems natural to treat them as a random variable. The Bayesian view of statistics uses this approach; unknown (or latent) parameters are treated as random variables and we want to learn a distribution of these parameters conditional on the what we can observe in the training data.

During the "learning" process of BNNs, unknown model weights are inferred based on what we do know or what we can observe. This is the problem of inverse probability, and is solved through the use of Bayes Theorem. The weights in our model ω are hidden or latent variables; we cannot immediately observe their true distribution. Bayes Theorem allows us to represent a distribution over these weights in terms of probabilities we can observe, resulting in the distribution of model parameters conditional on the data we have seen $p(\omega|\mathcal{D})$,[7] which we call the posterior distribution.

Before training, we can observe the joint distribution between our weights and our data $p(\omega, \mathcal{D})$. This joint distribution is defined by our prior beliefs over our latent variables $p(\omega)$ and our choice of model/likelihood $p(\mathcal{D}|\omega)$,

$$p(\omega, \mathcal{D}) = p(\omega)p(\mathcal{D}|\omega). \tag{3.6}$$

Our choice of network architecture and loss function is used to define the likelihood term in Eq. (3.6). For example, for a 1-D homoscedastic regression problem with a mean squared error loss and a known noise variance, the likelihood is a Gaussian distribution with the mean value specified by the output of the network,

$$p(\mathcal{D}|\omega) = \mathcal{N}\big(\mathbf{f}^{\omega}(\mathcal{D}), \sigma^2\big).$$

Under this modelling scheme, it is typically assumed that all samples from $\mathcal{D}$ are i.i.d., meaning that the likelihood can then be written as a product of the contribution from the N individual terms in the data set,

$$p(\mathcal{D}|\omega) = \prod_{i=1}^{N} \mathcal{N}\big(\mathbf{f}^{\omega}(\mathbf{x}_i), \sigma^2\big). \tag{3.7}$$

Our prior distribution should be specified to incorporate our belief as to how the weights should be distributed, prior to seeing any data. Due to the black-box nature of NNs, specifying a meaningful prior is challenging. In many practical NNs trained under the frequentist scheme, the weights of the trained network have a low magnitude, and are roughly centred around zero. Following this empirical observation, we may use a zero mean Gaussian with a small variance for our prior, or a spike-slab prior centred at zero to encourage sparsity in our model.

With the prior and likelihood specified, Bayes theorem is then applied to yield the posterior distribution over the model weights,

$$\pi(\omega|\mathcal{D}) = \frac{p(\omega)p(\mathcal{D}|\omega)}{\int p(\omega)p(\mathcal{D}|\omega)d\omega} = \frac{p(\omega)p(\mathcal{D}|\omega)}{p(\mathcal{D})}. \tag{3.8}$$

The denominator in the posterior distribution is called the marginal likelihood, or the evidence. This quantity is a constant with respect to the unknown model weights, and normalises the posterior to ensure it is a valid distribution.

---

[7]$\mathcal{D}$ is used here to denote the set of training data $(\mathbf{x}, \mathbf{y})$.

From this posterior distribution, we can perform predictions of any quantity of interest. Predictions are in the form of an expectation with respect to the posterior distribution,

$$\mathbb{E}_\pi[f] = \int f(\boldsymbol{\omega})\pi(\boldsymbol{\omega}|\mathcal{D})d\boldsymbol{\omega}. \tag{3.9}$$

All predictive quantities of interest will be an expectation of this form. Whether it be a predictive mean, variance or interval, the predictive quantity will be an expectation over the posterior. The only change will be in the function $f(\boldsymbol{\omega})$ with which the expectation is applied to. Prediction can then be viewed as an average of the function f weighted by the posterior $\pi(\boldsymbol{\omega})$.

We see that the Bayesian inference process revolves around marginalisation (integration) over our unknown model weights. By using this marginalisation approach, we are able to learn about the generative process of a model, as opposed to an optimisation scheme used in the frequentist setting. With access to this generative model, our predictions are represented in the form of valid conditional probabilities.

In this description, it was assumed that many parameters such as the noise variance σ or any prior parameters were known. This is rarely the case, and as such we need to perform inference for these unknown variables. The Bayesian framework allows us to perform inference over these variables similarly to how we perform inference over our weights; we treat these additional variables as latent variables, assign a prior distribution (or sometimes called a hyper-prior) and then marginalise over them to find our posterior. For more of a description of how this can be performed for BNNs, please refer to [33, 38].

For many models of interest, computation of the posterior (Eq. (3.8)) remains intractable. This is largely due to the computation of the marginal likelihood. For non-conjugate models or those that are non-linear in the latent variables (such as NNs), this quantity can be analytically intractable. For high dimensional models, a quadrature approximation of this integral can become computationally intractable. As a result, approximations for the posterior must be made. The following sections detail how approximate Bayesian inference can be achieved in BNNs.

3.2.3 Origin of Bayesian Neural Networks

From this survey and those conducted prior [70], the first instance of what could be considered a BNN was developed in [39]. This paper emphasises key statistical properties of NNs by developing a statistical interpretation of loss functions used. It was shown that minimisation of a squared error term is equivalent to finding the Maximum Likelihood Estimate (MLE) of a Gaussian. More importantly, it was shown that by specifying a prior over the network weights, Bayes Theorem can be used to obtain an appropriate posterior. Whilst this work provides key insights into the Bayesian perspective of NNs, no means for finding the marginal likelihood (evidence) is supplied, meaning that no practical means for inference is suggested. Denker and LeCun [40] extend on this work, offering a practical means for performing approximate inference using the Laplace approximation, though minimal experimental results are provided.

A NN is a generic function approximator. It is well known that as the limit of the number of parameters approaches infinity in a single hidden layer network, any arbitrary function can be represented [41–43]. This means that for the practical case, our finite training data set can be well approximated by a single layer NN as long as there are sufficient trainable parameters in the model. Similar to high-degree polynomial regression, although we can represent any function and even exactly match the training data in certain cases, as the number of parameters in a NN increases or the degree of the polynomial used increases, the model complexity increases leading to issues of overfitting. This leads to a fundamental challenge found in NN design; how complex should I make my model?

Building on the work of Gull and Skilling [44], MacKay demonstrates how a Bayesian framework naturally lends itself to handle the task of model design and comparison of generic statistical models [45]. In this work, two levels of inference are described: inference for fitting a model and inference for assessing the suitability of a model. The first level of inference is the typical application of Bayes rule for updating model parameters,

$$P(\omega|\mathcal{D}, \mathcal{H}_i) = \frac{P(\mathcal{D}|\omega, \mathcal{H}_i) P(\omega|\mathcal{H}_i)}{P(\mathcal{D}|\mathcal{H}_i)}, \tag{3.10}$$

where ω is the set of parameters in the generic statistical model, $\mathcal{D}$ is our data and $\mathcal{H}_i$ represents the i'th model used for this level of inference.[8] This is then described as,

$$\text{Posterior} = \frac{\text{Likelihood} \times \text{Prior}}{\text{Evidence}}.$$

It is important to note that the normalising constant in Eq. (3.10) is referred to as the evidence for the specific model of interest $\mathcal{H}_i$. Evaluation of the posterior remains intractable for most models of interest, so approximations must be made. In this work, the Laplace approximation is used.

Though computation of the posterior over parameters is required, the key aim of this work is to demonstrate methods of assessing the posterior over the model hypothesis $\mathcal{H}_i$. The posterior over model design is represented as,

$$P(\mathcal{H}_i|\mathcal{D}) \propto P(\mathcal{D}|\mathcal{H}_i) P(\mathcal{H}_i), \tag{3.11}$$

---

[8]$\mathcal{H}$ is used to refer to the model "hypothesis".

which translates to,

$$\text{Model Posterior} \propto \text{Evidence} \times \text{Model Prior}.$$

The data dependent term in Eq. (3.11) is the evidence for the model. Despite the promising interpretation of the posterior normalisation constant, as described earlier, evaluation of this distribution is intractable for most BNNs. Assuming a Gaussian distribution, the Laplace approximation of the evidence can be found as,

$$P(\mathcal{D}|\mathcal{H}_i) = \int P(\mathcal{D}|\omega, \mathcal{H}_i) P(\omega|\mathcal{H}_i) d\omega \tag{3.12}$$

$$\approx P(\mathcal{D}|\omega_{\text{MAP}}, \mathcal{H}_i) \Big[P(\omega_{\text{MAP}}|\mathcal{H}_i) \Delta\omega \Big] \tag{3.13}$$

$$= P(\mathcal{D}|\omega_{\text{MAP}}, \mathcal{H}_i) \Big[P(\omega_{\text{MAP}}|\mathcal{H}_i)(2\pi)^{\frac{k}{2}} \det^{-\frac{1}{2}} \mathbf{A} \Big] \tag{3.14}$$

$$= \text{Best Likelihood Fit} \times \text{Occam Factor}.$$

This can be interpreted as a single Riemann approximation to the model evidence with the best likelihood fit representing the peak of the evidence, and the Occam factor is the width that is characterised by the curvature around the peak of the Gaussian. The Occam factor can be interpreted as the ratio of the width of the posterior $\Delta\omega$ and the range of the prior $\Delta\omega_0$ for the given model $\mathcal{H}_i$,

$$\text{Occam Factor} = \frac{\Delta\omega}{\Delta\omega_0}, \tag{3.15}$$

meaning that the Occam factor is the ratio of change in plausible parameter space from the prior to the posterior. Figure 3.4 demonstrates this concept graphically.

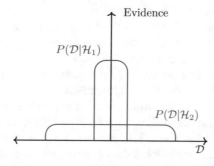

Fig. 3.4 Graphical illustration of how the evidence plays a role in investigating different model hypotheses. The simple model $\mathcal{H}_1$ is able to predict a small range of data with greater strength, while the more complex model $\mathcal{H}_2$ is able to represent a larger range of data, though with lower probability. Adapted from [45, 46]

With this representation, a complex model able to represent a large range of data will have a wider evidence, thus having a larger Occam factor. A simple model will have a lower capacity to capture a complex generative process, but a smaller range of data will be able to be modelled with greater certainty, resulting in a lower Occam Factor. This results in a natural regularisation for the complexity of a model. An unnecessarily complex model will typically result in a wide posterior, resulting in a large Occam factor and low evidence for the given model. Similarly, a wide or less informative prior will result in a reduced Occam factor, providing further intuition into the Bayesian setting of regularisation.

Using this evidence framework requires computation of the marginal likelihood, which is an expensive (and the key challenge) within Bayesian modelling. Given the large investment required to approximate the marginal likelihood, it may be infeasible to compare many different architectures. Despite this, the use of the evidence framework can used to assess solutions for BNNs. For most NN architectures of interest, the objective function is non-convex with many local minima. Each local minima can be regarded as a possible solution for the inference problem. MacKay uses this as motivation to compare the solutions from each local minimum using the corresponding evidence function [47]. This allows for assessment of model complexity at each solution without prohibitive computational requirements.

3.2.3.1   Early Variational Inference for BNNs

The machine learning community has continuously excelled at optimisation based problems. While many ML models, such as Support Vector Machines and Linear Gaussian Models result in a convex objective function, NNs have a highly non-convex objective function with many local minima. A difficult to locate global minimum motivates the use of gradient based optimisation schemes such as backpropagation [3]. This type of optimisation can be viewed in a Bayesian context through the lens of Variational Inference (VI).

VI is an approximate inference method that frames marginalisation required during Bayesian inference as an optimisation problem [48–50]. This is achieved by assuming the form of the posterior distribution and performing optimisation to find the assumed density that closest to the true posterior. This assumption simplifies computation and provides some level of tractability.

The assumed posterior distribution $q_\theta(\omega)$ is a suitable density over the set of parameters ω, that is restricted to a certain family of distributions parameterised by θ. The parameters for this variational distribution are then adjusted to reduce the dissimilarity between the variational distribution and the true posterior $p(\omega|\mathcal{D})$.[9]

---

[9]The model hypothesis $\mathcal{H}_i$ used previously will be omitted for further expressions, as little of the remaining key research items deal with model comparison and simply assume a single architecture and solution.

The means to measure similarity for VI is often the forward KL-Divergence between the variational and true distribution,

$$KL\Big(q_\theta(\omega)||p(\omega|\mathcal{D})\Big) = \int q_\theta(\omega) \log \frac{q_\theta(\omega)}{p(\omega|\mathcal{D})} d\omega. \tag{3.16}$$

For VI, Eq. (3.16) serves as the objective function we wish to minimise w.r.t variational parameters θ. This can be expanded out as,

$$\mathrm{KL}\Big(q_\theta(\omega)||p(\omega|\mathcal{D})\Big) = \mathbb{E}_q\Big[\log \frac{q_\theta(\omega)}{p(\omega)} - \log p(\mathcal{D}|\omega)\Big] + \log p(\mathcal{D}) \tag{3.17}$$

$$= \mathrm{KL}\Big(q_\theta(\omega)||p(\omega)\Big) - \mathbb{E}_q[\log p(\mathcal{D}|\omega)] + \log p(\mathcal{D}) \tag{3.18}$$

$$= -\mathcal{F}[q_\theta] + \log p(\mathcal{D}), \tag{3.19}$$

where $\mathcal{F}[q_\theta] = -\mathrm{KL}\Big(q_\theta(\omega)||p(\omega)\Big) + \mathbb{E}_q[\log p(\mathcal{D}|\omega)]$. The combination of terms into $\mathcal{F}[q]$ is to separate the tractable terms from the intractable log marginal likelihood. We can now optimise this function using backpropagation, and since the log marginal likelihood does not depend on variational parameters θ, it's derivative evaluates to zero. This leaves only term of containing variational parameters, which is $\mathcal{F}[q_\theta]$.

This notation used in Eq. (3.19), particularly the choice to include the negative of $\mathcal{F}[q_\theta]$ is deliberate to highlight a different but equivalent derivation to the identical result, and to remain consistent with existing literature. This result can be obtained by instead of minimising the KL-Divergence between the true and approximate distribution, but by approximating the intractable log marginal likelihood. Through application of Jensen's inequality, we can then find that $\mathcal{F}[q_\theta]$ forms a lower bound on the logarithm of the marginal likelihood [48, 51]. This can be seen by re-arranging Eq. (3.19) and noting that the KL divergence is strictly ≥ 0 and only equals zero when the two distributions are equal. The logarithm of the marginal likelihood is equal to the sum of the KL divergence between the approximate and true posterior and $\mathcal{F}[q_\theta]$. By minimising the KL divergence between the approximate and true posterior, the closer $\mathcal{F}[q_\theta]$ will be to the logarithm of the marginal likelihood. For this reason, $\mathcal{F}[q_\theta]$ is commonly referred to as the Evidence Lower Bound (ELBO). Figure 3.5 illustrates this graphically.

The first application of VI to BNNs was by Hinton and Van Camp [53], where they tried to address the problem of overfitting in NNs. They argued that by using a probabilistic perspective of model weights, the amount of information they could contain would be reduced and would simplify the network. Formulation of this problem was through an information theoretic basis, particularly the Minimum Descriptive Length (MDL) principle, though its application results in a framework equivalent to VI. As is common in VI, the mean-field approach was used. Mean-Field Variational Bayes (MFVB) assumes a posterior distribution that factorises

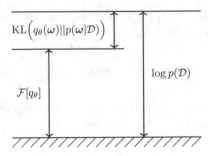

Fig. 3.5 Graphical illustration of how the minimisation of the KL divergence between the approximate and true posterior maximises the lower bound on the evidence. As the KL Divergence between our approximate and true posterior is minimised, the ELBO $\mathcal{F}[q_\theta]$ tightens to the log-evidence. Therefore maximising the ELBO is equivalent to minimising the KL divergence between the approximate and true posterior. Adapted from [52]

over parameters of interest. For the work in [53], the posterior distribution over model weights was assumed to be a factorisation of independent Gaussians,

$$q_\theta(\omega) = \prod_{i=1}^{P} \mathcal{N}(w_i|\mu_i, \sigma_i^2), \tag{3.20}$$

where P is the number of weights in the network. For a regression network with a single hidden layer, an analytic solution for this posterior is available. The ability to achieve an analytic solution to the approximation is an desirable property, as analytic solutions significantly reduce the time to perform inference.

There are a few issues with this work, though one of the most prominent issues is the assumption of a posterior that factorises over individual network weights. It is well known that strong correlation between parameters in a NN is present. A factorised distribution simplifies computation by sacrificing the rich correlation information between parameters. MacKay highlighted this limitation in an early survey of BNNs [32] and offers insight into how a preprocessing stage of inputs to hidden layers could allow for more comprehensive approximate posterior distributions.

Barber and Bishop [52] again highlight this limitation, and offer a VI based approach that extends on the work in [53] to allow for full correlation between the parameters to be captured by using a full rank Gaussian for the approximating posterior. For a single hidden layer regression network utilising a Sigmoid activation, analytic expressions for evaluating the ELBO is provided.[10] This is achieved by replacing the Sigmoid with the appropriately scaled error function.

---

[10]Numerical methods are required to evaluate certain terms in the analytic expression for the ELBO.

An issue with this modelling scheme is the increased number of parameters. For a full covariance model, the number of parameters scales quadratically with the number of weights in the network. To rectify this, Barber and Bishop propose a restricted form for the covariance often used in factor analysis, such that,

$$\mathbf{C} = \text{diag}(d_1^2, \ldots, d_n^2) + \sum_{i=1}^{s} \mathbf{s}_i \mathbf{s}_i^T, \qquad (3.21)$$

where the diag operator creates a diagonal matrix from the vector $\mathbf{d}$ of size n, where n is the number of weights in the model. This form then scales linearly with the number of hidden units in the network.

These bodies of work provide important insight into how the prominent back-propagation method can be applied to challenging Bayesian problems. This allows for properties of the two areas of research to be merged and offer the benefits nominally seen in isolation. Complex regression tasks for large bodies of data sets could now be handled in a probabilistic sense using NNs.

Despite the insight offered by these methods, there are limitations to these methods. Both the work of Hinton and Van Camp and Barber and Bishop focus on development of a closed form representation of the networks.[11] This analytic tractability imposes many restrictions on the networks. As discussed previously, [53] assume a factorised posterior over individual weights which is unable to capture any correlation in parameters. Covariance structure is captured in [52], though the authors limit their analysis to the use of a Sigmoid activation function (which is well approximated by the error function), which is seldom used in modern networks due to the low magnitude in the gradient.[12] A key limitation common to both of these approaches is the restriction of a single hidden layer network.

As stated previously, a NN can approximate any function arbitrarily well by adding additional hidden units. For modern networks, empirical results have shown that similarly complex functions can be represented with fewer hidden units by increasing the number of hidden layers in the network. This has lead to the term "deep learning", where depth refers to the number of hidden layers. The reduction in number of weight variables is especially important for when trying to approximate the full covariance structure between layers. For example, correlation between hidden units within a single layer may be captured, while assuming that parameters between the different layers are independent. An assumption such as this can significantly reduce the number of correlation parameters. With modern networks having hundreds of millions of weights across many layers (with these networks only being able to offer point estimates), the need to develop practical probabilistic interpretations beyond a single layer is essential.

---

[11] Although there are a large number of benefits to such an approach, as illustrated earlier.

[12] Analytic results may be achievable using other activation functions, such as TanH, which suffer less from such an issue.

3.2.3.2 Hybrid Monte Carlo for BNNs

It is worthwhile at this point to reflect on the actual quantities of interest. So far the emphasis has been placed on finding good approximations for the posterior, though the accurate representation of the posterior is usually not the end design requirement. The main quantities of interest are predictive moments and intervals. We want to make good predictions accompanied by confidence information. The reason we emphasise computation of the posterior is that predictive moments and intervals are all computed as expectations of the posterior $\pi(\omega|\mathcal{D})$.[13] This expectation is listed in Eq. (3.9), and is repeated here for convenience,

$$\mathbb{E}_\pi[f] = \int f(\omega)\pi(\omega|\mathcal{D})d\omega.$$

This is why computation of the posterior is emphasised; accurate predictions rely on accurate approximations of the intractable posterior.

The previous methods employed optimisation based schemes such as VI or Laplace approximations of the posterior. In doing so, strong assumptions and restrictions on the form of posterior are enforced. The restrictions placed are often credited with inaccuracies induced in predictions, though this is not the only limitation.

As highlighted by Betancourt [54] and Betancourt et al. [55], the expectation computed for predictive quantities not just a probability mass, it the product of the probability mass and a volume. The probability mass is our posterior distribution $\pi(\omega|\mathcal{D})$, and the volume $d\omega$ over which we are integrating. It is likely that for all models of interest, the contribution of the expectation from this product of the density and volume will not be at the maximum for the mass. Therefore optimisation based schemes which consider only the mass can deliver inaccurate predictive quantities. To make accurate predictions with finite computational resources, we need to evaluate this expectation not just when the mass is greatest, but when the product of the mass and volume is largest. The most promising way to achieve this is with Markov Chain Monte Carlo (MCMC).

MCMC algorithms remains at the forefront of Bayesian research and applied statistics.[14] MCMC is a general approach for sampling from arbitrary and intractable distributions. The ability to sample from a distribution enables the use of Monte Carlo integration for prediction,

$$\mathbb{E}_\pi[f] = \int f(\omega)\pi(\omega|\mathcal{D})d\omega \approx \frac{1}{N}\sum_{i=1}^{N} f(\omega_i), \tag{3.22}$$

---

[13]Note that π is used to represent the true posterior distribution here, as appose to q used previously to denote an approximation of the posterior.

[14]MCMC is regarded as one of the most influential algorithms of the twenty-first century [56].

where ω_i represents an independent sample from the posterior distribution. MCMC enables sampling from our posterior distribution, with the samples converging to when the product of the probability density and volume are greatest [54].

Assumptions previously made in VI methods, such as a factorised posterior are not required in the MCMC context. MCMC provides convergence to the true posterior as the number of samples approaches infinity. By avoiding such restrictions, with enough time and computing resources we can yield a solution that is closer to the true predictive quantities. This is an important challenge for BNNs, as the posterior distributions is typically quite complex.

Traditional MCMC methods demonstrate a random-walk behaviour, in that new proposals in the sequence are generated randomly. Due to the complexity and high dimension of the posterior in BNNs, this random-walk behaviour makes these methods unsuitable for performing inference in any reasonable time. To avoid the random-walk behaviour, Hybrid/Hamiltonian Monte Carlo (HMC) can be employed to incorporate gradient information into the iterative behaviour. While HMC was initially proposed for statistical physics [57], Neal highlighted the potential for HMC to address Bayesian inference and specifically researched the applications to BNNs and the wider statistics community as a whole [38].

Given that HMC was initially proposed for physical dynamics, it is appropriate to build intuition for applied statistics through a physical analogy. Treat our parameters of interest ω as a position variable. An auxiliary variable is then introduced to model the momentum $\mathbf{v}$ of our current position. This auxiliary variable is not of statistical interest, and is only introduced to aid in development of the system dynamics. With a position and momentum variable, we can represent the potential energy $U(\omega)$ and the kinetic energy $K(\mathbf{v})$ of our system. The total energy of a system is then represented as,

$$H(\omega, \mathbf{v}) = U(\omega) + K(\mathbf{v}). \tag{3.23}$$

We now consider the case of a lossless system, in that the total energy $H(\omega, \mathbf{v})$ is constant.[15] This is described as a Hamiltonian system, and is represented as the following system of differential equations [58],

$$\frac{dw_i}{dt} = \frac{\partial H}{\partial v_i}, \tag{3.24}$$

$$\frac{dv_i}{dt} = -\frac{\partial H}{\partial w_i}, \tag{3.25}$$

where t represents time and the i denotes the individual elements in ω and $\mathbf{v}$.

With the dynamics of the system defined, we wish to relate the physical interpretation to a probabilistic interpretation. This can be achieved through the

---

[15]The values for ω and $\mathbf{v}$ will change, though the total energy of the system will remain constant.

canonical distribution,[16]

$$P(\omega, \mathbf{v}) = \frac{1}{Z} \exp\left(-H(\omega, \mathbf{v})\right) = \frac{1}{Z} \exp\left(-U(\omega)\right) \exp\left(-K(\mathbf{v})\right), \qquad (3.26)$$

where Z is a normalising constant and $H(\omega, \mathbf{v})$ is our total energy as defined in Eq. (3.23). From this joint distribution, we see that our position and momentum variable are independent.

Our end goal is to find predictive moments and intervals. For a Bayesian this makes the key quantity of interest the posterior distribution. Therefore, we can set the potential energy which we wish to sample from to,

$$U(\omega) = -\log\left(p(\omega)p(\mathcal{D}|\omega)\right). \qquad (3.27)$$

Within HMC, the kinetic energy can be freely selected from a wide range of suitable functions, though is typically chosen such that it's marginal distribution of $\mathbf{v}$ is a diagonal Gaussian centred at the origin.

$$K(\mathbf{v}) = \mathbf{v}^T M^{-1} \mathbf{v}, \qquad (3.28)$$

where M is a diagonal matrix referring to the "mass" of our variables in this physical interpretation. Although this is the most common kinetic energy function used, it may not be the most suitable. Betancourt [54] surveys the selection the design of other Gaussian kinetic energies with an emphasis on the geometric interpretations. It is also highlighted that selection of appropriate kinetic energy functions remains an open research topic, particularly in the case of non-Gaussian functions.

Since Hamiltonian dynamics leaves the total energy invariant, when implemented with infinite precision, the dynamics proposed are reversible. Reversibility is a sufficient property to satisfy the condition of detailed balance, which is required to ensure that the target distribution (the posterior we are trying to sample from) remains invariant. For practical implementations, numerical errors arise due to discretisation of variables. The discretisation method most commonly employed is the leapfrog method. The leapfrog method specifies a step size ϵ and a number of steps L to be used before possibly accepting the new update. The leapfrog method first performs a half update of the momentum variable v, followed by a full update

---

[16]As is commonly done, we assume the temperature variable included in physical representations of the canonical distribution is set to one. For more information, see [58, p. 11], [59, p. 123].

of the position w and then the remaining half update of the momentum [58],

$$v_i\left(t + \frac{\epsilon}{2}\right) = v_i(t) + \frac{\epsilon}{2}\frac{dv_i}{dt}(v(t)), \tag{3.29}$$

$$w_i(t + \epsilon) = w_i(t) + \epsilon\frac{dw_i}{dt}(w(t)), \tag{3.30}$$

$$v_i(t + \epsilon) = v_i\left(t + \frac{\epsilon}{2}\right) + \frac{\epsilon}{2}\frac{dv_i}{dt}\left(v\left(t + \frac{\epsilon}{2}\right)\right). \tag{3.31}$$

If the value of ϵ is chosen such that this dynamical system remains stable, it can be shown that this leapfrog method preserves the volume (total energy) of the Hamiltonian.

For expectations to be approximated using (3.22), we require each sample ω_i to be independent from subsequent samples. We can achieve practical independence[17] by using multiple leapfrog steps L. In this way, after L leapfrog steps of size ϵ, the new position is proposed. This reduces correlation between samples and can allow for faster exploration of the posterior space. A Metropolis step is then applied to determine whether this new proposal is accepted as the newest state in the Markov Chain [58].

For the BNN proposed by Neal [38], a hyper-prior $p(\gamma)$ is induced to model the variance over prior parameter precision and likelihood precision. A Gaussian prior is used for the prior over-parameters and the likelihood is set to be Gaussian. Therefore, the prior over the γ was Gamma distributed, such that it was conditionally conjugate. This allows for Gibbs sampling to be used for performing inference over hyperparameters. HMC is then used to update the posterior parameters. Sampling from the joint posterior $P(\omega, \gamma|\mathcal{D})$ then involves alternating between the Gibbs sampling step for the hyperparameters and Hamiltonian dynamics for the model parameters. Superior performance of HMC for simple BNN models was then demonstrated and compared with random walk MCMC and Langevin methods [38].

3.2.4 Modern BNNs

Considerably less research was conducted into BNNs following early work of Neal, MacKay and Bishop proposed in the 90s. This relative stagnation was seen throughout the majority of NN research, and was largely due to the high computational demand for training NNs. NNs are parametric models that are able to capture any function with arbitrary accuracy, but to capture complex functions accurately requires large networks with many parameters. Training of such large networks became infeasible even for the traditional frequentist perspective, and the

---

[17]Where for all practical purposes each sample can be viewed as independent.

computational demand significantly increases to investigate the more informative Bayesian counterpart.

Once it was shown that general purpose GPUs could accelerate and allow training of large models, interest and research into NNs saw a resurgence. GPUs enabled large scale parallelism of the linear algebra performed during back propagation. This accelerated computation has allowed for training of deeper networks, where successive concatenation of hidden layers is used. With the proficiency of GPUs for optimising complex networks and the great empirical success seen by such models, interest into BNNs resumed.

Modern research into BNNs has largely focused on the VI approach, given that these problems can be optimised using a similar backpropagation approach used for point estimate networks. Given that the networks offering the most promising results use multiple layers, the original VI approaches shown in [52, 53], which focus on analytical approximations for regression networks utilising a single hidden layer became unsuitable. Modern NNs now exhibit considerably different architectures with varying dimensions, hidden layers, activations and applications. More general approaches for viewing networks in a probabilistic sense was required.

Given the large scale of modern networks, large data sets are typically required for robust inference.[18] For these large data sets, evaluation of the complete log-likelihood becomes infeasible for training purposes. To combat this, a Stochastic Gradient Descent (SGD) approach is used, where mini-batches of the data are used to approximate the likelihood term, such that our variational objective becomes,

$$\mathcal{L}(\boldsymbol{\omega}, \boldsymbol{\theta}) = -\frac{N}{M} \sum_{i=1}^{N} \mathbb{E}_q[\log\left(p(\mathcal{D}_i|\boldsymbol{\omega})\right)] + \mathrm{KL}\left(q_{\boldsymbol{\theta}}(\boldsymbol{\omega})||p(\boldsymbol{\omega})\right), \qquad (3.32)$$

where $\mathcal{D}_i \subset \mathcal{D}$, and each subset is of size M. This provides an efficient way to utilise large data sets during training. After passing a single subset $\mathcal{D}_i$, backpropagation is applied to update the model parameters. This sub-sampling of the likelihood induces noise into our inference process, hence the name SGD. This noise that is induced is expected to average out over evaluation of each individual subset [60]. SGD is the most common method for training NNs and BNNs utilising a VI approach.

A key paper in the resurgence of BNN research was published by Graves [61]. This work proposes a MFVB treatment using a factorised Gaussian approximate posterior. The key contribution of this work is the computation of the derivatives. The VI objective (ELBO) can be viewed as a sum of two expectations,

$$\mathcal{F}[q_{\boldsymbol{\theta}}] = \mathbb{E}_q[\log\left(p(\mathcal{D}|\boldsymbol{\omega})\right)] - \mathbb{E}_q[\log q_{\boldsymbol{\theta}}(\boldsymbol{\omega}) - \log p(\boldsymbol{\omega})] \qquad (3.33)$$

---

[18]Neal [38] argues that this not true for Bayesian modelling; claims that if suitable prior information is available, complexity of a model should only be limited by computational resources.

It is these two expectations that we need to optimise w.r.t model parameters, meaning that we require the gradient of expectations. This work shows how using the gradient properties of a Gaussian proposed in [62] can be used to perform parameter updates,

$$\nabla_\mu \, \mathbb{E}_{p(\omega)}[f(\omega)] = \mathbb{E}_{p(\omega)}[\nabla_\omega f(\omega)], \tag{3.34}$$

$$\nabla_\Sigma \, \mathbb{E}_{p(\omega)}[f(\omega)] = \frac{1}{2}\mathbb{E}_{p(\omega)}[\nabla_\omega \nabla_\omega f(\omega)]. \tag{3.35}$$

MC integration could be applied to Eqs. (3.34) and (3.35) to approximate the gradient of the mean and variance parameters. This framework allows for optimisation of the ELBO to generalise to any log-loss parametric model.

Whilst addressing the problem of applying VI to complex BNNs with more hidden layers, practical implementations have shown inadequate performance which is attributed to large variance in the MC approximations of the gradient computations [63]. Developing gradient estimates with reduced variance has become an integral research topic in VI [64]. Two of the most common methods for deriving gradient approximations rely on the use of score functions and path-wise derivative estimators.

Score function estimators rely on the use of the log-derivative property, such that,

$$\frac{\partial}{\partial \theta} p(x|\theta) = p(x|\theta)\frac{\partial}{\partial \theta} \log p(x|\theta). \tag{3.36}$$

Using this property, we can form Monte Carlo estimates of the derivatives of an expectation, which is often required in VI,

$$\nabla_\theta \mathbb{E}_q[f(\omega)] = \int f(\omega)\nabla_\theta q_\theta(\omega)\partial\omega$$

$$= \int f(\omega)q_\theta(\omega)\nabla_\theta \log\left(q_\theta(\omega)\right)\partial\omega$$

$$\approx \frac{1}{L}\sum_{i=1}^{L} f(\omega_i)\nabla_\theta \log\left(q_\theta(\omega_i)\right). \tag{3.37}$$

A common problem with score function gradient estimators is that they exhibit considerable variance [64]. One of the most common methods to reduce the variance in Monte Carlo estimates is the introduction of control variates [65].

The second type of gradient estimator commonly used in the VI literature is the pathwise derivative estimator. This work builds on the "reparameterisation trick" [66–68], where a random variable is represented as a deterministic and differentiable expression. For example, for a Gaussian with $\theta = \{\mu, \sigma\}$,

$$\omega \sim \mathcal{N}(\mu, \sigma^2)$$

$$\omega = g(\theta, \epsilon) = \mu + \sigma \odot \epsilon \tag{3.38}$$

where $\epsilon \sim \mathcal{N}(0, I)$ and $\odot$ represents the Hadamard product. Using this method allows for efficient sampling for Monte Carlo estimates of expectations. This is shown in [67], that with $\omega = g(\theta, \epsilon)$, we know that $q(\omega|\theta)d\omega = p(\epsilon)d\epsilon$. Therefore, we can show that,

$$\int q_\theta(\omega) f(\omega) d\omega = \int p(\epsilon) f(\omega) d\epsilon$$

$$= \int p(\epsilon) f(g(\theta, \epsilon)) d\epsilon$$

$$\approx \frac{1}{M} \sum_{i=1}^{M} f(g(\theta, \epsilon_i)) = \frac{1}{M} \sum_{i=1}^{M} f(\mu + \sigma \odot \epsilon_i) \tag{3.39}$$

Since Eq. (3.39) is differentiable w.r.t θ, gradient descent methods can be used to optimise this expectation approximation. This is an important property in VI, since the VI objective contains expectations of the log-likelihood that are often intractable. The reparameterisation trick serves as the basis for pathwise-gradient estimators. Pathwise estimators are favourable for their reduced variance over score function estimators [64, 67].

A key benefit of having a Bayesian treatment of NNs is the ability to extract uncertainty in our models and their predictions. This has been a recent research topic of high interest in the context of NNs. Promising developments regarding uncertainty estimation in NNs has been found by relating existing regularisation techniques such as Dropout [69] to approximate inference. Dropout is a Stochastic Regularisation Technique (SRT) that was proposed to address overfitting commonly seen in point-estimate networks. During training, Dropout introduces an independent random variable that is Bernoulli distributed, and multiplies each individual weight element-wise by a sample from this distribution. For example, a simple MLP implementing Dropout is of the form,

$$\rho_u \sim \text{Bernoulli}(p),$$

$$\phi_j = \theta\left(\sum_{i=1}^{N_1} (x_i \rho_u) w_{ij}\right). \tag{3.40}$$

Looking at Eq. (3.40), it can be seen that the application of Dropout introduces stochasticity into the network parameters in a similar manner as to that of the reparameterisation trick shown in Eq. (3.38). A key difference is that in the case of

Dropout, stochasticity is introduced into the input space, as appose to the parameter space required for Bayesian inference. Yarin Gal [70] identified this similarity, and demonstrated how noise introduced through the application of Dropout can be transferred to the networks weights efficiently as,

$$\mathbf{W}_\rho^1 = \text{diag}(\rho)\mathbf{W}^1 \qquad (3.41)$$

$$\mathbf{\Phi}_\rho = a(\mathbf{X}^T\mathbf{W}_\rho^1). \qquad (3.42)$$

Where ρ is a vector sampled from the Bernoulli distribution, and the diag($\cdot$) operator creates a square diagonal matrix from a vector. In doing this it can be seen that a single dropout variable is shared amongst each row of the weight matrix, allowing some correlation within rows to be maintained. By viewing the stochastic component in terms of network weights, the formulation becomes suitable for approximate inference using the VI framework. In this work, the approximate posterior is of the form of a Bernoulli distribution multiplied by the network weights.

The reparameterisation trick is then applied to allow for partial derivatives w.r.t. network parameters to be found. The ELBO is then formed and backpropagation is performed to maximise the lower bound. MC integration is used to approximate the analytically intractable expected log-likelihood. The KL divergence between the approximate posterior and the prior distribution in the ELBO is then found by approximating the scaled Bernoulli approximate posterior as a mixture of two Gaussians with very small variance.

In parallel to this work, Kingma et al. [71] identified this same similarity between Dropout and it's potential for use within a VI framework. As appose to the typical Bernoulli distributed r.v. introduced in Dropout, Kingma et al. [71] focuses attention to the case when the introduced r.v. is Gaussian [72]. It is also shown how with selection of an appropriate prior that is independent of parameters, current applications of NNs using dropout can be viewed as approximate inference.

Kingma et al. also aims to reduce the variance in the stochastic gradients using a refined, local reparameterisation. This is done by instead of sampling from the weight distribution before applying the affine transformation, the sampling is performed afterwards. For example, consider a MFVB case where each weight is assumed to be an independent Gaussian $w_{ij} \sim \mathcal{N}(\mu_{ij}, \sigma_{ij}^2)$. After the affine transformation $\phi_j = \sum_{i=1}^{N_1}(x_i\rho_i)w_{ij}$, the posterior distribution of ϕ_j conditional on the inputs will also be a factorised Gaussian,

$$q(\psi_j|\mathbf{x}) = \mathcal{N}(\gamma_j, \delta_j^2), \qquad (3.43)$$

$$\gamma_j = \sum_{i=1}^{N} x_i\mu_{i,j}, \qquad (3.44)$$

$$\delta_j^2 = \sum_{i=1}^{N} x_i^2\sigma_{i,j}^2. \qquad (3.45)$$

It is advantageous to sample from this distribution for ϕ as appose to the distribution of the weights w themselves, as this results in a gradient estimator whose variance scales linearly with the number of mini-batches used during training.[19]

These few bodies of work are important in addressing the serious lack of rigour seen in ML research. For example, the initial Dropout paper [69] lacks any significant theoretical foundation. Instead, the method cites a theory for sexual reproduction [73] as motivation for the method, and relies heavily on the empirical results given. These empirical results have been further demonstrated throughout many high impact[20] research items which utilise this technique merely as a regularisation method. The work in [70] and [71] show that there is theoretical justification for such an approach. In attempts to reduce the effect of overfitting in a network, the frequentist methodology relied on the application of a weakly justified technique that shows empirical success, while Bayesian analysis provides a rich body of theory that naturally leads to a meaningful understanding of this powerful approximation.

Whilst addressing the problem of applying VI to complex BNNs with more hidden layers, practical implementations have shown inadequate performance which is attributed to large variance in the MC approximations of the gradient computations. Hernandez et al. [63] acknowledge this limitation and propose a new method for practical inference of BNNs titled Probabilistic Back Propagation (PBP). PBP deviates from the typical VI approach, and instead employs an Assumed Density Filtering (ADF) method [74]. In this format, the posterior is updated in an iterative fashion through application of Bayes rule,

$$p(\omega_{t+1}|\mathcal{D}_{t+1}) = \frac{p(\omega_t|\mathcal{D}_t)p(\mathcal{D}_{t+1}|\omega_t)}{p(\mathcal{D}_{t+1})}. \tag{3.46}$$

As opposed to traditional network training where the predicted error is the objective function, PBP uses a forward pass to compute the log-marginal probability of a target and updates the posterior distribution of network parameters. The moment matching method defined in [75] updates the posterior using a variant of backpropagation, whilst maintaining equivalent mean and variance between the approximate and variational distribution,

$$\mu_{t+1} = \mu_t + \sigma_t \frac{\partial \log p(\mathcal{D}_{t+1})}{\partial \mu} \tag{3.47}$$

$$\sigma_{t+1} = \sigma_t + \sigma_t^2 \Big[\Big(\frac{\partial p(\mathcal{D}_{t+1})}{\partial \mu_t} \Big)^2 - 2 \frac{\partial p(\mathcal{D}_{t+1})}{\partial \sigma} \Big]. \tag{3.48}$$

---

[19]This method also has computational advantages, as the dimension of ϕ is typically much lower than that of ω.

[20]At the time of writing, [69] has over ten thousand citations.

Experimental results on multiple small data-sets illustrate reasonable performance in terms of predicted accuracy and uncertainty estimation when compared with HMC methods for simple regression problems [63]. A key limitation of this method is the computational bottleneck introduced by the online training method. This approach may be suitable for some applications, or for updating existing BNNs with additional data as it becomes available, though for performing inference on large data sets the method is computationally prohibitive.

A promising method for approximate inference in BNNs was proposed by Blundell et al., titled "Bayes by Backprop" [76]. The method utilises the reparameterisation trick to show how unbiased estimates of the derivative of an expectation can be found. For a random variable $\omega \sim q_\theta(\omega)$ that can be reparameterised as deterministic and differentiable function $\omega = g(\epsilon, \theta)$, the derivative of the expectation of an arbitrary function $f(\omega, \theta)$ can be expressed as,

$$\frac{\partial}{\partial \theta} \mathbb{E}_q[f(\omega, \theta)] = \frac{\partial}{\partial \theta} \int q_\theta(\omega) f(\omega, \theta) d\omega \qquad (3.49)$$

$$= \frac{\partial}{\partial \theta} \int p(\epsilon) f(\omega, \theta) d\epsilon \qquad (3.50)$$

$$= \mathbb{E}_{q(\epsilon)} \left[\frac{\partial f(\omega, \theta)}{\partial \omega} \frac{\partial \omega}{\partial \theta} + \frac{\partial f(\omega, \theta)}{\partial \theta} \right]. \qquad (3.51)$$

In the Bayes by Backprop algorithm, the function $f(\omega, \theta)$ is set as,

$$f(\omega, \theta) = \log \frac{q_\theta(\omega)}{p(\omega)} - \log p(\mathbf{X}|\omega). \qquad (3.52)$$

This $f(\omega, \theta)$ can be seen as the argument for the expectation performed in Eq. (3.17), which is part of the lower bound.

Combining Eqs. (3.51) and (3.52),

$$\mathcal{L}(\omega, \theta) = \mathbb{E}_q[f(\omega, \theta)] = e_q \left[\log \frac{q_\theta(\omega)}{p(\omega)} - \log p(\mathcal{D}|\omega) \right] = -\mathcal{F}[q_\theta] \qquad (3.53)$$

which is shown to be the negative of the ELBO, meaning that Bayes by Backprop aims to minimise the KL divergence between the approximate and true posterior. Monte Carlo integration is used[21] to approximate the cost in Eq. (3.53),

$$\mathcal{F}[q_\theta] \approx \sum_{i=1}^{N} \log \frac{q_\theta(\omega_i)}{p(\omega_i)} - \log p(\mathbf{X}|\omega_i) \qquad (3.54)$$

---

[21] Some terms may be tractable in this integrand, depending on the form of the prior and posterior approximation. MC integration allows for arbitrary distributions to be approximated.

where ω_i is the ith sample from $q_\theta(\omega)$. With the approximation in Eq. (3.54), the unbiased gradients can be found using the result shown in Eq. (3.51).

For the Bayes by Backprop algorithm, a fully factorised Gaussian posterior is assumed such that $\theta = \{\mu, \rho\}$, where $\sigma = \text{softplus}(\rho)$ is used to ensure the standard deviation parameter is positive. With this, the distribution of weights $\omega \sim \mathcal{N}(\mu, \text{softplus}(\rho)^2)$ in the network are reparameterised as,

$$\omega = g(\theta, \epsilon) = \mu + \text{softplus}(\rho) \odot \epsilon. \tag{3.55}$$

In this BNN, the trainable parameters are μ and ρ. Since a fully factorised distribution is used, following from Eq. (3.20), the logarithm of the approximate posterior can be represented as,

$$\log q_\theta(\omega) = \sum_{l,j,k} \log \left(\mathcal{N}(w_{ljk}; \mu_{ljk}, \sigma_{ljk}^2) \right). \tag{3.56}$$

The complete Bayes by Backprop algorithm is described in Algorithm 1.

Algorithm 1 Bayes by Backprop (BbB) algorithm [76]

1: **procedure** BBB(θ, **X**, α)
2:     **repeat**
3:         $\mathcal{F}[q_\theta] \leftarrow 0$ ▷ Initialise cost
4:         **for** i in $[1, \ldots, N]$ **do**   ▷ Number of samples for MC estimate
5:             Sample $\epsilon_i \sim \mathcal{N}(0, 1)$
6:             $\omega \leftarrow \mu + \text{softplus}(\rho) \cdot \epsilon_i$
7:             $\mathcal{L} \leftarrow \log q(\omega|\theta) - \log p(\omega) - \log p(\mathbf{X}|\omega)$
8:             $\mathcal{F}[q_\theta]+ = \text{sum}(\mathcal{L})/N$ ▷ Sum across all log of weights in set ω
9:         **end for**
10:         $\theta \leftarrow \theta - \alpha \nabla_\theta \mathcal{F}[q_\theta]$ ▷ Update parameters
11:     **until** convergence
12: **end procedure**

3.2.5 Gaussian Process Properties of BNNs

Neal [38] also provided derivation and experimentation results to illustrate that for a network with a single hidden layer, a Gaussian Process (GP) prior over the network output arises when the number of hidden units approaches infinity, and a Gaussian prior is placed over parameters.[22] Figure 3.6 illustrates this result.

---

[22]For a regression model with no non-linear activation function placed on the output units.

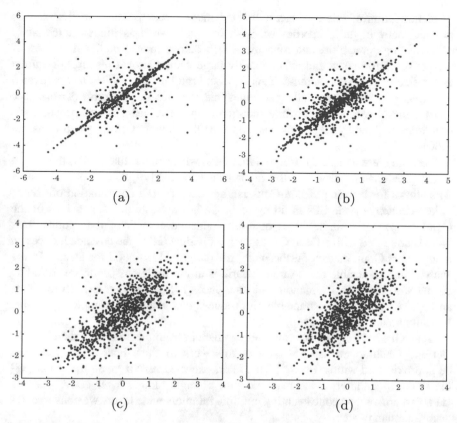

Fig. 3.6 Illustration of GP prior induced on output when placing a Gaussian prior over parameters as the network size increases. Experimentation replicated from [38, p. 33]. Each dot corresponds to the output of a network with parameters sampled from the prior, with the x-axis as $f(0.2)$ and the y-axis as $f(-0.4)$. For each network, the number of hidden units are (**a**) 1, (**b**) 3, (**c**) 10, (**d**) 100

This important link between NNs and GPs can be seen from Eqs. (3.1) and (3.2). From these expressions, it can be seen that a NN with a single hidden layer is a sum of N parametric basis functions applied to the input data. If the parameters for each basis function in Eq. (3.1) are r.v.'s, Eq. (3.2) becomes the sum of r.v.'s. Under the central limit theorem, as the number of hidden layers $N \rightarrow \infty$, the output becomes Gaussian. Since the output is then described as an infinite sum of basis functions, the output can be seen to become a GP. Following from a full derivation of this result and the illustrations show in Fig. 3.6, Neal [38] shows how an approximate Gaussian nature is achieved for finite computing resources and how the magnitude of this sum can be maintained. Williams then demonstrated how the form of the covariance function could be analysed for different activation functions [77]. The relation between GPs and infinitely wide networks with a single hidden layer work has recently been extended to the case of deep networks [78].

Identification of this link has motivated many research works in BNNs. GPs provide many of the properties we wish to obtain, such as reliable uncertainty estimates, interpretability and robustness. GPs deliver these benefits at the cost of predictive performance and exponentially large computational resources required as the size of data sets increase. This link between GPs and BNNs has motivated the merging of the two modelling schemes; maintaining the predictive performance and flexibility seen in NNs while incorporating the robustness and probabilistic properties enabled by GPs. This has led to the development of the Deep Gaussian Process.

Deep GPs are a cascade of individual GPs, where much like a NN, the output of the previous GP serves as the input to a new GP [79, 80]. This stacking of GPs allows for learning of non-Gaussian densities from a combination of GPs.[23] A key challenge with GPs is fitting to large data sets, as the dimensions of the Gram matrix for a single GP is quadratic with the number of data points. This issue is amplified with a Deep GP, as each individual GP in the cascade induces an independent Gram matrix. Furthermore, the marginal likelihood for Deep GPs are analytically intractable due to non-linearities in the functions produced. Building on the work in [82], Damianou and Lawrence [79] use a VI approach to create an approximation that is tractable and reduces computational complexity to that typically seen in sparse GPs [83].

Deep GPs have shown how the GPs can benefit from methodology seen in NNs. Gal and Ghahramani [84–86] built of this work to show how a Deep GP can be approximated with a BNN.[24] This is an expected result; given that Neal [38] identified an infinitely wide network with a single hidden layer converges to a Gaussian process, by concatenating multiple infinitely wide layers we converge to a deep Gaussian process.

Alongside this analysis of deep Gaussian processes, [84–86] build on the work in [77] to analyse the relationship between the modern non-linear activation used within BNNs and the covariance function for a GP. This is promising work that could allow for more principled selection of activation functions in NNs, similar to that of GPs. Which activation functions will yield a stationary process? What is the expected length scale for our process? These questions may be able to be addressed using the rich theory existing for GPs.

The GP properties are not restricted to MLP BNNs. Recent research has identified certain relationships and conditions that induce GP properties in convolutional BNNs [87, 88]. This result is expected since CNNs can be implemented as MLPs with structure enforced in the weights. What this work identifies is how the GP is constructed when this structure is enforced. Van der Wilk et al. [89] proposed the Convolutional Gaussian Process, which implements a patch based operation similar to that seen in CNNs to define the GP prior over functions. Practical implementation

---

[23] A complete introduction to Deep GPs, along with code and lectures has been offered by Neil Lawrence [81].

[24] Approximation becomes a Deep GP as the number of hidden units in each layer approaches ∞.

of this method requires the use of approximation methods, due to the prohibitive cost of evaluating large data sets, and even evaluation at each patch. Inducing points are formed with a VI framework to reduce the number of data points to evaluate and the number of patches evaluated.

3.2.6 Limitations in Current BNNs

Whilst great effort has been put into developing Bayesian methods for performing inference in NNs, there are significant limitations to these methods and many gaps remaining in the literature. A key limitation is the heavy reliance on VI methods. Within the VI framework, the most common approach is the Mean Field approach. MFVB provides a convenient way to represent an approximate posterior distribution by enforcing strong assumptions of independence between parameters. This assumption allows for factorised distributions to be used to approximate the posterior. This assumption of independence significantly reduces the computational complexity of approximate inference at the cost of probabilistic accuracy.

A common finding with VI approaches is that resulting models are overconfident, in that predictive means can be accurate while variance is considerably under estimated [50, 90–93]. This phenomenon is described in Section 10.1.2 of [2] and Section 21.2.2 of [35], both of which are accompanied by examples and intuitive figures to illustrate this property. This property of under-estimated variance is present within much of the current research in BNNs [70]. Recent work has aimed to address these issues through the use of noise contrastive priors [94] and through use of calibration data sets [95]. The authors in [96] employ the use of the concrete distribution [97] to approximate the Bernoulli parameter in the MC Dropout method [85], allowing for it to be optimised, resulting in posterior variances that are better calibrated. Despite these efforts, the task of formulating reliable and calibrated uncertainty estimates within a VI framework for BNNs remains unsolved.

It is reasonable to consider that perhaps the limitations of the current VI approaches are influenced by the choice of approximate distribution used, particularly the usual MFVB approach of independent Gaussians. If more comprehensive approximate distributions are used, will our predictions be more consistent with the data we have and haven't seen? Mixture based approximations have been proposed for the general VI approach [48, 98], though introduction of N mixtures increases the number of variational parameters by N. Matrix-Normal approximate posteriors have been introduced to the case of BNNs [99], which reduces the number of variational parameters in the model when compared with a full rank Gaussian, though this work still factorises over individual weights, meaning no covariance structure is modelled.[25] MCDropout is able to maintain some correlation

---

[25]Though this work highlights that even with a fully factorised distribution over weights, the outputs of each layer will be correlated.

information within the rows of weight matrix, at the compromise of a low entropy approximate posterior.

A recent approach for VI has been proposed to capture more complex posterior distributions through the use of normalising flows [100, 101]. Within a normalising flow, the initial distribution "flows" through a sequence of invertible functions to produce a more complex distribution. This can be applied within the VI framework using amortized inference [102]. Amortized inference introduces an inference network which maps input data to the variational parameters of generative model. These parameters are then used to sample from the posterior of the generative process. The use of normalising flows has been extended to the case of BNNs [103]. Issues arise with this approach relating to the computational complexity, along with limitations of amortized inference. Normalising flows requires the calculation of the determinant of the Jacobian for applying the change of variables used for each invertible function, which can be computationally expensive for certain models. Computational complexity can be reduced by restricting the normalising flow to contain invertible operations that are numerically stable [102, 104]. These restrictions have been shown to severely limit the flexibility of the inference process, and the complexity of the resulting posterior approximation [105].

As stated previously, in the VI framework, an approximate distribution is selected and the ELBO is then maximised. This ELBO arises from the applying the KL divergence between the true and approximate posterior, but this begs the question, why use the KL? The KL divergence is a well known measure to assess the similarity of between two distributions, and satisfies all the key properties of a divergence (i.e. is positive and only zero when the two distributions are equal). A divergence allows us to know whether our approximation is approaching the true distribution, but not how close we are to it. Why not use of a well defined distance as appose to a divergence?

The KL divergence is used as it allows us to separate the intractable quantity (the marginal likelihood) out of our objective function (the ELBO) which we can optimise. Our goal with our Bayesian inference is to identify the parameters that best fit our model under prior knowledge and the distribution of the observed data. The VI framework poses inference as an optimisation problem, where we optimise our parameters to minimise the KL divergence between our approximate and true distribution (which maximises our ELBO). Since we are optimising our parameters, by separating the marginal likelihood from our objective function, we are able to compute derivatives with respect to the tractable quantities. Since the marginal likelihood is independent of the parameters, this component vanishes when the derivative is taken. This is the key reason why the KL divergence is used, as it allows us to separate the intractable quantity out of our objective function, which will then be evaluated as zero when using gradient information to perform optimisation.

The KL divergence has been shown to be part of a generic family of divergences known as α-divergences [106, 107]. The α-divergence is represented as,

$$D_\alpha[p(\omega)||q(\omega)] = \frac{1}{\alpha(1-\alpha)}\left(1 - \int p(\omega)^\alpha q(\omega)^{1-\alpha}d\omega\right). \qquad (3.57)$$

The forward KL divergence used in VI is found from Eq. (3.57) in the limit that $\alpha \to -1$, and the reverse KL divergence $\text{KL}(p||q)$ occurs in the limit of $\alpha \to 1$, which is used during expectation propagation. While the use of the forward KL divergence used in VI typically results in an under-estimated variance, the use of the reverse KL will often over-estimate variance [2]. Similarly, the Hellinger distance arises from (3.57) when $\alpha = 0$,

$$D_H(p(\omega)||q(\omega))^2 = \int \left(p(\omega)^{\frac{1}{2}} - q(\omega)^{\frac{1}{2}}\right)^2 d\omega. \qquad (3.58)$$

This is a valid distance, in that it satisfies the triangle inequality and is symmetric. Minimisation of the Hellinger distance has shown to provide reasonable compromise in variance estimate when compared with the two KL divergences [107]. Though these measures may provide desirable qualities, they are not suitable for direct use within VI, as the intractable marginal likelihood cannot be separated from the other terms of interest.[26] While these measures cannot be immediately used, it illustrates how a change in the objective measure can result in different approximations. It is possible that more accurate posterior expectations can be found by utilising a different measure for the objective function.

The vast majority of modern works have revolved around the notion of VI. This is largely due to its amenability to SGD. Sophisticated tools now exist to simplify and accelerate the implementation of automatic differentiation and back-propagation [108–114]. Another benefit of VI is it's acceptance of sub-sampling in the likelihood. Sub-sampling reduces the computational expense for performing inference required to train over large data sets currently available. It is this key reason that more traditional MCMC based methods have received significantly less attention in the BNN community.

MCMC serves as the gold standard for performing Bayesian inference due to it's rich theoretical development, asymptotic guarantees and practical convergence diagnostics. Traditional MCMC based methods require sampling from the full joint likelihood to perform updates, requiring all training data to be seen before any new proposal can be made. Sub-sampling MCMC, or Stochastic Gradient MCMC (SG-MCMC) approaches have been proposed in [60, 115, 116], which have since been applied to BNNs [117]. It has since been shown that the naive sub-sampling within MCMC will bias the trajectory of the stochastic updates away

---

[26]This may be easy to see for the Hellinger distance, but less so for the reverse KL divergence. Enthusiastic readers are encouraged to not take my word for it, and to put pen and paper to prove this for themselves!

from the posterior [118]. This bias removes the theoretical advantages gained from a traditional MCMC approach, making them less desirable than a VI approach which is often less computationally expensive. For sampling methods to become feasible, sub-sampling methods need to be developed that assure convergence to the posterior distribution.

3.3 Comparison of Modern BNNs

From the literature survey presented within, two prominent methods for approximate inference in BNNs was Bayes by Backprop and MC Dropout [85]. These methods have found to be the most promising and highest impact methods for approximate inference in BNNs. These are both VI methods that are flexible enough to permit the use of SGD, making deployment to large and practical data sets feasible. Given their prominence, it is worthwhile to compare the methods to see how well they perform.

To compare these methods, a series of simple homoskedastic regression tasks were conducted. For these regression models, the likelihood is represented as Gaussian. With this we can write that the un-normalised posterior is,

$$p(\omega|\mathcal{D}) \propto p(\omega)\mathcal{N}(\mathbf{f}^{\omega}(\mathcal{D}), \sigma^2\mathbf{I}), \tag{3.59}$$

where $\mathbf{f}^{\omega}(\mathcal{D})$ is the function represented by the BNNs. A mixture of Gaussians was used to model a spike-slab prior for both models. The approximate posterior $q_\theta(\omega)$ was then found for each model using the respective methods proposed. For Bayes by Backprop, the approximate posterior is a fully factorised Gaussian, and for MC Dropout is a scaled Bernoulli distribution. With the approximate posterior for each model, predictive quantities can be found using MC Integration. The first two moments can be approximated as [70],

$$\mathbb{E}_q[\mathbf{y}^*] \approx \frac{1}{N} \sum_{i=1}^{N} \mathbf{f}^{\omega_i}(\mathbf{x}^*) \tag{3.60}$$

$$\mathbb{E}_q[\mathbf{y}^{*T}\mathbf{y}^*] \approx \sigma^2\mathbf{I} + \frac{1}{N} \sum_{i=1}^{N} \mathbf{f}^{\omega_i}(\mathbf{x}^*)^T\mathbf{f}^{\omega_i}(\mathbf{x}^*) \tag{3.61}$$

where the star superscript denotes the new input and output sample $\mathbf{x}^*, \mathbf{y}^*$ from the test set.

The data sets used to evaluate these models were simple toy data sets from high impact papers, where similar experimentation was provided as empirical evidence [76, 119]. Both BNN methods were then compared with a GP model. Figure 3.7 illustrates these results.

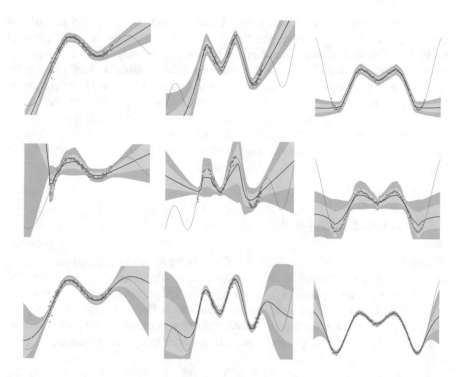

Fig. 3.7 Comparison of BNNs with GP for a regression task over three toy data sets. The top row is a BNN trained with Bayes By Backprop [76], the centre row is trained with MC dropout [70], and the bottom a GP with a Mattern52 kernel fitted with the GPflow package [120]. The two BNNs consisted of two hidden layers utilising ReLU activation. Training data is shown with the dark grey scatter, the mean is shown in purple, the true test function is shown in blue, and the shaded regions representing ± one and two std. from the mean. Best viewed on a computer screen

Analysis of the regression results shown in Fig. 3.7 shows contrasting performance in terms of bias and variance in predictions. Models trained with Bayes by Backprop and a factorised Gaussian approximate posterior show reasonable predictive results within the distribution of training data, though variance outside the region of training data is significantly under estimated when compared with the GP. MC Dropout with a scaled Bernoulli approximate posterior typically exhibits greater variance for out of distribution data, though maintains unnecessarily high variance within the distribution of training data. Little tuning of hyperparameters was done to these models. Better results may be achieved, particularly for MC Dropout, with better selection of hyperparameters. Alternatively, a more complete Bayesian approach can be used, where hyperparameters are treated as latent variables and marginalisation is performed over these variables.

It is worthwhile noting the computational and practical difficulties encountered with these methods. The MC Dropout method is incredibly versatile, in that it was less sensitive to the choice of prior distribution. It also managed to fit to more complex distributions with fewer samples and training iterations. On top all this is the significant savings in computational resources. Given that training a model using MC Dropout is often identical to how many existing deep networks are trained, inference is performed in the same time as traditional vanilla networks. It also offers no increase in the number of parameters to a network, where Bayes by Backprop requires twice as many. These factors should be taken into account for practical scenarios. If the data being modelled is smooth, is in sufficient quantity and additional time for inference is permitted, Bayes by Backprop may be preferable. For large networks with complex functions, sparse data and more stringent time requirements, MC Dropout may be more suitable.

3.3.1 Convolutional BNNs

Whilst the MLP serves as the basis for NNs, the most prominent NN architecture is the Convolutional Neural Network (CNN) [121]. These networks have excelled at challenging image classification tasks, with predictive performance far exceeding prior kernel based or feature engineered methods. A CNN differs from a typical MLP through it's application a convolution-like operator as oppose to inner products.[27] The output of a single convolutional layer can be expressed as,

$$\boldsymbol{\Phi} = u(\mathbf{X}^T * \mathbf{W}) \tag{3.62}$$

where $u(\cdot)$ is a non-linear activation and $*$ represents the convolution-like operation. Here the input $\mathbf{X}$ and the weight matrix $\mathbf{W}$ are no longer restricted to either vectors or matrices, and can instead be multi-dimensional arrays. It can be shown that CNNs can be written to have an equivalent MLP model, allowing for optimised linear algebra packages to be used for training with back-propagation [122].

Extending on the current research methods, a new type of Bayesian Convolutional Neural Network (BCNN) can be developed. This is achieved here by extending on the Bayes by Backprop method [76] to the case of models suitable for image classification. Each weight in the convolutional layers is assumed to be independent, allowing for factorisation over each individual parameter.

Experimentation was conducted to investigate the predictive performance of BCNNs, and the quality of their uncertainty estimates. These networks were configured for classification of the MNIST hand digit dataset [123].

---

[27]Emphasis is placed on "convolution like", as it is not equivalent to the mathematical operation of linear or circular convolution.

Since this task is a classification task, the likelihood for the BCNN was set to a Softmax function,

$$\text{softmax}(\mathbf{f}_i^\omega) = \frac{\mathbf{f}_i^\omega(\mathcal{D})}{\sum_j \exp\left(\mathbf{f}_j^\omega(\mathcal{D})\right)}. \tag{3.63}$$

The un-normalised posterior can then be represented as,

$$p(\omega|\mathcal{D}) \propto p(\omega) \times \text{softmax}(\mathbf{f}^\omega(\mathcal{D})). \tag{3.64}$$

The approximate posterior is then found using Bayes by Backprop. Predictive mean for test samples can be found using Eq. (3.60), and MC integration is used to approximate credible intervals [35].

Comparison with a vanilla CNN was made to evaluate the predictive performance of the BCNN. For both the vanilla and BCNN, the popular LeNet architecture [123] was used. Classification was conducted using the mean output of the BCNN, with credible intervals being used to assess the models uncertainty. Overall predictive performance for both networks on the 10,000 test images in the MNIST dataset showed comparative performance. The BCNN showed a test prediction accuracy of 98.99%, while the vanilla network showed a slight improvement with a prediction accuracy of 99.92%. Whilst the competitive predictive performance is essential, the main benefit of the BCNN is that we yield valuable information about the uncertainty of our predictions. Examples of difficult to classify digits are shown in the Appendix, accompanied by plots of the mean prediction and 95% credible intervals for each class. From these examples, we can see the large amount of predictive uncertainty for these challenging images, which could be used to make more informed decisions in practical scenarios.

This uncertainty information is invaluable for many scenarios of interest. As statistical models are increasingly employed for complex tasks containing human interaction, it is crucial that many of these systems make responsible decisions based on their perceived model of the world. For example, NNs are largely used within the development of autonomous vehicles. Development of autonomous vehicles is an incredibly challenging feat, due to the high degree of variability in scenarios and the complexity relating to human interaction. Current technologies are insufficient for safely enabling this task, and as discussed earlier, the use of these technologies have been involved in multiple deaths [24, 25]. It is not possible to model all variables within such a highly complex system. This accompanied by imperfect models and reliance on approximate inference, it is important that our models can communicate any uncertainty relating to decisions made. It is crucial that we acknowledge that in essence, our models are wrong. This is why probabilistic models are favoured for such scenarios; there is an underlying theory to help us deal with heterogeneity in our data and to account for uncertainty induced by variables not included in the model. It is vital that models used for such complex scenarios can communicate their uncertainty when used in such complex and high risk scenarios.

3.4   Conclusion

Throughout this report, the problems that arise with overconfident predictions from typical NNs and ad hoc model design have been illustrated. Bayesian analysis has been shown to provide a rich body of theory to address these challenges, though exact computation remains analytically and computationally intractable for any BNN of interest. In practice, approximate inference must be relied upon to yield accurate approximations to the posterior.

Many of the approximate methods for inference within BNNs have revolved around the MFVB approach. This provides a tractable lower bound to optimise w.r.t variational parameters. These methods are attractive due to their relative ease of use, accuracy of predictive mean values and acceptable number of induced parameters. Despite this, it was shown through the literature survey and experimentation results that the assumptions made within a fully factorised MFVB approach result in over-confident predictions. It was shown that these MFVB approaches can be extended upon to more complex models such as CNNs. Experimental results indicate comparable predictive performance to point estimate CNNs for image classification tasks. The Bayesian CNN was able to provide credible intervals on the predictions, which were found to be highly informative and intuitive measure of uncertainty for difficult to classify data points.

This survey and these experiments highlight the capabilities of Bayesian analysis to address common challenges seen in the machine learning community. These results also highlight how current approximate inference methods for BNNs are insufficient and can provide inaccurate variance information. Additional research is required to not only determine how these networks operate, but how accurate inference can be achieved with modern large networks. Methods to scale exact inference methods such as MCMC to large data sets would allow for a more principled method of performing inference. MCMC offers diagnostic methods to assess convergence and quality of inference. Similar diagnostics for VI would allow researchers and practitioners to evaluate the quality of their assumed posterior, and inform them with ways to improve on this assumption. Achieving these goals will allow us to obtain accurate posterior approximations. From this we will be able to sufficiently determine what our models know, but also what they don't know.

Appendix

See Fig. 3.8.

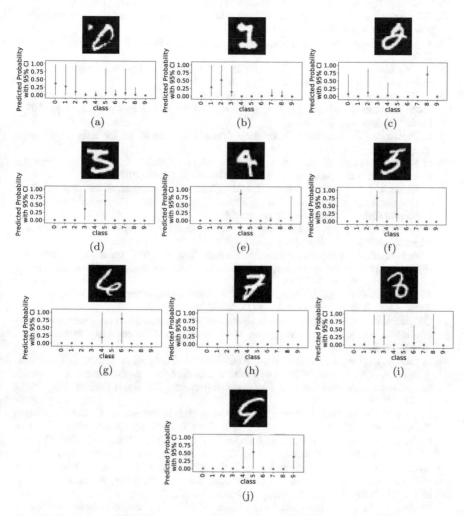

Fig. 3.8 Examples of difficult to classify images from each class in MNIST. True class for each image is 0–9 (**a–j**) arranged in alphabetical order. The bottom plot illustrates the 95% credible intervals for these predictions. Best viewed on a computer screen

References

1. F. Rosenblatt, The perceptron: a probabilistic model for information storage and organization in the brain. Psychol. Rev. **65**(6), 386–408 (1958)
2. C. Bishop, *Pattern Recognition and Machine Learning* (Springer, New York, 2006)
3. D.E. Rumelhart, G.E. Hinton, R.J. Williams, Learning representations by back-propagating errors. Nature **323**(6088), 533 (1986)
4. K.-S. Oh, K. Jung, Gpu implementation of neural networks. Pattern Recog. **37**(6), 1311–1314 (2004)
5. D.C. Ciresan, U. Meier, L.M. Gambardella, J. Schmidhuber, Deep big simple neural nets excel on handwritten digit recognition. CoRR (2010)
6. A. Krizhevsky, I. Sutskever, G.E. Hinton, Imagenet classification with deep convolutional neural networks, in *Advances in Neural Information Processing Systems* (2012), pp. 1097–1105
7. K. Simonyan, A. Zisserman, Very deep convolutional networks for large-scale image recognition. CoRR (2014)
8. C. Szegedy, W. Liu, Y. Jia, P. Sermanet, S. Reed, D. Anguelov, D. Erhan, V. Vanhoucke, A. Rabinovich et al., Going deeper with convolutions, in *CVPR* (2015)
9. R. Girshick, J. Donahue, T. Darrell, J. Malik, Rich feature hierarchies for accurate object detection and semantic segmentation, in *Proceedings of the IEEE Conference on Computer Vision and Pattern Recognition* (2014), pp. 580–587
10. S. Ren, K. He, R. Girshick, J. Sun, Faster r-cnn: towards real-time object detection with region proposal networks, in *Advances in Neural Information Processing Systems* (2015), pp. 91–99
11. J. Redmon, S. Divvala, R. Girshick, A. Farhadi, You only look once: unified, real-time object detection, in *Proceedings of the IEEE Conference on Computer Vision and Pattern Recognition* (2016), pp. 779–788
12. A. Mohamed, G.E. Dahl, G. Hinton, Acoustic modeling using deep belief networks. IEEE Trans. Audio Speech Lang. Process. **20**(1), 14–22 (2012)
13. G.E. Dahl, D. Yu, L. Deng, A. Acero, Context-dependent pre-trained deep neural networks for large-vocabulary speech recognition. IEEE Trans. Audio Speech Lang. Process. **20**(1), 30–42 (2012)
14. G. Hinton, L. Deng, D. Yu, G.E. Dahl, A.-r. Mohamed, N. Jaitly, A. Senior, V. Vanhoucke, P. Nguyen, T.N. Sainath et al., Deep neural networks for acoustic modeling in speech recognition: The shared views of four research groups. IEEE Signal Process. Mag. **29**(6), 82–97 (2012)
15. D. Amodei, S. Ananthanarayanan, R. Anubhai, J. Bai, E. Battenberg, C. Case, J. Casper, B. Catanzaro, Q. Cheng, G. Chen, J. Chen, J. Chen, Z. Chen, M. Chrzanowski, A. Coates, G. Diamos, K. Ding, N. Du, E. Elsen, J. Engel, W. Fang, L. Fan, C. Fougner, L. Gao, C. Gong, A. Hannun, T. Han, L. Johannes, B. Jiang, C. Ju, B. Jun, P. LeGresley, L. Lin, J. Liu, Y. Liu, W. Li, X. Li, D. Ma, S. Narang, A. Ng, S. Ozair, Y. Peng, R. Prenger, S. Qian, Z. Quan, J. Raiman, V. Rao, S. Satheesh, D. Seetapun, S. Sengupta, K. Srinet, A. Sriram, H. Tang, L. Tang, C. Wang, J. Wang, K. Wang, Y. Wang, Z. Wang, Z. Wang, S. Wu, L. Wei, B. Xiao, W. Xie, Y. Xie, D. Yogatama, B. Yuan, J. Zhan, Z. Zhu, Deep speech 2: end-to-end speech recognition in English and Mandarin, in *Proceedings of The 33rd International Conference on Machine Learning*. Proceedings of Machine Learning Research, New York, 20–22 Jun 2016, vol. 48, ed. by M.F. Balcan, K.Q. Weinberger (2016), pp. 173–182.
16. D. Silver, J. Schrittwieser, K. Simonyan, I. Antonoglou, A. Huang, A. Guez, T. Hubert, L. Baker, M. Lai, A. Bolton et al., Mastering the game of go without human knowledge. Nature **550**(7676), 354 (2017)
17. McKinsey & Company, Inc., Smartening up with artificial intelligence (ai) - what's in it for Germany and its industrial sector? McKinsey & Company, Inc, Tech. Rep., April 2017. Available: https://www.mckinsey.de/files/170419_mckinsey_ki_final_m.pdf

18. E.V.T.V. Serooskerken, Artificial intelligence in wealth and asset management. Pictet on Robot Advisors, Tech. Rep., Jan 2017. Available: https://perspectives.pictet.com/wp-content/uploads/2016/12/Edgar-van-Tuyll-van-Serooskerken-Pictet-Report-winter-2016-2.pdf
19. A. van den Oord, T. Walters, T. Strohman, Wavenet launches in the google assistant. Available: https://deepmind.com/blog/wavenet-launches-google-assistant/
20. Siri Team, Deep learning for siri's voice: on-device deep mixture density networks for hybrid unit selection synthesis, Aug 2017. Available: https://machinelearning.apple.com/2017/08/06/siri-voices.html
21. J. Wakefield, Microsoft chatbot is taught to swear on twitter. Available: www.bbc.com/news/technology-35890188
22. J. Guynn, Google photos labeled black people 'gorillas'. Available: https://www.usatoday.com/story/tech/2015/07/01/google-apologizes-after-photos-identify-black-people-as-gorillas/29567465/
23. J. Buolamwini, T. Gebru, Gender shades: intersectional accuracy disparities in commercial gender classification, in *Conference on Fairness, Accountability and Transparency* (2018), pp. 77–91
24. Tesla Team, A tragic loss. Available: https://www.tesla.com/en_GB/blog/tragic-loss
25. ABC News, Uber suspends self-driving car tests after vehicle hits and kills woman crossing the street in Arizona, 2018. Available: http://www.abc.net.au/news/2018-03-20/uber-suspends-self-driving-car-tests-after-fatal-crash/9565586
26. Council of European Union, Regulation (EU) 2016/679 of the European Parliment and of the council (2016)
27. B. Goodman, S. Flaxman, European union regulations on algorithmic decision-making and a "right to explanation" *AI magazine*, **38**(3), 50–57 (2017)
28. M. Vu, T. Adali, D. Ba, G. Buzsaki, D. Carlson, K. Heller, C. Liston, C. Rudin, V. Sohal, A. Widge, H. Mayberg, G. Sapiro, K. Dzirasa, A shared vision for machine learning in neuroscience. J. Neurosci. **38**(7), 1601–1607 (2018)
29. A. Holzinger, C. Biemann, C.S. Pattichis, D.B. Kell, What do we need to build explainable ai systems for the medical domain. Preprint, arXiv: 1712.09923 (2017)
30. R. Caruana, Y. Lou, J. Gehrke, P. Koch, M. Sturm, N. Elhadad, Intelligible models for healthcare: predicting pneumonia risk and hospital 30-day readmission, in *Proceedings of the 21th ACM SIGKDD International Conference on Knowledge Discovery and Data Mining* (ACM, New York, 2015), pp. 1721–1730
31. D. Gunning, Explainable artificial intelligence (XAI). Defense Advanced Research Projects Agency (DARPA), nd Web (2017)
32. D.J. MacKay, Probable networks and plausible predictions—a review of practical Bayesian methods for supervised neural networks. Netw. Comput. Neural Syst. **6**(3), 469–505 (1995)
33. J. Lampinen, A. Vehtari, Bayesian approach for neural networks—review and case studies. Neural Netw. **14**(3), 257–274 (2001)
34. H. Wang, D.-Y. Yeung, Towards Bayesian deep learning: a survey. Preprint, arXiv:1604.01662 (2016)
35. K. Murphey, *Machine Learning, A Probabilistic Perspective* (MIT Press, Cambridge, 2012)
36. X. Glorot, A. Bordes, Y. Bengio, Deep sparse rectifier neural networks, in *AISTATS* (2011), pp. 315–323
37. A.L. Maas, A.Y. Hannun, A.Y. Ng, Rectifier nonlinearities improve neural network acoustic models, in *ICML*, vol. 30 (2013), p. 3
38. R.M. Neal, *Bayesian Learning for Neural Networks*, vol. 118 (Springer Science & Business Media, New York, 1996)
39. N. Tishby, E. Levin, S.A. Solla, Consistent inference of probabilities in layered networks: predictions and generalizations, in *International 1989 Joint Conference on Neural Networks*, vol. 2 (1989), pp. 403–409
40. J.S. Denker, Y. Lecun, Transforming neural-net output levels to probability distributions, in *NeurIPS* (1991), pp. 853–859

41. G. Cybenko, Approximation by superpositions of a sigmoidal function. Math. Control Signals Syst. **2**(4), 303–314 (1989)
42. K.-I. Funahashi, On the approximate realization of continuous mappings by neural networks. Neural Netw. **2**(3), 183–192 (1989)
43. K. Hornik, Approximation capabilities of multilayer feedforward networks. Neural Netw. **4**(2), 251–257 (1991)
44. S.F. Gull, J. Skilling, *Quantified Maximum Entropy MemSys5 Users' Manual*, vol. 33 (Maximum Entropy Data Consultants Ltd, Cambridge, 1991)
45. D.J. MacKay, Bayesian interpolation. Neural Comput. **4**(3), 415–447 (1992)
46. D.J. MacKay, Bayesian methods for adaptive models. Ph.D. dissertation, California Institute of Technology, 1992
47. D.J. MacKay, A practical Bayesian framework for backpropagation networks. Neural Comput. **4**(3), 448–472 (1992)
48. M.I. Jordan, Z. Ghahramani, T.S. Jaakkola, L.K. Saul, An introduction to variational methods for graphical models. Mach. Learn. **37**(2), 183–233 (1999)
49. M.J. Wainwright, M.I. Jordan et al., Graphical models, exponential families, and variational inference. Found. Trends Mach. Learn. **1**(1–2), 1–305 (2008)
50. D.M. Blei, A. Kucukelbir, J.D. McAuliffe, Variational inference: a review for statisticians. J. Am. Stat. Assoc. **112**(518), 859–877 (2017)
51. M.D. Hoffman, D.M. Blei, C. Wang, J. Paisley, Stochastic variational inference. J. Mach. Learn. Res. **14**(1), 1303–1347 (2013)
52. D. Barber, C.M. Bishop, Ensemble learning in Bayesian neural networks, in *NATO ASI Series F Computer and Systems Sciences*, vol. 168 (Springer, New York, 1998), pp. 215–238
53. G.E. Hinton, D. Van Camp, Keeping the neural networks simple by minimizing the description length of the weights, in *Proceedings of the Sixth Annual Conference on Computational Learning Theory* (ACM, New York, 1993), pp. 5–13
54. M. Betancourt, A conceptual introduction to Hamiltonian Monte Carlo. Preprint, arXiv:1701.02434 (2017)
55. M. Betancourt, S. Byrne, S. Livingstone, M. Girolami et al., The geometric foundations of Hamiltonian Monte Carlo. Bernoulli **23**(4A), 2257–2298 (2017)
56. G. Madey, X. Xiang, S.E. Cabaniss, Y. Huang, Agent-based scientific simulation. Comput. Sci. Eng. **2**(01), 22–29 (2005)
57. S. Duane, A.D. Kennedy, B.J. Pendleton, D. Roweth, Hybrid monte carlo. Phys. Lett. B **195**(2), 216–222 (1987)
58. R.M. Neal et al., MCMC using Hamiltonian dynamics, in *Handbook of Markov Chain Monte Carlo*, vol. 2, no. 11 (CRC Press, Boca Raton, 2011), p. 2
59. S. Brooks, A. Gelman, G. Jones, X.-L. Meng, *Handbook of Markov Chain Monte Carlo* (CRC Press, Boca Raton, 2011)
60. M. Welling, Y. Teh, Bayesian learning via stochastic gradient Langevin dynamics, in *Proceedings of the 28th International Conference on Machine Learning, ICML 2011* (2011), pp. 681–688
61. A. Graves, Practical variational inference for neural networks, in *Advances in Neural Information Processing Systems 24*, ed. by J. Shawe-Taylor, R.S. Zemel, P.L. Bartlett, F. Pereira, K.Q. Weinberger (Curran Associates, Inc., Granada, 2011), pp. 2348–2356
62. M. Opper, C. Archambeau, The variational Gaussian approximation revisited. Neural Comput. **21**(3), 786–792 (2009)
63. J.M. Hernández-Lobato, R. Adams, Probabilistic backpropagation for scalable learning of Bayesian neural networks, in *International Conference on Machine Learning* (2015), pp. 1861–1869
64. J. Paisley, D. Blei, M. Jordan, Variational Bayesian inference with stochastic search. Preprint, arXiv:1206.6430 (2012)
65. J.R. Wilson, Variance reduction techniques for digital simulation. Am. J. Math. Manag. Sci. **4**(3–4), 277–312 (1984)

66. M. Opper, C. Archambeau, The variational Gaussian approximation revisited. Neural Comput. **21**(3), 786–792 (2009)
67. D.P. Kingma, M. Welling, Auto-encoding variational Bayes. Preprint, arXiv:1312.6114 (2013)
68. D.J. Rezende, S. Mohamed, D. Wierstra, Stochastic backpropagation and approximate inference in deep generative models, in *Proceedings of the 31st International Conference on Machine Learning (ICML)* (2014), pp. 1278–1286.
69. N. Srivastava, G. Hinton, A. Krizhevsky, I. Sutskever, R. Salakhutdinov, Dropout: a simple way to prevent neural networks from overfitting. J. Mach. Learn. Res. **15**(1), 1929–1958 (2014)
70. Y. Gal, Uncertainty in deep learning. University of Cambridge, 2016
71. D.P. Kingma, T. Salimans, M. Welling, Variational dropout and the local reparameterization trick, in *Advances in Neural Information Processing Systems* (2015), pp. 2575–2583
72. S. Wang, C. Manning, Fast dropout training, in *International Conference on Machine Learning* (2013), pp. 118–126
73. A. Livnat, C. Papadimitriou, N. Pippenger, M.W. Feldman, Sex, mixability, and modularity. Proc. Natl. Acad. Sci. **107**(4), 1452–1457 (2010)
74. M. Opper, O. Winther, A Bayesian approach to on-line learning, in *On-Line Learning in Neural Networks* (1998), pp. 363–378
75. T.P. Minka, A family of algorithms for approximate Bayesian inference. Ph.D. dissertation, Massachusetts Institute of Technology, 2001
76. C. Blundell, J. Cornebise, K. Kavukcuoglu, D. Wierstra, Weight uncertainty in neural networks. Preprint, arXiv:1505.05424 (2015)
77. C.K. Williams, Computing with infinite networks, in *Advances in Neural Information Processing Systems* (1997), pp. 295–301
78. J. Lee, J. Sohl-Dickstein, J. Pennington, R. Novak, S. Schoenholz, Y. Bahri, Deep neural networks as Gaussian processes, in *International Conference on Learning Representations* (2018)
79. A. Damianou, N. Lawrence, Deep Gaussian processes, in *AISTATS* (2013), pp. 207–215
80. A. Damianou, Deep Gaussian processes and variational propagation of uncertainty. Ph.D. dissertation, University of Sheffield, 2015
81. N. Lawrence, Deep Gaussian processes, 2019. Available: http://inverseprobability.com/talks/notes/deep-gaussian-processes.html
82. A. Damianou, M.K. Titsias, N.D. Lawrence, Variational Gaussian process dynamical systems, in *NeurIPS* (2011), pp. 2510–2518
83. M. Titsias, Variational learning of inducing variables in sparse Gaussian processes, in *Proceedings of the Twelfth International Conference on Artificial Intelligence and Statistics.* Proceedings of Machine Learning Research, 16–18 April 2009, vol. 5, ed. by D. van Dyk, M. Welling (Hilton Clearwater Beach Resort, Clearwater, 2009), pp. 567–574
84. Y. Gal, Z. Ghahramani, Dropout as a Bayesian approximation: insights and applications, in *Deep Learning Workshop, ICML*, vol. 1 (2015), p. 2
85. Y. Gal, Z. Ghahramani, Dropout as a Bayesian approximation: representing model uncertainty in deep learning, in *ICML* (2016), pp. 1050–1059
86. Y. Gal, Z. Ghahramani, Dropout as a Bayesian approximation: appendix. Preprint, arXiv:1506.02157 (2015)
87. A. Garriga-Alonso, L. Aitchison, C.E. Rasmussen, Deep convolutional networks as shallow Gaussian processes. Preprint, arXiv:1808.05587 (2018)
88. R. Novak, L. Xiao, Y. Bahri, J. Lee, G. Yang, D.A. Abolafia, J. Pennington, J. Sohl-Dickstein, Bayesian deep convolutional networks with many channels are Gaussian processes, in *International Conference on Learning Representations* (2019)
89. M. Van der Wilk, C.E. Rasmussen, J. Hensman, Convolutional Gaussian processes, in *Advances in Neural Information Processing Systems* (2017), pp. 2849–2858
90. D.J. MacKay, D.J. Mac Kay, *Information Theory, Inference and Learning Algorithms* (Cambridge University Press, Cambridge, 2003)

91. B. Wang, D. Titterington, Inadequacy of interval estimates corresponding to variational Bayesian approximations, in *AISTATS*, Barbados (2005)
92. R.E. Turner, M. Sahani, Two problems with variational expectation maximisation for time-series models, in *Bayesian Time Series Models*, ed. by D. Barber, A.T. Cemgil, S. Chiappa (Cambridge University Press, Cambridge, 2011)
93. R. Giordano, T. Broderick, M.I. Jordan, Covariances, robustness, and variational bayes. J. Mach. Learn. Res. **19** (51), 1–49 (2018)
94. D. Hafner, D. Tran, A. Irpan, T. Lillicrap, J. Davidson, Reliable uncertainty estimates in deep neural networks using noise contrastive priors. Preprint, arXiv:1807.09289 (2018)
95. V. Kuleshov, N. Fenner, S. Ermon, Accurate uncertainties for deep learning using calibrated regression. Preprint, arXiv:1807.00263 (2018)
96. Y. Gal, J. Hron, A. Kendall, Concrete dropout, in *Advances in Neural Information Processing Systems* (2017), pp. 3581–3590
97. C.J. Maddison, A. Mnih, Y.W. Teh, The concrete distribution: a continuous relaxation of discrete random variables. Preprint, arXiv:1611.00712 (2016)
98. T.S. Jaakkola, M.I. Jordan, Improving the mean field approximation via the use of mixture distributions, in *Learning in Graphical Models* (Springer, Berlin, 1998), pp. 163–173
99. C. Louizos, M. Welling, Structured and efficient variational deep learning with matrix Gaussian posteriors, in *International Conference on Machine Learning* (2016), pp. 1708–1716
100. E.G. Tabak, E. Vanden-Eijnden, Density estimation by dual ascent of the log-likelihood. Commun. Math. Sci. **8**(1), 217–233 (2010)
101. E.G. Tabak, C.V. Turner, A family of nonparametric density estimation algorithms. Commun. Pure Appl. Math. **66**(2), 145–164 (2013)
102. D.J. Rezende, S. Mohamed, Variational inference with normalizing flows. Preprint, arXiv:1505.05770 (2015)
103. C. Louizos, M. Welling, Multiplicative normalizing flows for variational Bayesian neural networks, in *Proceedings of the 34th International Conference on Machine Learning - Volume 70*, ICML'17, JMLR.org (2017), pp. 2218–2227
104. L. Dinh, J. Sohl-Dickstein, S. Bengio, Density estimation using real NVP. CoRR, vol. abs/1605.08803 (2016)
105. C. Cremer, X. Li, D.K. Duvenaud, Inference suboptimality in variational autoencoders. CoRR, vol. abs/1801.03558 (2018)
106. S.-i. Amari, *Differential-Geometrical Methods in Statistics*, vol. 28 (Springer Science & Business Media, New York, 2012)
107. T. Minka et al., Divergence measures and message passing. Technical report, Microsoft Research (2005)
108. Y. Jia, E. Shelhamer, J. Donahue, S. Karayev, J. Long, R. Girshick, S. Guadarrama, T. Darrell, Caffe: convolutional architecture for fast feature embedding. Preprint, arXiv:1408.5093 (2014)
109. F. Chollet, keras (2015). https://github.com/fchollet/keras
110. M. Abadi, A. Agarwal, P. Barham, E. Brevdo, Z. Chen, C. Citro, G.S. Corrado, A. Davis, J. Dean, M. Devin, S. Ghemawat, I.J. Goodfellow, A. Harp, G. Irving, M. Isard, Y. Jia, R. Józefowicz, L. Kaiser, M. Kudlur, J. Levenberg, D. Mané, R. Monga, S. Moore, D.G. Murray, C. Olah, M. Schuster, J. Shlens, B. Steiner, I. Sutskever, K. Talwar, P.A. Tucker, V. Vanhoucke, V. Vasudevan, F.B. Viégas, O. Vinyals, P. Warden, M. Wattenberg, M. Wicke, Y. Yu, X. Zheng, Tensorflow: large-scale machine learning on heterogeneous distributed systems. CoRR, vol. abs/1603.04467 (2016)
111. J.V. Dillon, I. Langmore, D. Tran, E. Brevdo, S. Vasudevan, D. Moore, B. Patton, A. Alemi, M.D. Hoffman, R.A. Saurous, Tensorflow distributions. CoRR, vol. abs/1711.10604 (2017)
112. A. Paszke, S. Gross, S. Chintala, G. Chanan, E. Yang, Z. DeVito, Z. Lin, A. Desmaison, L. Antiga, A. Lerer, Automatic differentiation in pytorch, in *Proceedings of Neural Information Processing Systems* (2017)

113. T. Chen, M. Li, Y. Li, M. Lin, N. Wang, M. Wang, T. Xiao, B. Xu, C. Zhang, Z. Zhang, Mxnet: a flexible and efficient machine learning library for heterogeneous distributed systems. CoRR, vol. abs/1512.01274 (2015)
114. A. Kucukelbir, D. Tran, R. Ranganath, A. Gelman, D.M. Blei, Automatic differentiation variational inference. e-prints, arXiv:1603.00788 (2016)
115. S. Patterson, Y.W. Teh, Stochastic gradient Riemannian Langevin dynamics on the probability simplex, in *Advances in Neural Information Processing Systems 26*, ed. by C.J.C. Burges, L. Bottou, M. Welling, Z. Ghahramani, K.Q. Weinberger (Curran Associates, Inc., Granada, 2013), pp. 3102–3110
116. T. Chen, E. Fox, C. Guestrin, Stochastic gradient Hamiltonian monte carlo, in *Proceedings of the 31st International Conference on Machine Learning*. Proceedings of Machine Learning Research, 22–24 Jun 2014, vol. 32, ed. by E.P. Xing, T. Jebara (2014), pp. 1683–1691
117. C. Li, C. Chen, D. Carlson, L. Carin, Preconditioned stochastic gradient Langevin dynamics for deep neural networks. arXiv e-prints, Dec 2015
118. M. Betancourt, The fundamental incompatibility of scalable Hamiltonian monte carlo and naive data subsampling, in *Proceedings of the 32Nd International Conference on International Conference on Machine Learning - Volume 37*, ICML'15, JMLR.org (2015), pp. 533–540
119. I. Osband, C. Blundell, A. Pritzel, B.V. Roy, Deep exploration via bootstrapped DQN. CoRR, vol. abs/1602.04621 (2016)
120. A.G.d.G. Matthews, M. van der Wilk, T. Nickson, K. Fujii, A. Boukouvalas, P. León-Villagrá, Z. Ghahramani, J. Hensman, GPflow: a Gaussian process library using TensorFlow. J. Mach. Learn. Res. **18**(40), 1–6 (2017)
121. Y. LeCun, B. Boser, J.S. Denker, D. Henderson, R.E. Howard, W. Hubbard, L.D. Jackel, Backpropagation applied to handwritten zip code recognition. Neural Comput. **1**(4), 541–551 (1989)
122. I. Goodfellow, Y. Bengio, A. Courville, *Deep Learning* (MIT Press, Cambridge, 2016)
123. Y. Lecun, L. Bottou, Y. Bengio, P. Haffner, Gradient-based learning applied to document recognition. Proc. IEEE **86**(11), 2278–2324 (1998)

Chapter 4
Markov Chain Monte Carlo Algorithms for Bayesian Computation, a Survey and Some Generalisation

Wu Changye and Christian P. Robert

Abstract This chapter briefly recalls the major simulation based methods for conducting Bayesian computation, before focusing on partly deterministic Markov processes and a novel modification of the bouncy particle sampler that offers an interesting alternative when dealing with large datasets.

Keywords Monte Carlo methods · MCMC algorithms · Bouncy particle sampler · PDMP · Big Data

4.1 Bayesian Statistics

In statistical analysis, the statistician frames observations, $(X_{1:n}) \subset \mathcal{X}$, within a model that belongs to a class of probability distributions $\mathcal{P} = \{P_\theta, \theta \in \Theta\}$ over the sample space $(\mathcal{X}, \mathcal{A})$, where θ is called the model parameter and Θ is an arbitrary set. In this book, we mostly focus on parametric cases—that is, $\Theta \subset \mathbb{R}^d$, and suppose the distributions to be dominated by some measure $\mu(dx)$. In both frequentist statistics and Bayesian statistics, the likelihood plays a crucial role in inference, which encompasses the plausibility of parameter values, given the observed data.

The source of this chapter appears in the PhD thesis of the first author, which he defended in 2018 at Université Paris Dauphine.

W. Changye
Université Paris Dauphine PSL, Paris, France

C. P. Robert (✉)
Université Paris Dauphine PSL, Paris, France

University of Warwick, Coventry, UK
e-mail: xian@ceremade.dauphine.fr

K. L. Mengersen et al. (eds.), *Case Studies in Applied Bayesian Data Science*, Lecture Notes in Mathematics 2259, https://doi.org/10.1007/978-3-030-42553-1_4

Definition 4.1.1 The likelihood function is defined as a function of the parameter θ associated with the probability mass function or density function of the observations $(X_{1:n})$ conditioned on the parameter θ and is denoted by $\mathcal{L}(X_{1:n}|\theta)$.

While there exists a true, fixed parameter θ_0 such that P_{θ_0} is the distribution that spanned the observations, Bayesian statistics models the parameter θ as a random variable, associated with a prior distribution that describes our beliefs about the parameter and is independent of the data. As a result, the parameter space is equipped with a probability structure $(\Theta, \mathcal{B}, \pi_0)$ and Bayesian analysis extracts information about θ by combining these prior beliefs, π_0, and the information provided by the observed data. In the Bayesian paradigm [5, 42], once the prior and the likelihood have been chosen, the information about the parameter is modelled by the posterior distribution, which is defined as follows.

Definition 4.1.2 The posterior distribution is the probability distribution of the parameter θ, given the observations $(X_{1:n})$, over the parameter space $(\Theta, \mathcal{B})$. According to Bayes' Theorem, it has the following form,

$$\pi(d\theta|X_{1:n}) = \frac{\mathcal{L}(X_{1:n}|\theta)\pi_0(d\theta)}{\int_\Theta \mathcal{L}(X_{1:n}|\theta')\pi_0(d\theta')}$$

The integral, $\int_\Theta \mathcal{L}(X_{1:n}|\theta')\pi_0(d\theta')$, in the denominator is called the evidence, or the marginal likelihood and is denoted by $m_{\pi_0}(X_{1:n})$.

Compared with frequentist statistics, Bayesian methods explicitly use probability tools as a way to quantify uncertainties about the unknown quantities. While Bayesian inference is by nature uniquely defined and unique, in practice, problems usually arise as the posterior has no closed or interpretable form. For instance, the evidence may be not explicitly available. In the early days of Bayesian analysis, it was confined to problems where the posteriors are explicitly available, such as conjugate priors.

Definition 4.1.3 If the posterior distribution $\pi(\cdot|X_{1:n})$ is in the same family of probability distributions as the prior π_0, the prior and the likelihood are then said to be conjugate distributions, and the associated family of prior distributions is called a conjugate family for the likelihood function $\mathcal{L}(X_{1:n}|\theta)$.

Unfortunately, even though conjugacy is definitely useful in many applications, it is far too restrictive a notion and thus does not offer a universal modelling solution.

Approaches to overcoming this restriction can be separated into two main groups: approximation methods and Monte Carlo methods. Approximation methods, such as Laplace's, expectation propagation and variational Bayes, project the exact posterior of interest into a tractable family of probability distributions and approximate it with the closest element of this family. Unless the posterior belongs to the chosen family, there is an intrinsic gap between the resulting approximation and the posterior of interest. Monte Carlo methods, which are also the focus of this chapter, have different behaviours and can produce solutions that approximate the distributions or quantities of interest in any degree of precision when the computation effort grows to infinity.

4.2   Monte Carlo Methods

The concept of Monte Carlo approximations is based on the Law of Large Numbers (LLN) to approximate the integrals of interest. Consider the instance when we are interested in computing an integral of the form

$$I_h := \mathbb{E}_P(h(X))$$

assuming its existence. The LLN says that if $X_1, X_2, \cdots$, is an infinite sequence of independent and identically distributed (i.i.d.) random variables according to the probability distribution P, then

$$\frac{1}{N} \sum_{i=1}^{N} h(X_i) \xrightarrow[N \to \infty]{P} I_h$$

Based on an additional assumption that $\sigma^2 := \mathbb{E}_P(h^2(X)) - I_h^2 < \infty$, the central limit theorem (CLT) gives a stronger result,

$$\sqrt{N} \left(\sum_{i=1}^{N} h(X_i) - I_h \right) \xrightarrow[N \to \infty]{\mathcal{L}} \mathcal{N}(0, \sigma^2)$$

By the CLT, the Monte Carlo estimator converges to I_h at a rate $\mathcal{O}(N^{-1/2})$, regardless of the dimensionality of X_i's.

4.2.1   The Inverse Transform

When the distribution to sample from is one-dimensional with c.d.f F, we can sample $U \sim \mathcal{U}[0, 1]$ and compute $X = F^{-1}(U)$, where F^{-1} is the generalized inverse function of F. It is easy to verify that $X \sim F$.

Definition 4.2.1 For a non-decreasing function F on $\mathbb{R}$, the generalized inverse of F, F^{-1}, is the function defined by

$$F^{-1}(u) = \inf\{x | F(x) \geq u\}$$

This method is only applicable to one-dimensional distributions and requires deep knowledge about the generalized inverse function F^{-1}, which restricts its applicability in practice. General transformation methods extend this inverse transform by taking advantage of the links between the target distribution and some tractable distributions. Unfortunately, only a few distributions can be expressed as a transformation of other easier distributions. Even worse, posterior distributions in

Bayesian statistics, which are the main target distributions of this chapter, usually cannot be sampled by this method. See [18, 43] for more details about general transformation methods and some examples.

4.2.2 Accept-Reject Sampling

Accept-Reject sampling only requires that the target p of interest is known up to a multiplicative constant and is based on the *Fundamental Theorem of Simulation*, which says that sampling uniformly over the subgraph of a probability density function results in samples marginally distributed according to the distribution.

Theorem 4.2.1 (Fundamental Theorem of Simulation) *Sampling from*

$$X \sim p$$

is equivalent to sample from

$$(X, U) \sim \mathcal{U}(\{(x, u)|0 < u < p(x)\})$$

and marginalise with respect to U.

In light of the simple fact that

$$p(x) = \int_0^{p(x)} du,$$

it is easy to show that if we can sample (X, U) from $\mathcal{F} = \{(x, u)|0 < u < f(x)\}$ uniformly, then the marginal X is distributed to our desired target p. In practice, sampling (X, U) is not always feasible or too expensive, we can sample from an bigger set and discard the pairs which are outside of $\mathcal{F}$ to bypass this difficulty. Generally, such an auxiliary set is constructed by the subgraph of Mq, where q is a probability density function, which it is easy to sample from, and M is a known constant such that $p \leq Mq$ on the support of p. Of course the choice of the auxiliary distribution q will impact the efficiency of the algorithm.

Algorithm 1 Accept-reject algorithm: sample from p

Sample $X \sim q$ and $U \sim \mathcal{U}[0, Mq(X)]$
if $u \leq f(X)$ **then**
    Accept X
else
    Reject X
end if

4.2.3   Importance Sampling

As in accept-reject sampling, importance sampling (IS) introduces an auxiliary distribution, Q, and is based on the identity

$$\mathbb{E}_P(h(X)) = \mathbb{E}_Q(h(X)w(X))$$

where Q dominates P and is called the *importance distribution*, and $w(x) = \frac{dP}{dQ}(x)$ is called the *weight function*. By the LLN, the integral of interest, I_h, can be approximated by

$$\hat{I}_h^N = \frac{1}{N} \sum_{i=1}^{N} h(X_i)w(X_i), \quad X_i \overset{iid}{\sim} Q$$

The remarkable advantage of importance sampling is that the weight function w can be known up to a multiplicative constant, which is extremely valuable for sampling from posterior in Bayesian inference. In fact, the multiplicative constant can be estimated by $\frac{1}{N}\sum_{i=1}^{N} w(X_i)$ and one can show that the normalized (and biased) estimator

$$\frac{\sum_{i=1}^{N} h(X_i)w(X_i)}{\sum_{i=1}^{N} w(X_i)}$$

converges to the integral of interest.

As in accept-reject sampling, the importance distribution Q has a large influence over the efficiency of IS. The optimal Q, which minimises the variance of the estimator $\hat{I}_h^N$, not only depends of the target distribution P, but also on the integrand h. However, it always satisfies

$$\mathrm{Var}_Q[h(X)w(X)] \leq \mathrm{Var}_P[h(X)],$$

where Q is optimal. For more details of the choice of optimal Q's see [43].

So far, all of the above requires a sequence of i.i.d. proposals and the obtained samples are thus independent to each other. In the next section, we describe a class of sampling algorithms, based on Markov chains, which produce correlated samples to approximate the target distribution or the integrals of interest.

4.3   Markov Chain Monte Carlo Methods

Markov chain Monte Carlo (MCMC) algorithms have been used for nearly 60 years, becoming a reference method for analysing Bayesian complex models in the early 1990s [21]. The strength of this method is that it guarantees convergence to the quantity (or quantities) of interest with minimal requirements on the targeted distribution (also called *target*) behind such quantities. In that sense, MCMC algorithms are robust or universal, as opposed to the most standard Monte Carlo methods (see, e.g., [43, 44]) that require direct simulations from the target distribution.

MCMC methods have a history (see, e.g. [12]) that starts at approximately the same time as the Monte Carlo methods, in conjunction with the conception of the first computers. They have been devised to handle the simulation of complex target distributions, when complexity stems from the shape of the target density, the size of the associated data, the dimension of the object to be simulated, or from time requirements. For instance, the target density $p(x)$ may happen to be expressed in terms of multiple integrals that cannot be solved analytically,

$$p(x) = \int \omega(x, \xi) d\xi,$$

which requires the simulation of the entire vector (x, ξ). In cases when ξ is of the same dimension as the data, as for instance in latent variable models, this significant increase in the dimension of the object to be simulated creates computational difficulties for standard Monte Carlo methods, from managing the new target $\omega(x, \xi)$, to devising a new and efficient simulation algorithm. A Markov chain Monte Carlo (MCMC) algorithm allows for an alternative resolution of this computational challenge by simulating a Markov chain that explores the space of interest (and possibly supplementary spaces of auxiliary variables) without requiring a deep preliminary knowledge on the density p, besides the ability to compute $p(x_0)$ for a given parameter value x_0 (if up to a normalising constant) and possibly the gradient $\nabla \log p(x_0)$.

The validation of the method (e.g., [43]) is that the Markov chain is *ergodic* (e.g., [34]), namely that it converges in distribution to the distribution with density π, no matter where the Markov chain is started at time $t = 0$. As the basic Monte Carlo method, MCMC enjoys corresponding LLN and CLT.

Theorem 4.3.1 (Ergodic Theorem, [43]) *If* $(X_n)_{n \geq 0}$ *is a positive Harris recurrent Markov chain with invariant measure P, then for every* $h \in L_1(P)$, *we have*

$$\frac{1}{N} \sum_{i=1}^{N} h(X_i) \xrightarrow[N \to \infty]{} \int h(x) P(dx)$$

Theorem 4.3.2 (Markov Central Limit Theorem, [43]) *If* $(X_n)_{n \geq 0}$ *is a positive Harris recurrent and irreducible Markov chain, geometrically ergodic with invariant measure P, and if the function h satisfies* $\mathbb{E}_P[h(X)] = 0$ *and* $\mathbb{E}_P[|h(X)|^{2+\epsilon}] < \infty$ *for some* $\epsilon > 0$, *then*

$$\frac{1}{N} \sum_{i=1}^{N} h(X_i) \xrightarrow[N \to \infty]{\mathcal{L}} \mathcal{N}(0, \sigma_h^2)$$

for some finite $\sigma_h^2 = \mathbb{E}_P[h(X_0)^2] + 2\sum_{k=1}^{\infty} \mathbb{E}_P[h(X_0)h(X_k)] < \infty.$

See [43] for a comprehensive survey of the Markov chain theory used with MCMC algorithms.

4.3.1   Metropolis-Hastings Algorithms

The Metropolis–Hastings algorithm is a generic illustration of the principle of MCMC and is named after Nicholas Metropolis, who first proposed the algorithm by using symmetric proposal distribution in [33], and Keith Hastings, who extended it to the more general case in [23]. The basic algorithm is constructed by choosing a *proposal*, that is, a conditional density $q(x'|x)$ (also known as a *Markov kernel*), the Markov chain $\{X_n\}_{n=1}^{\infty}$ being then derived by successive simulations of the transition

$$X_{n+1} = \begin{cases} X' \sim q(X'|X_n) & \text{with probability } \left\{ \dfrac{p(X')}{p(X_n)} \times \dfrac{q(X_n|X')}{q(X'|X_n)} \right\} \wedge 1, \\ X_n & \text{otherwise.} \end{cases}$$

This acceptance-rejection feature of the algorithm makes it appropriate for targeting p as its stationary distribution if the resulting Markov chain $(X_n)_n$ is irreducible, i.e., has a positive probability of visiting any region of the support of p in a finite number of iterations. (Stationarity can easily be shown, e.g., by using the so-called *detailed balance property* that makes the chain time-reversible, see, e.g., [43].) The most widely used Markov kernel in Metropolis–Hastings might be the random walk proposals, in which $q(x|y) = q(y|x)$ and the density of proposal cancels in the acceptance ratio. Considering the initial goal of simulating samples from the target distribution p, the performances of MCMC methods like the Metropolis–Hastings algorithm above often vary quite a lot, depending primarily on the adequacy between the proposal q and the target p. For instance, if $q(\cdot|X_n) = p(\cdot)$, the Metropolis–Hastings algorithm reduces to i.i.d. sampling from the target, which is of course a formal option when i.i.d. sampling from p proves impossible. Although there exist rare instances when the Markov chain (X_n) leads to negative correlations between the successive terms of the chain, making it *more efficient* than regular i.i.d. sampling [30], the most common occurrence is one of positive correlation between the simulated values (sometimes uniformly, see [29]). This feature implies a reduced efficiency of the algorithm and hence requires a larger number of simulations to achieve the same precision as an approximation based on i.i.d. simulations (without accounting for differences in computing time). More generally, an MCMC algorithm may require a large number of iterations to escape the attraction of its starting point X_0 and to reach stationarity, to the extent that some versions of such algorithms fail to converge in the time available (i.e., in practice if not in theory).

Algorithm 2 Metropolis-Hastings algorithm

Input: the starting point X_0, the proposal distribution q and the number of iterations N.
for $n = 1, 2, \cdots, N$ **do**
   Sample $X' \sim q(\cdot|X_{n-1})$
   Compute the acceptance probability $\alpha(X_{n-1}, X')$, where

$$\alpha(X_{n-1}, X') = \min\left\{1, \frac{p(X')q(X_{n-1}|X')}{p(X_{n-1})q(X'|X_{n-1})}\right\}$$

   Sample $U \sim \mathcal{U}[0, 1]$;
   if $U < \alpha(X_{n-1}, X')$ **then**
      $X_n \to X'$
   else
      $X_n \to X_{n-1}$
   end if
end for

4.3.2 Hamiltonian Monte Carlo

Hamiltonian (or hybrid) Monte Carlo (HMC) is an auxiliary variable technique that takes advantage of a continuous time Markov process to sample from the target p. This approach comes from physics [20] and was popularised in statistics by Neal [36, 37] and MacKay [32]. Given a target $p(x)$, where $x \in \mathbb{R}^d$, an artificial auxiliary variable $v \in \mathbb{R}^d$ is introduced along with a density $\varphi(v|x)$ so that the joint distribution of (x, v) enjoys $p(x)$ as its marginal. While there is complete freedom in this representation, the HMC literature often calls v the *momentum* of a particle located at x by analogy with physics. Based on the representation of the joint distribution

$$\rho(x, v) = p(x)\varphi(v|x) \propto \exp\{-H(x, v)\},$$

where $H(\cdot)$ is called the *Hamiltonian*, Hamiltonian Monte Carlo (HMC) is associated with the continuous time process (x_t, v_t) generated by the so-called *Hamiltonian equations*

$$\frac{\mathrm{d}x_t}{\mathrm{d}t} = \frac{\partial H}{\partial v}(x_t, v_t) \qquad \frac{\mathrm{d}v_t}{\mathrm{d}t} = -\frac{\partial H}{\partial x}(x_t, v_t),$$

which keep the Hamiltonian target stable over time, as

$$\frac{\mathrm{d}H(x_t, v_t)}{\mathrm{d}t} = \frac{\partial H}{\partial v}(x_t, v_t)\frac{\mathrm{d}v_t}{\mathrm{d}t} + \frac{\partial H}{\partial x}(x_t, v_t)\frac{\mathrm{d}x_t}{\mathrm{d}t} = 0.$$

Obviously, the above continuous time Markov process is deterministic and only explores a given level set,

$$\{(x, v) : H(x, v) = H(x_0, v_0)\},$$

instead of the whole augmented state space $\mathbb{R}^{2d}$, which induces an issue with irreducibility. An acceptable solution to this problem is to refresh the momentum, $v_t \sim \varphi(v|x_{t-})$, at random times $\{\tau_n\}_{n=1}^{\infty}$, where x_{t-} denotes the location of x immediately prior to time t, and the random durations $\{\tau_n - \tau_{n-1}\}_{n=2}^{\infty}$ follow an exponential distribution. By construction, continuous-time Hamiltonian Markov chain can be regarded as a specific piecewise deterministic Markov process (PDMP) using Hamiltonian dynamics [10, 15, 16] and our target, π, is the marginal of its associated invariant distribution.

Before moving to the practical implementation of the concept, let us point out that the free cog in the machinery is the conditional density $\varphi(v|x)$, which is usually chosen as a Gaussian density with either a constant covariance matrix M corresponding to the target covariance or as a local curvature depending on x in Riemannian Hamiltonian Monte Carlo [22]. Betancourt [6] argues in favour of these two cases against non-Gaussian alternatives and [31] analyses how different choices of kinetic energy in Hamiltonian Monte Carlo impact algorithm performance. For a fixed covariance matrix, the Hamilton equations become

$$\frac{dx_t}{dt} = M^{-1}v_t \qquad \frac{dv_t}{dt} = \nabla \log p(x_t),$$

which is the score function. The velocity (or momentum) of the process is thus driven by this score function, gradient of the log-target.

The above description remains quite conceptual in that there is no generic methodology for producing this continuous time process, since Hamilton equations cannot be solved exactly in most cases. Furthermore, standard numerical solvers like Euler's method create an unstable approximation that induces a bias as the process drifts away from its true trajectory. There exists however a discretisation simulation technique that produces a Markov chain and which is well-suited to the Hamiltonian equations in that it preserves the stationary distribution [6]. It is called the *symplectic integrator*, and one version in the independent case with constant covariance consists in the following (so-called *leapfrog*) steps

$$v_{t+\epsilon/2} = v_t + \epsilon \nabla \log p(x_t)/2,$$

$$x_{t+\epsilon} = x_t + \epsilon M^{-1} v_{t+\epsilon/2},$$

$$v_{t+\epsilon} = v_{t+\epsilon/2} + \epsilon \nabla \log p(x_{t+\epsilon})/2,$$

where ϵ is the time-discretisation step. Using a proposal on v_0 drawn from the Gaussian auxiliary target and deciding on the acceptance of the value of $(x_{T\epsilon}, v_{T\epsilon})$ by a Metropolis–Hastings step can limit the danger of missing the target. Note that the first two leapfrog steps induce a Langevin move on x_t:

$$x_{t+\epsilon} = x_t + \epsilon^2 M^{-1} \nabla \log p(x_t)/2 + \epsilon M^{-1} v_t,$$

thus connecting with the MALA algorithm. In practice, it is important to note that discretising Hamiltonian dynamics introduces two free parameters, the step size ϵ and the trajectory length T, both to be calibrated. As an empirically successful and popular variant of HMC, the "no-U-turn sampler" (NUTS) of [24] adapts the value of ϵ based on primal-dual averaging. It also eliminates the need to choose the trajectory length T via a recursive algorithm that builds a set of candidate proposals for a number of forward and backward leapfrog steps and stops automatically when the simulated path retraces.

Algorithm 3 Leapfrog(x_0, v_0, ϵ, L)

Input: the starting position x_0, the starting momentum v_0, the step-size ϵ, the steps L
for $\ell = 0, 1, \cdots, L - 1$ **do**
  $v_{\ell+1/2} = v_\ell + \epsilon \nabla \log p(x_\ell)$
  $x_{\ell+1} = x_\ell + \epsilon M^{-1} v_{\ell+1/2}$
  $v_{\ell+1} = v_{\ell+1/2} + \epsilon \nabla \log p(x_{\ell+1})$
end for
Output: (x_L, v_L)

Algorithm 4 Hamiltonian Monte Carlo algorithm

Input: the step-size ϵ, the steps of leapfrog integrator L, starting position x_0, the desired number of iterations N.
for $n = 1, \cdots, N$ **do**
  Sample $v_{n-1} \sim \varphi(v)$;
  Compute $(x^*, v^*) \leftarrow$ Leapfrog($x_{n-1}, v_{n-1}, \epsilon, L$);
  Compute the acceptance ratio α, where

$$\alpha = \min\left\{1, \frac{\exp(-H(x^*, -v^*))}{\exp(-H(x_{n-1}, v_{n-1}))}\right\};$$

  Sample $u \sim \mathcal{U}[0, 1]$;
  if $u < \alpha$ **then**
    $x_n \leftarrow x^*$
  else
    $x_n \leftarrow x_{n-1}$
  end if
end for

4.3.3   Scalable MCMC

The explosion in the collection and analysis of "big" datasets in recent years[1] has brought new challenges to the MCMC algorithms that are used for Bayesian inference. When examining whether or not a new proposed sample is accepted at the accept-reject step, an MCMC algorithm such as the Metropolis-Hastings version needs to sweep over the whole data set, at each and every iteration, for the evaluation of the likelihood function. MCMC algorithms are then difficult to scale up, which in turn strongly hinders their application in big data settings. In some cases, the datasets may be too large to fit on a single machine. It may also be that confidentiality measures impose different databases to stand on separate networks, with the possible added burden of encrypted data [2]. Communication between the separate machines may prove impossible on an MCMC scale that involves thousands or hundreds of thousands iterations.

In the recent years, efforts have been made to design *scalable* algorithms, namely, solutions that manage to handle large scale targets by breaking the problem into manageable or scalable pieces. Roughly speaking, these methods can be classified into two categories [4]: divide-and-conquer approaches and sub-sampling approaches.

Divide-and-conquer approaches partition the whole data set, denoted $\mathcal{D}$, into batches, $\{\mathcal{D}_1, \cdots, \mathcal{D}_k\}$, and run separate MCMC algorithms on each data batch, independently, as if they were independent Bayesian inference problems.[2] These methods then combine the simulated parameter outcomes together to approximate the original posterior distribution. Depending on the treatments of the batches selected in the MCMC stages, these approaches can be further subdivided into two finer groups: sub-posterior methods and boosted sub-posterior methods. Sub-posterior methods are motivated by the independent product equation:

$$\pi(\theta|\mathcal{D}) \propto \prod_{i=1}^{k} \left(\pi_0(\theta)^{1/k} \prod_{\ell \in \mathcal{X}_i} p(x_\ell|\theta) \right) = \prod_{i=1}^{k} \pi_i(\theta), \qquad (4.1)$$

and they target the densities $\pi_i(\theta)$ (up to a constant) in their respective MCMC steps. They thus bypass communication costs [45], by running MCMC samplers independently on each batch, and they most often increase MCMC mixing rates (in effective samples sizes produced by second), given that the sub-posterior distributions $\pi_i(\theta)$ are based on smaller datasets. For instance, [45] combine the samples

---

[1] Some of the material in this section was also used in the paper "Accelerating MCMC Algorithms", written by Christian P. Robert, Víctor Elvira, Nick Tawn, and Wu Changye and published in WIRES (2018).

[2] In order to keep the notations consistent, we still denote the target density by π, with the prior density denoted as π_0 and the sampling distribution of one observation x as $p(x|\theta)$. The dependence on the sample $\mathcal{D}$ is not reported unless necessary.

from the sub-posteriors, $\pi_i(\theta)$, by a Gaussian reweighting. Neiswanger et al. [38] estimate the sub-posteriors $\pi_i(\theta)$ by non-parametric and semi-parametric methods, and they run additional MCMC samplers on the product of these estimators towards approximating the true posterior $\pi(\theta)$. Wang and Dunson [49] refine this product estimator with an additional Weierstrass sampler, while [50] estimate the posterior by partitioning the space of samples with step functions.

As an alternative to sampling from the sub-posteriors, boosted sub-posterior methods target instead the components

$$\tilde{\pi}_i(\theta) \propto \pi_0(\theta) \left(\prod_{\ell \in \mathcal{X}_i} p(x_\ell | \theta) \right)^k \tag{4.2}$$

in separate MCMC runs. Since they formally amount to repeating each batch k times towards producing pseudo data sets with the same size as the true one, the resulting boosted sub-posteriors, $\tilde{\pi}_1(\theta), \cdots, \tilde{\pi}_k(\theta)$, have the same scale in variance of each component of the parameters, θ, as the true posterior, and can thus be treated as a group of estimators of the true posterior. In the subsequent combining stage, these sub-posteriors are merged together to construct a better approximation of the target distribution. For instance, [35] approximate the posterior with the geometric median of the boosted sub-posteriors, embedding them into associated reproducing kernel Hilbert spaces (rkhs), while [46] achieve this goal using the barycentres of $\tilde{\pi}_1, \cdots, \tilde{\pi}_k$, these barycentres being computed with respect to a Wasserstein distance.

In a perspective different from the above parallel scheme of divide-and-conquer approaches, sub-sampling approaches aim at reducing the number of individual datapoint likelihood evaluations operated at each iteration towards accelerating MCMC algorithms. From a general perspective, these approaches can be further classified into two finer classes: exact subsampling methods and approximate subsampling methods, depending on their resulting outputs. Exact subsampling approaches typically require subsets of data of random size at each iteration. One solution to this effect is taking advantage of pseudo-marginal MCMC via constructing unbiased estimators of the target density evaluated on subsets of the data [1]. Quiroz et al. [40] follow this direction by combining the debiasing technique of [41] and the correlated pseudo-marginal MCMC approach of [17]. Another direction is to use piecewise deterministic Markov processes (PDMP) [15, 16], which enjoy the target distribution as the marginal of their invariant distribution. This PDMP version requires unbiased estimators of the gradients of the logarithm of the likelihood function, instead of the likelihood itself. By using a tight enough bound on the event rate function of the associated Poisson processes PDMP can produce super-efficient scalable MCMC algorithms. The bouncy particle sampler [11] and the zig-zag sampler [8] are two competing PDMP algorithms, while [9] unifies and extends these two methods. Besides, one should note that PDMP produces a non-reversible Markov chain, which means that the algorithm should be more efficient in terms of mixing rate and asymptotic variance, when compared with reversible MCMC algorithms, such as MH, HMC and MALA, as observed in some theoretical and experimental works [7, 13, 25, 47].

Approximate subsampling approaches aim at constructing an approximation of the target distribution. One direction is to approximate the acceptance probability with high accuracy by using subsets of the data [3, 4]. Another solution is based on a direct modification of exact methods. The seminal work [51] in this direction, SGLD, is to exploit the Langevin diffusion

$$d\boldsymbol{\theta}_t = \frac{1}{2}\boldsymbol{\Lambda}\nabla \log \pi(\boldsymbol{\theta}_t)dt + \boldsymbol{\Lambda}^{1/2}d\mathbf{B}_t, \quad \boldsymbol{\theta}_0 \in \mathbb{R}^d, t \in [0, \infty) \qquad (4.3)$$

where $\boldsymbol{\Lambda}$ is a user-specified matrix, π is the target distribution and $\mathbf{B}_t$ is a d-dimensional Brownian process. By virtue of the Euler-Maruyama discretisation and using unbiased estimators of the gradient of the log-target density, SGLD and its variants [14, 19] often produce fast and accurate results in practice when compared with MCMC algorithms using MH steps.

4.4 Continuous-Time MCMC Samplers

All the above MCMC samplers are based on discrete-time, reversible Markov chains, however, continuous-time, non-reversible MCMC samplers have been drawing the attention of computational statisticians, which are based on piecewise deterministic Markov processes (PDMP). Even though PDMP was proposed as early as 1984 by Davis [15], its prevalence in statistics for sampling problems began the remarkable applications by Peters et al. [39], Bouchard-Côté et al. [11], and Bierkens et al. [8].

Suppose p be the continuous target distribution over $\mathbb{R}^d$ and for convenience sake, we also use $p(x)$ for the probability density function of p. Like HMC, an auxiliary variable, $v \in \mathcal{V}$ is introduced in PDMP framework and PDMP-based sampler explores the augmented space $\mathbb{R}^d \times \mathcal{V}$, targeting a variable $z = (x, v)$ with distribution $\rho(dx, dv)$ over $\mathbb{R}^d \times \mathcal{V}$ as its invariant distribution. By construction, the distribution ρ enjoys p as its marginal distribution in x. In practice, the distribution of v is often chosen to be independent to x and we denote it $\varphi(v)$. A piecewise deterministic Markov process $z_t = (x_t, v_t)$ consists of three distinct components: its deterministic dynamic between events, an event occurrence rate and a transition dynamic at event time. Specifically,

1. **Deterministic dynamic**: between two events, the Markov process evolves deterministically, according to some ordinary differential equation:

$$\frac{dz_t}{dt} = \Phi(z_t).$$

2. **Event occurrence rate**: an event occurs at time t with rate $\lambda(z_t)$.

3. **Transition dynamic**: At an event time, τ, the state prior to τ is denoted by $z_{\tau-}$, with the new state being generated by $z_\tau \sim Q(\cdot|z_{\tau-})$.

Here, an "event" refers to an occurrence of a time-inhomogeneous Poisson process with rate $\lambda(\cdot)$ [26]. The powerful tool to analyse PDMP is the extended generator, which is defined as follows.

Algorithm 5 Simulation of PDMP

Initialize the starting point z_0, $\tau_0 \leftarrow 0$.
for $k = 1, 2, 3, \cdots$ **do**
    Sample inter-event time η_k from following distribution

$$\mathbb{P}(\eta_k > t) = \exp\left\{-\int_0^t \lambda(z_{\tau_{k-1}+s})ds\right\}.$$

$\tau_k \leftarrow \tau_{k-1} + \eta_k$, $z_{\tau_{k-1}+s} \leftarrow \Psi_s(z_{\tau_{k-1}})$, for $s \in (0, \eta_k)$, where Ψ is the ODE flow of Φ.
$z_{\tau_k-} \leftarrow \Psi_{\eta_k}(z_{\tau_{k-1}})$, $z_{\tau_k} \sim Q(\cdot|z_{\tau_k-})$.
end for

Definition 4.4.1 ([16]) Let $\mathcal{D}(\mathcal{L})$ denote the set of measurable functions $f : \mathcal{Z} \to \mathbb{R}$ with the following property: there exists a measurable function $h : \mathcal{Z} \to \mathbb{R}$ such that the function $t \to h(z_t)$ is integrable P_z-a.s. for each $z \in \mathcal{Z}$ and the process

$$C_t^f = f(z_t) - f(z_0) - \int_0^t h(z_s)ds$$

is a local martingale. Then we write $h = \mathcal{L}f$ and call $(\mathcal{L}, \mathcal{D}(\mathcal{L}))$ the extended generator of the process $\{z_t\}_{t \geq 0}$.

Actually, we can compute the generator of above PDMP explicitly, by Davis [16, Theorem 26.14].

Theorem 4.4.1 ([16]) *The generator, $\mathcal{L}$, of above PDMP is, for $f \in \mathcal{D}(\mathcal{L})$*

$$\mathcal{L}f(z) = \nabla f(z) \cdot \Phi(z) + \lambda(z) \int_{z'} [f(z') - f(z)] Q(dz'|z).$$

Furthermore, $\rho(dz)$ is an invariant distribution of above PDMP, if

$$\int \mathcal{L}f(z)\rho(dz) = 0, \quad \text{for all } f \in \mathcal{D}(\mathcal{L}).$$

By choosing appropriate deterministic dynamic, rate function and transition dynamic, it is easy to make sure the specific PDMP admit the desired distribution, ρ, as its invariant distribution. However, there are two main difficulties in implementing such PDMP-based MCMC sampler. The first one is the computation of the ODE flow, Ψ. Almost all existing PDMP-based samplers adopt the linear dynamic, which means that

$$\frac{dx_t}{dt} = v_t, \quad \frac{dv_t}{dt} = 0.$$

Vanetti et al. [48] uses an approximation, $\hat{p}$, of the target p and adopts the Hamiltonian dynamic of $\hat{p} \times \varphi$ in HMC-BPS. The second difficulty comes from the generation of inter-event time, which corresponds to the first occurrence time of inhomogeneous Poisson process. The comment techniques to overcome such difficulty are based on the following two theorems.

Theorem 4.4.2 (Superposition Theorem, [26]) *Let $\Pi_1, \Pi_2, \cdots,$ be a countable collection of independent Poisson processes on state space $\mathbb{R}^+$ and let Π_n have rate $\lambda_n(\cdot)$ for each n. If $\sum_{n=1}^{\infty} \lambda_n(t) < \infty$ for all t, then the superposition*

$$\Pi = \bigcup_{n=1}^{\infty} \Pi_n$$

is a Poisson process with rate

$$\lambda(t) = \sum_{n=1}^{\infty} \lambda_n(t)$$

Theorem 4.4.3 (Thinning Theorem, [28]) *Let $\lambda : \mathbb{R}^+ \to \mathbb{R}^+$ and $\Lambda : \mathbb{R}^+ \to \mathbb{R}^+$ be continuous functions such that $\lambda(t) \leq \Lambda(t)$ for all $t \geq 0$. Let $\tau_1, \tau_2, \cdots,$ be the increasing finite or infinite sequence of a Poisson process with rate $\Lambda(\cdot)$. If, for all i, the point τ_i is removed from the sequence with probability $1 - \lambda(t)/\Lambda(t)$, then the remaining points $\tilde{\tau}_1, \tilde{\tau}_2, \cdots$ form a non-homogeneous Poisson process with rate $\lambda(\cdot)$.*

In order to estimate the integral of interest $I_h = \int h(x)P(dx)$, there are two approaches used to construct the estimator by PDMP-based samplers. Given a simulated path $(x_t, v_t)_{t=0}^{T}$, the first estimator is

$$\tilde{I}_h^T = \frac{1}{T} \int_0^T h(x_t)dt.$$

By discretising the path uniformly with respect to t, we can construct another estimator as

$$\hat{I}_h^T = \frac{1}{N} \sum_{n=1}^{N} h(x_{nT/N}).$$

4.4.1 Bouncy Particle Sampler

Bouncy particle sampler (BPS) is a specific piecewise deterministic Markov process, which admits $\pi(\mathbf{x})d\mathbf{x} \otimes d\mathbf{v}$ over the state space $\mathbb{R}^d \times S_{d-1}$ as its invariant distribution, by specifying the event rate $\lambda(\mathbf{z})$ and the transition dynamic $Q(d\mathbf{z}'; \mathbf{z})$.

1. **The deterministic dynamic**:

$$\frac{dx_t^{(i)}}{dt} = v_t^{(i)}, \quad \frac{dv_t^{(i)}}{dt} = 0, \quad i = 1, \cdots, d$$

2. **The event occurrence**: $\lambda(\mathbf{z}_t) = \max\{0, -\mathbf{v}_t \cdot \nabla \log \pi(\mathbf{x}_t)\}$.
3. **The transition dynamic**: $Q(\cdot|\mathbf{x}, \mathbf{v}) = \delta_{(\mathbf{x}, P_\mathbf{x}\mathbf{v})}(\cdot)$, where

$$P_\mathbf{x}\mathbf{v} = \mathbf{v} - 2\frac{\langle \mathbf{v}, \nabla \log \pi(\mathbf{x})\rangle}{\langle \nabla \log \pi(\mathbf{x}), \nabla \log \pi(\mathbf{x})\rangle}\nabla \log \pi(\mathbf{x})$$

Bouchard-Côté et al. [11] has shown that BPS admits $\pi(\mathbf{x})d\mathbf{x} \otimes d\mathbf{v}$ as its invariant distribution. However, the authors also find that pure BPS (specified above) meets with a reducibility problem and add a reference Poisson process into BPS to overcome it. The workflow of BPS with refreshment is shown in Algorithm 6.

Algorithm 6 Bouncy particle sampler

Initialize: $\mathbf{x}_0, \mathbf{v}_0, T_0 = 0$.
for $i = 1, 2, 3, \cdots$ **do**
    Generate $\tau \sim PP(\lambda(\mathbf{x}_t, \mathbf{v}_t))$
    Generate $\tau^{\text{ref}} \sim PP(\lambda^{\text{ref}})$
    if $\tau \leq \tau^{\text{ref}}$ **then**
        $T_i \leftarrow T_{i-1} + \tau$
        $\mathbf{x}_i \leftarrow \mathbf{x}_{i-1} + \tau\mathbf{v}_{i-1}$
        $\mathbf{v}_i \leftarrow \mathbf{v}_{i-1} - 2\frac{\langle \mathbf{v}_{i-1}, \nabla \log \pi(\mathbf{x}_i)\rangle}{\langle \nabla \log \pi(\mathbf{x}_i), \nabla \log \pi(\mathbf{x}_i)\rangle}\nabla \log \pi(\mathbf{x}_i)$
    else
        $T_i \leftarrow T_{i-1} + \tau^{\text{ref}}$
        $\mathbf{x}_i \leftarrow \mathbf{x}_{i-1} + \tau^{\text{ref}}\mathbf{v}_{i-1}$
        $\mathbf{v}_i \sim \mathcal{U}(S_{d-1})$
    end if
end for

4.4.2 Generalized Bouncy Particle Sampler

In BPS, at event time, the velocity changes deterministically. However, we found that the velocity can be changed into other directions, according to some distribution, at event time, which incorporates the randomness of the reference Poisson process in BPS to overcome reducibility. In this section, we generalize the BPS as follows: prior to event time, we decompose the velocity according to the gradient of $\log \pi (\mathbf{x})$, flip the parallel subvector and resample the orthogonal subvector with respect to some distribution. The details in the same PDMP framework are as follows:

1. **The deterministic dynamic**:

$$\frac{dx_t^{(i)}}{dt} = v_t^{(i)}, \quad \frac{dv_t^{(i)}}{dt} = 0, \quad i = 1, \cdots, d$$

2. **The event occurrence**: $\lambda(\mathbf{z}_t) = \max\{0, -\langle \mathbf{v}_t, \nabla \log \pi (\mathbf{x}_t)\rangle\}$.
3. **The transition dynamic**: $Q(d\mathbf{x}', d\mathbf{v}'|\mathbf{x}, \mathbf{v}) = \delta_{\{\mathbf{x}\}}(d\mathbf{x}')\delta_{\{-\mathbf{v}_1\}}(d\mathbf{v}_1')\mathcal{N}_{\mathbf{v}_1^\perp}(d\mathbf{v}_2')$, where

$$\mathbf{v}_1 = \frac{\langle \mathbf{v}, \nabla \log \pi (\mathbf{x})\rangle}{\langle \nabla \log \pi (\mathbf{x}), \nabla \log \pi (\mathbf{x})\rangle}\nabla \log \pi (\mathbf{x}), \quad \mathbf{v}_2 = \mathbf{v} - \mathbf{v}_1$$

$$\mathbf{v}_1' = \frac{\langle \mathbf{v}', \nabla \log \pi (\mathbf{x})\rangle}{\langle \nabla \log \pi (\mathbf{x}), \nabla \log \pi (\mathbf{x})\rangle}\nabla \log \pi (\mathbf{x}), \quad \mathbf{v}_2' = \mathbf{v}' - \mathbf{v}_1'$$

$$\mathbf{v}_1^\perp = \left\{\mathbf{u} \in \mathbb{R}^d : \langle \mathbf{u}, \mathbf{v}_1\rangle = 0\right\}$$

where $\mathcal{N}_{\mathbf{v}_1^\perp}$ denotes the $(d-1)$-dimensional standard normal distribution over the subspace $\mathbf{v}_1^\perp$, that is, the hyperplane orthogonal to $\mathbf{v}_1$.

We summarize the GBPS in Algorithm 7.

Algorithm 7 Generalized bouncy particle sampler

Initialize: x_0, v_0, $T_0 = 0$.
  for $i = 1, 2, 3, \cdots$ **do**
    Generate $\tau \sim PP(\lambda(\mathbf{x}_t, \mathbf{v}_t))$
    $T_i \leftarrow T_{i-1} + \tau$
    $x_i \leftarrow x_{i-1} + \tau v_{i-1}$
    $v_i \leftarrow Q(d\mathbf{v}|x_i, v_{i-1})$
  end for

Theorem 4.4.4 *The above piecewise deterministic Markov chain admits $\pi (\mathbf{x})d\mathbf{x} \otimes \psi_d(\mathbf{v})d\mathbf{v}$ over $\mathbb{R}^{2d}$ as its invariant distribution, where $\psi_d(\mathbf{v})$ is the density function of d-dimensional standard normal distribution.*

Proof In order to prove $\pi (\mathbf{x})d\mathbf{x} \otimes \psi_d(\mathbf{v})d\mathbf{v}$ is the invariant distribution of generator $\mathcal{A}$ of the above Markov chain, we just need to prove the following equation is satisfied by appropriate functions f:

$$\int_{\mathbb{R}^d} \int_{\mathbb{R}^d} \mathcal{A}f(\mathbf{x}, \mathbf{v})\pi(\mathbf{x})\psi_d(\mathbf{v})d\mathbf{x}d\mathbf{v} = 0$$

where by Theorem 26.14 [16]

$$\mathcal{A}f(\mathbf{z}) = \nabla f(\mathbf{z}) \cdot \Psi(\mathbf{z}) + \lambda(\mathbf{x}, \mathbf{v})\int_{\mathbf{v}' \in \mathbb{R}^d} f(\mathbf{x}, \mathbf{v}')Q(d\mathbf{v}'|\mathbf{x}, \mathbf{v}) - \lambda(\mathbf{x}, \mathbf{v})f(\mathbf{x}, \mathbf{v})$$

This is established in [52]. In order to establish the ergodicity theorem of GBPS, we introduce a specific assumption on the target distribution $\pi(\mathbf{x})$.

Assumption 4.4.1 For any two points $\mathbf{x}_1, \mathbf{x}_2 \in \mathbb{R}^d$ and any velocity $\mathbf{v} \in \mathbb{R}^d$, $\|\mathbf{v}\|_2 = 1$, there exists $t > 0$, such that

$$\mathbf{x}_2 \in S^\perp(\mathbf{x}_1 + t\mathbf{v}, \mathbf{v})$$

Theorem 4.4.5 *Under Assumption 4.4.1, the Markov chain $\mathbf{z}'_t = (\mathbf{x}_t, \frac{\mathbf{v}_t}{\|\mathbf{v}_t\|})$ induced by GBPS admits $\pi(\mathbf{x}) \times \mathcal{U}(S_{d-1})$ as its unique invariant distribution.*

The proof of Theorem 4.4.5 and the definitions of notations in Assumption 4.4.1 can be found in [52]. Whether or not Theorem 4.4.5 remains correct without Assumption 4.4.1 is an open question.

4.4.3 Construction of the Estimator and Implementation

While constructing an unbiased estimator of $I = \int h(\mathbf{x})\pi(d\mathbf{x})$, we cannot use the skeleton of the simulated GBPS path directly. In fact, such an estimator is biased. Suppose $\{\mathbf{x}_i, \mathbf{v}_i, T_i\}_{i=0}^M$ be the skeleton of a simulated trajectory, which means that at event time T_i, the state is $(\mathbf{x}_i, \mathbf{v}_i)$. Then, the whole trajectory $\mathbf{x}_{[0,T_M]}$ is filled up with

$$\mathbf{x}_t = \mathbf{x}_i + (t - T_i)\mathbf{v}_i, \quad T_i \le t < T_{i+1}$$

Let n be the number of data points selected from this trajectory, then an estimator of I is constructed as

$$\hat{I} = \frac{1}{n}\sum_{i=1}^n h(\mathbf{x}_{\frac{iT_M}{n}})$$

The main difficulty in implementing BPS and GBPS is to simulate the event times, which follows a Poisson process. The common techniques are based on the thinning and superposition theorems of Poisson process recalled above [27, 28]. In GBPS, from a given state $(\mathbf{x}, \mathbf{v})$, the associated Poisson process $\Pi_{\mathbf{x}, \mathbf{v}}$ has a rate function $\lambda(t) = \lambda(\mathbf{x} + t\mathbf{v}, \mathbf{v})$. With the help of the above two theorems, we can truly simulate a sample from $\Pi_{\mathbf{x}, \mathbf{v}}$.

Let $\eta(t) = \int_0^t \lambda(s)ds$, then the first event time, τ, of Poisson process Π, whose rate function is $\lambda(t)$, satisfies

$$\mathbb{P}(\tau > u) = \mathbb{P}(\Pi \cap [0, u] = \emptyset) = \exp(-\eta(u))$$

By the inverse theorem, τ can be simulated with the help of a uniform variate $V \sim \mathcal{U}(0, 1)$ via:

$$\tau = \eta^{-1}(-\log(V))$$

If we can compute η^{-1} analytically, it is easy to simulate the event times. Otherwise, the simulations commonly depend on the superposition and thinning theorems.

4.4.4 GBPS with Sub-sampling in Big Data

In Bayesian analysis, we suppose the observations $\{y_1, y_2, \cdots, y_N\}$ are i.i.d. samples from some distribution in the family $\{\mathbb{P}_{\mathbf{x}}, \mathbf{x} \in \mathbb{R}^d\}$ and let $\mathbb{P}_{\mathbf{x}}$ admit the density $p_{\mathbf{x}}$ with respect to the Lebesgue measure on $\mathbb{R}^d$. Given a prior $\pi_0(\mathbf{x})$ over the parameter $\mathbf{x}$, the posterior is

$$\pi(\mathbf{x}) \stackrel{\text{def}}{=} \pi(\mathbf{x}|y_1, \cdots, y_N) \propto \pi_0(\mathbf{x}) \prod_{n=1}^N p_{\mathbf{x}}(y_n)$$

Traditional MCMC algorithms (with MH step) are difficult to scale for large data sets, since each MH step needs to sweep over the whole data set. However, as indicated in [8], PDMP may be super-efficient by using sub-sampling to simulate samples from the target distribution if we can give a tight upper bound of the rate function. In GBPS, we only use the gradient of the logarithm of the target distribution, which means we can simulate the posterior by knowing it up to a constant. Besides, we can give an unbiased estimator of the gradient of the logarithm of the posterior by using its sum structure to simulate the posterior exactly:

$$\widehat{\nabla \log \pi}(\mathbf{x}) = N\nabla \log \pi_I(\mathbf{x}) = \nabla \log \pi_0(\mathbf{x}) + N\nabla_{\mathbf{x}} \log p_{\mathbf{x}}(y_I), \quad I \sim \mathcal{U}\{1, 2, \cdots, N\}$$

In Algorithm 8, we show the workflow of the implementation of subsampling in GBPS. Notice that $\lambda(\Delta, \mathbf{v}_{i-1})$ equals to $\lambda(\mathbf{x}, \mathbf{v}_{i-1})$ in which $\nabla \log \pi(\mathbf{x})$ is replaced by Δ. $\Lambda(t)$ is an upper bound of $\lambda(\mathbf{x}, \mathbf{v})$.

Algorithm 8 Subsampling version

Initialize: $\mathbf{x}_0, \mathbf{v}_0, T_0 = 0$.
for $i = 1, 2, 3, \cdots$ **do**
    Generate $\tau \sim PP(\Lambda(t))$
    $T_i \leftarrow T_{i-1} + \tau$
    $\mathbf{x}_i \leftarrow \mathbf{x}_{i-1} + \tau \mathbf{v}_{i-1}$
    $I \sim \mathcal{U}(\{1, \cdots, N\})$
    $\Delta \leftarrow N \nabla \log \pi_I(\mathbf{x}_i)$
    $q \leftarrow \lambda(\Delta, \mathbf{v}_{i-1})/\Lambda(\tau)$
    $u \sim \mathcal{U}(0, 1)$
    if $u \leq q$ **then**
        $\mathbf{v}_i \leftarrow Q(d\mathbf{v}|\Delta, \mathbf{v}_{i-1})$
    else
        $\mathbf{v}_i \leftarrow \mathbf{v}_{i-1}$
    end if
end for

4.4.5 Numerical Simulations

In this section, we apply GBPS algorithm on three numerical experiments. Example 4.4.1 shows that reducibility problem appears in isotropic Gaussian distribution for BPS without refreshment but is not encountered by GBPS. In Example 4.4.2, we can find that GBPS works well on multimode distributions and with similar performance with BPS. Finally, we present the GBPS with sub-sampling on Bayesian logistic model.

Example 4.4.1 (Isotropic Gaussian Distribution) In this example, we show the reducibility problem of BPS without refreshment. The target distribution is

$$\pi(\mathbf{x}) = \frac{1}{2\pi} \exp\left\{-\frac{x_1^2 + x_2^2}{2}\right\}$$

First we apply the BPS without reference Poisson process and show its reducibility in Fig. 4.1. Compared with BPS without refreshment, GBPS is irreducible, shown in Fig. 4.2.

Second, we compare the performances of GBPS and of BPS with refreshment. For BPS, we set $\lambda^{\text{ref}} = \{0.01, 0.1, 0.2, 0.5, 1\}$. Each method is run 50 times and each sampled path has length 10^4. For each path, we sample 10^4 points with length gap 1. Figure 4.3 shows the errors of the first and second moments of each component and Fig. 4.4 presents the errors in terms of Wasserstein-2 distance with respect to the target distribution and the effective sample size of each method.

For BPS, we need to tune the rate of reference Poisson process to balance the efficiency and accuracy. Even though BPS is ergodic for every positive refreshment rate λ^{ref} in theory, the value of λ^{ref} matters in implementation. The smaller the refreshment rate, the larger the effective sample size (high efficiency), the more

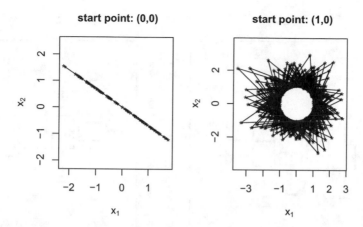

Fig. 4.1 Reducibility problem in isotropic Gaussian distributions: (left) the first 50 segments of a BPS path without refreshment which starts from the center of the Gaussian distribution, the trajectory is on a line; (right) the first 500 segments of another BPS path with $\lambda^{\text{ref}} = 0$ starting from an point except the center, the trajectory cannot explore the center area

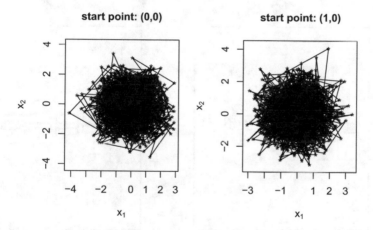

Fig. 4.2 GBPS is irreducible in isotropic Gaussian distribution: (left) the first 1000 segments of a GBPS path which starts from the center of the Gaussian distribution; (right) the first 1000 segments of another GBPS path starting from an point except the center

slowly the chain mixes. The larger the refreshment rate, the smaller the effective sample size (low efficiency), the faster the chain mixes. However, when the refreshment rate is extremely large or small, BPS will produce chains approximating the target distribution poorly. On the other hand, there is no hyper-parameter to tune in GBPS, which incorporates the randomness of BPS in refreshment into transition dynamics. Compared with BPS with different refreshment rates, GBPS performs better when considering the first and second moments of each component. In terms of Wasserstein-2 distance, as well as ESS, GBPS outperforms BPS.

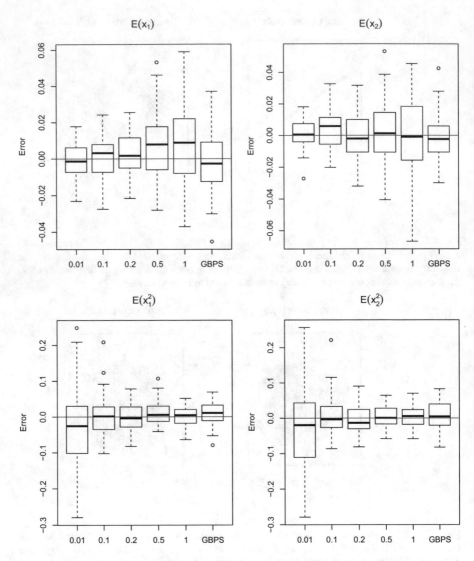

Fig. 4.3 Comparison between BPS and GBPS in isotropic Gaussian distribution. For each graph, the first five boxes represent the BPS method with different refreshment rates $\lambda^{\text{ref}} = \{0.01, 0.1, 0.2, 0.5, 1\}$

Example 4.4.2 (Mixture of Gaussian Model) In this example, we show how to simulate the event time by using superposition and thinning theorems. The target is a mixture of Gaussian distributions:

$$\pi(x_1, x_2) = \frac{p}{2\pi\sigma_1\sigma_2} \exp\left\{-\frac{(x_1-3)^2}{2\sigma_1^2} - \frac{x_2^2}{2\sigma_2^2}\right\} + \frac{1-p}{2\pi\sigma_3\sigma_4} \exp\left\{-\frac{x_1^2}{2\sigma_3^2} - \frac{(x_2-3)^2}{2\sigma_4^2}\right\}$$

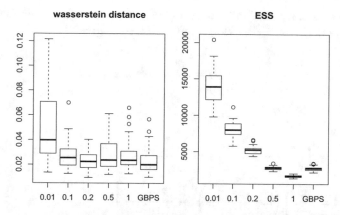

Fig. 4.4 Comparison between BPS and GBPS in isotropic Gaussian distribution in terms of Wasserstein distance and effective sample size. For each graph, the first five boxes represent the BPS method with different refreshment rates $\lambda^{\text{ref}} = \{0.01, 0.1, 0.2, 0.5, 1\}$

In our experiment, we set $p = 0.5$, $(\sigma_1, \sigma_2, \sigma_3, \sigma_4) = (1, 1.5, 2, 1)$. The gradient is

$$
\frac{\partial \pi(x_1, x_2)}{\partial x_1} = \frac{p}{2\pi \sigma_1 \sigma_2} \exp\left\{ -\frac{(x_1 - 3)^2}{2\sigma_1^2} - \frac{x_2^2}{2\sigma_2^2} \right\} \left(-\frac{(x_1 - 3)}{\sigma_1^2} \right)
$$
$$
+ \frac{1 - p}{2\pi \sigma_3 \sigma_4} \exp\left\{ -\frac{x_1^2}{2\sigma_3^2} - \frac{(x_2 - 3)^2}{2\sigma_4^2} \right\} \left(-\frac{x_1}{\sigma_3^2} \right)
$$

$$
\frac{\partial \pi(x_1, x_2)}{\partial x_2} = \frac{p}{2\pi \sigma_1 \sigma_2} \exp\left\{ -\frac{(x_1 - 3)^2}{2\sigma_1^2} - \frac{x_2^2}{2\sigma_2^2} \right\} \left(-\frac{x_2}{\sigma_2^2} \right)
$$
$$
+ \frac{1 - p}{2\pi \sigma_3 \sigma_4} \exp\left\{ -\frac{x_1^2}{2\sigma_3^2} - \frac{(x_2 - 3)^2}{2\sigma_4^2} \right\} \left(-\frac{(x_2 - 3)}{\sigma_4^2} \right)
$$

We can give an upper bound for the norm of the gradient of the logarithm of the target density function:

$$
\|\nabla \log \pi(x_1, x_2)\|_2 \leq \frac{|x_1 - 3|}{\sigma_1^2} + \frac{|x_1|}{\sigma_3^2} + \frac{|x_2|}{\sigma_2^2} + \frac{|x_2 - 3|}{\sigma_4^2}
$$

Then an upper bound for $\lambda(\mathbf{x}, \mathbf{v})$ is given as

$$
\lambda(\mathbf{x}, \mathbf{v}) \leq \left(\frac{|x_1 - 3|}{\sigma_1^2} + \frac{|x_1|}{\sigma_3^2} + \frac{|x_2|}{\sigma_2^2} + \frac{|x_2 - 3|}{\sigma_4^2} \right) * \|\mathbf{v}\|_2
$$

By superposition, we need only focus on Poisson process whose rate function has such form: $\lambda(x, v) = \frac{|x-\mu|}{\sigma^2}$. Let $\lambda_s(x, v) = \lambda(x + sv, v) = \frac{|x+sv-\mu|}{\sigma^2}$. Define

$$\eta(t) = \int_0^t \lambda_s(x, v)ds$$

i): If $x > \mu, v > 0$,

$$\eta(t) = \int_0^t \frac{(x - \mu) + sv}{\sigma^2}ds = \frac{\frac{1}{2}vt^2 + (x - \mu)t}{\sigma^2} = \frac{v}{2\sigma^2}\left(t^2 + \frac{2(x - \mu)}{v}t\right)$$

$$= \frac{v}{2\sigma^2}\left[\left(t + \frac{(x - \mu)}{v}\right)^2 - \frac{(x - \mu)^2}{v^2}\right]$$

$$\eta^{-1}(z) = \sqrt{\frac{2\sigma^2 z}{v} + \frac{(x - \mu)^2}{v^2}} - \frac{(x - \mu)}{v}$$

ii) : If $x < \mu, v < 0$, then

$$\eta(t) = \int_0^t \frac{-(x - \mu) - sv}{\sigma^2}ds = \frac{-\frac{1}{2}vt^2 - (x - \mu)t}{\sigma^2} = -\frac{v}{2\sigma^2}\left(t^2 + \frac{2(x - \mu)}{v}t\right)$$

$$= -\frac{v}{2\sigma^2}\left[\left(t + \frac{(x - \mu)}{v}\right)^2 - \frac{(x - \mu)^2}{v^2}\right]$$

$$\eta^{-1}(z) = \sqrt{-\frac{2\sigma^2 z}{v} + \frac{(x - \mu)^2}{v^2}} - \frac{(x - \mu)}{v}$$

iii) : If $x > \mu, v \leq 0$:

$$\eta\left(-\frac{x - \mu}{v}\right) = \int_0^{-\frac{x-\mu}{v}} \frac{sv + (x - \mu)}{\sigma^2}ds = -\frac{(x - \mu)^2}{2v\sigma^2}$$

1. If $z > -\frac{(x-\mu)^2}{2v\sigma^2}$: $t_0 = -\frac{x-\mu}{v}$

$$-\frac{(x - \mu)^2}{2v\sigma^2} + \int_0^t -\frac{sv}{\sigma^2}ds = z, \quad t = \sqrt{-\frac{2\sigma^2 z}{v} - \frac{(x - \mu)^2}{v^2}}$$

$$\eta^{-1}(z) = \sqrt{-\frac{2\sigma^2 z}{v} - \frac{(x - \mu)^2}{v^2}} + \left(-\frac{x - \mu}{v}\right)$$

2. If $z \leq -\frac{(x-\mu)^2}{2v\sigma^2}$:

$$\int_0^t \frac{sv + (x - \mu)}{\sigma^2} ds = \frac{v}{2\sigma^2} \left(t^2 + \frac{2(x - \mu)}{v} t \right) = z$$

$$\eta^{-1}(z) = -\sqrt{\frac{2\sigma^2 z}{v} + \frac{(x - \mu)^2}{v^2}} + \left(-\frac{x - \mu}{v} \right)$$

$iv)$: If $x \leq \mu, v > 0$:

$$\eta \left(-\frac{x - \mu}{v} \right) = \int_0^{-\frac{x-\mu}{v}} \frac{-sv - (x - \mu)}{\sigma^2} ds = \frac{(x - \mu)^2}{2v\sigma^2}$$

1. If $z > \frac{(x-\mu)^2}{2v\sigma^2}$: $t_0 = -\frac{x-\mu}{v}$

$$\frac{(x - \mu)^2}{2v\sigma^2} + \int_0^t \frac{sv}{\sigma^2} ds = z, \quad t = \sqrt{\frac{2\sigma^2 z}{v} - \frac{(x - \mu)^2}{v^2}}$$

$$\eta^{-1}(z) = \sqrt{\frac{2\sigma^2 z}{v} - \frac{(x - \mu)^2}{v^2}} + \left(-\frac{x - \mu}{v} \right)$$

2. If $z \leq \frac{(x-\mu)^2}{2v\sigma^2}$:

$$\int_0^t \frac{-sv - (x - \mu)}{\sigma^2} ds = -\frac{v}{2\sigma^2} \left(t^2 + \frac{2(x - \mu)}{v} t \right) = z$$

$$\eta^{-1}(z) = -\sqrt{-\frac{2\sigma^2 z}{v} + \frac{(x - \mu)^2}{v^2}} + \left(-\frac{x - \mu}{v} \right)$$

In Fig. 4.5, we show the trajectory of the simulated GBPS path and associated samples. Figure 4.6 shows the marginal density functions of the target distribution. In Fig. 4.7, we compare the performance of BPS and GBPS. For BPS, we set $\lambda^{\text{ref}} = 0.01, 0.1, 1$. We sample 50 paths with length 10,000 for each method and take 10,000 points from each path with gap 1 to form samples. Empirically, BPS is ergodic over this example. With the increase of λ^{ref}, the refreshment occurs more frequently, which reduces the performance of BPS. Even though GBPS has worse performance, compared to BPS with some refreshment rates, it is quite reliable and has no parameter to tune.

Example 4.4.3 (Bayesian Logistic Model) For the Bayesian logistic model, we suppose $\mathbf{x} \in \mathbb{R}^d$ be the parameters and (y_i, z_i), for $i = 1, 2, \cdots, N$ be the

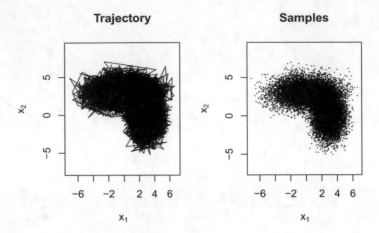

Fig. 4.5 The trajectory and samples from a GBPS path

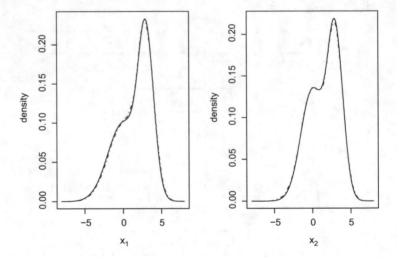

Fig. 4.6 Marginal density functions: the black solid lines are true marginal density, the red dotted lines are from a GBPS path

observations, where $y_i \in \mathbb{R}^d$, $z_i \in \{0, 1\}$, then

$$\mathbb{P}(z_i = 1 | y_i, \mathbf{x}) = \frac{1}{1 + \exp\{-\sum_{\ell=1}^{d} y_i^\ell x_\ell\}}$$

Choosing the improper prior, then the posterior is

$$\pi(\mathbf{x}) \propto \prod_{j=1}^{N} \frac{\exp\{z_j \sum_{\ell=1}^{d} y_j^\ell x_\ell\}}{1 + \exp\{\sum_{\ell=1}^{d} y_j^\ell x_\ell\}}$$

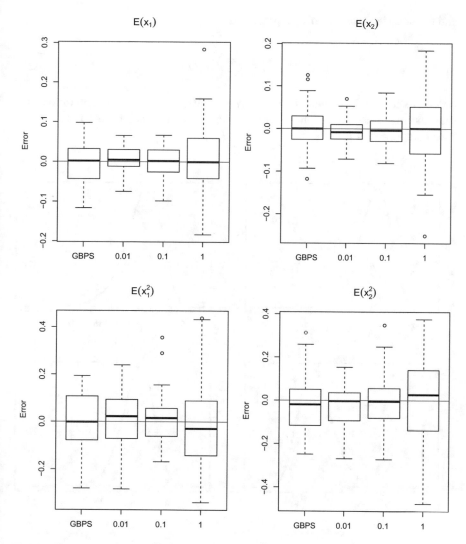

Fig. 4.7 Comparison between GBPS and BPS: for each graph, the last three boxes represent BPS with $\lambda^{\text{ref}} = 0.01, 0.1, 1$

for $k = 1, \cdots, d$, the partial derivative is

$$\frac{\partial}{\partial x_k} \log \pi(\mathbf{x}) = \sum_{j=1}^{N} \left[z_j - \frac{\exp\{z_j \sum_{\ell=1}^{d} y_j^\ell x_\ell\}}{1 + \exp\{\sum_{\ell=1}^{d} y_j^\ell x_\ell\}} \right] y_j^k$$

Then, they are bounded by

$$\left| \frac{\partial \log \pi(\mathbf{x})}{\partial x_k} \right| \leq \sum_{j=1}^{N} \left| y_j^k \right|$$

and the bounded rate for Poisson process is

$$\lambda^+ = \max_{1 \leq k \leq d} \sum_{j=1}^{N} \left| y_j^k \right|$$

In our experiment, we set $d = 5$, $N = 100$ and use 10 observations for subsampling at each iteration. Figure 4.8 shows the marginal density functions for each component of parameters.

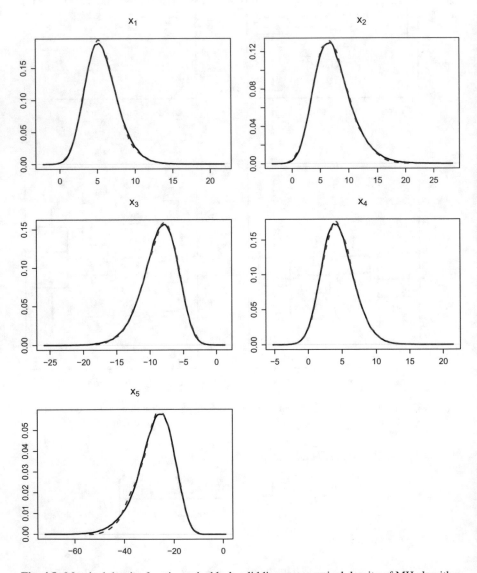

Fig. 4.8 Marginal density functions: the black solid lines are marginal density of MH algorithm, which are used as benchmark. The red dotted lines are from a GBPS path

4.5 Conclusion

In this chapter, we have generalized the bouncy particle sampler in terms of its transition dynamics. Our method—the generalized bouncy particle sampler—can be regarded as a bridge between bouncy particle and zig-zag process samplers. Compared with bouncy particle samplers, GBPS changes the direction velocity according to some distribution at event time. However, compared with zig-zag process samplers, GBPS can be regarded as attaching a moving coordinate system on the state space of $(\mathbf{x}, \mathbf{v})$, instead of using a fixed one as in a zig-zag process sampler. One main advantage of GBPS, compared to BPS, is that it has no parameter to tune.

Throughout this chapter, we have supposed that the parameter space has no restrictions. In practice, it is often the case one encounters restricted parameter space problems. In such cases, we may transfer the restricted region into the whole Euclidean space by reparameterization techniques. Besides, [9] experiments with some methods to simulate over restricted space. Another problem when implementing these methods is to figure out how to simulate event time from the associated Poisson process in a efficient manner. In general, simulations are based on superposition and thinning theorems. The upper bound of the rate function is however crucial. The tighter the upper bound is, the more efficient the simulation is. In Bayesian analysis, for large data sets, if the upper bound is $\mathcal{O}(N^{\alpha})$, then the effective sample size per likelihood computation is $\mathcal{O}(N^{-(1/2+\alpha)})$. If $\alpha < 1/2$, then both BPS and GBPS will be more efficient than traditional MCMC methods.

Exploring several simulation settings, we find that reducibility problem just appears in isotropic Gaussian distributions or in distributions who admit isotropic Gaussian distributions as their component for BPS. However, it is still an open question.

References

1. C. Andrieu, G. Roberts, The pseudo-marginal approach for efficient Monte Carlo computations. Ann. Stat. **37**, 697–725 (2009)
2. L.J. Aslett, P.M. Esperança, C.C. Holmes, A review of homomorphic encryption and software tools for encrypted statistical machine learning. Preprint, arXiv:1508.06574 (2015)
3. R. Bardenet, A. Doucet, C. Holmes, Towards scaling up Markov chain Monte Carlo: an adaptive subsampling approach, in *Proceedings of the 31st International Conference on Machine Learning (ICML-14)* (2014), pp. 405–413
4. R. Bardenet, A. Doucet, C. Holmes, On Markov chain Monte Carlo methods for tall data. J. Mach. Learn. Res. **18**(1), 1515–1557 (2017)
5. J. Berger, *Statistical Decision Theory and Bayesian Analysis*, 2nd edn. (Springer, New York, 1985)
6. M. Betancourt, A conceptual introduction to Hamiltonian Monte Carlo. Preprint, arXiv:1701.02434 (2017)
7. J. Bierkens, Non-reversible Metropolis–Hastings. Stat. Comput. **26**(6), 1213–1228 (2016)

8. J. Bierkens, P. Fearnhead, G. Roberts, The zig-zag process and super-efficient sampling for Bayesian analysis of big data. Preprint, arXiv:1607.03188 (2016)
9. J. Bierkens, A. Bouchard-Côté, A. Doucet, A.B. Duncan, P. Fearnhead, T. Lienart, G. Roberts, S.J. Vollmer, Piecewise deterministic Markov processes for scalable Monte Carlo on restricted domains. Stat. Probab. Lett. **136**, 148–154 (2018)
10. N. Bou-Rabee, J.M. Sanz-Serna et al., Randomized Hamiltonian Monte Carlo. Ann. Appl. Probab. **27**(4), 2159–2194
11. A. Bouchard-Côté, S.J. Vollmer, A. Doucet, The bouncy particle sampler: a nonreversible rejection-free Markov chain Monte Carlo method. J. Am. Stat. Assoc. 1–13 (2018). https://doi.org/10.1080/01621459.2017.1294075
12. O. Cappé, C. Robert, Ten years and still running! J. Am. Stat. Assoc. **95**(4), 1282–1286 (2000)
13. T. Chen, C. Hwang, Accelerating reversible Markov chains. Stat. Probab. Lett. **83**(9), 1956–1962 (2013)
14. T. Chen, E. Fox, C. Guestrin, Stochastic gradient Hamiltonian Monte Carlo, in *International Conference on Machine Learning* (2014), pp. 1683–1691
15. M.H. Davis, Piecewise-deterministic Markov processes: a general class of non-diffusion stochastic models. J. R. Stat. Soc. Ser. B (Methodol.) **46**, 353–388 (1984)
16. M.H. Davis, *Markov Models & Optimization*, vol. 49 (CRC Press, Boca Raton, 1993)
17. G. Deligiannidis, A. Doucet, M.K. Pitt, The correlated pseudo-marginal method. Preprint, arXiv:1511.04992 (2015)
18. L. Devroye, *Non-uniform Random Variate Generation* (Springer, New York, 1985)
19. N. Ding, Y. Fang, R. Babbush, C. Chen, R.D. Skeel, H. Neven, Bayesian sampling using stochastic gradient thermostats, in *Advances in Neural Information Processing Systems* (2014), pp. 3203–3211
20. S. Duane, A.D. Kennedy, B.J. Pendleton, D. Roweth, Hybrid Monte Carlo. Phys. Lett. B **195**, 216–222 (1987)
21. A. Gelfand, A. Smith, Sampling based approaches to calculating marginal densities. J. Am. Stat. Assoc. **85**, 398–409 (1990)
22. M. Girolami, B. Calderhead, Riemann manifold Langevin and Hamiltonian Monte Carlo methods. J. R. Stat. Soc. Ser. B (Stat. Methodol.) **73**, 123–214 (2011)
23. W.K. Hastings, Monte Carlo sampling methods using Markov chains and their applications. Biometrika **57**(1), 97–109 (1970)
24. M.D. Hoffman, A. Gelman, The No-U-turn sampler: adaptively setting path lengths in Hamiltonian Monte Carlo. Ann. Appl. Probab. **27**(4), 2159–2194 (2014)
25. C.-R. Hwang, S.-Y. Hwang-Ma, S.-J. Sheu, Accelerating Gaussian diffusions. Ann. Appl. Probab. **3**, 897–913 (1993)
26. J.F.C. Kingman, *Poisson Processes*, vol. 3 (Clarendon Press, Oxford, 1992)
27. J.F.C. Kingman, *Poisson Processes*. Wiley Online Library (1993)
28. P.A. Lewis, G.S. Shedler, Simulation of nonhomogeneous Poisson processes by thinning. Naval Res. Logist. **26**(3), 403–413 (1979)
29. J. Liu, W. Wong, A. Kong, Covariance structure of the Gibbs sampler with application to the comparison of estimators and augmentation schemes. Biometrika **81**, 27–40 (1994)
30. J. Liu, W. Wong, A. Kong, Covariance structure and convergence rates of the Gibbs sampler with various scans. J. R. Stat. Soc. B **57**, 157–169 (1995)
31. S. Livingstone, M.F. Faulkner, G.O. Roberts, Kinetic energy choice in Hamiltonian/hybrid Monte Carlo. Preprint, arXiv:1706.02649 (2017)
32. D.J.C. MacKay, *Information Theory, Inference & Learning Algorithms* (Cambridge University Press, Cambridge, 2002)
33. N. Metropolis, A.W. Rosenbluth, M.N. Rosenbluth, A.H. Teller, E. Teller, Equations of state calculations by fast computing machines. J. Chem. Phys. **21**, 1087–1092 (1953)
34. S. Meyn, R. Tweedie, *Markov Chains and Stochastic Stability* (Springer, New York, 1993)
35. S. Minsker, S. Srivastava, L. Lin, D. Dunson, Scalable and robust Bayesian inference via the median posterior. *International Conference on Machine Learning* (2014), pp. 1656–1664

36. R. Neal, *Bayesian Learning for Neural Networks*, vol. 118. Lecture Notes (Springer, New York, 1999)
37. R. Neal, MCMC using Hamiltonian dynamics, in *Handbook of Markov Chain Monte Carlo*, ed. by S. Brooks, A. Gelman, G.L. Jones, X.-L. Meng (CRC Press, New York, 2011)
38. W. Neiswanger, C. Wang, E. Xing, Asymptotically exact, embarrassingly parallel MCMC. Preprint, arXiv:1311.4780 (2013)
39. E.A. Peters et al., Rejection-free Monte Carlo sampling for general potentials. Phys. Rev. E **85**(2), 026703 (2012)
40. M. Quiroz, M. Villani, R. Kohn, Exact subsampling MCMC. Preprint, arXiv:1603.08232 (2016)
41. C.-h. Rhee, P.W. Glynn, Unbiased estimation with square root convergence for SDE models. Oper. Res. **63**(5), 1026–1043 (2015)
42. C. Robert, *The Bayesian Choice*, 2nd edn. (Springer, New York, 2001)
43. C. Robert, G. Casella, *Monte Carlo Statistical Methods*, 2nd edn. (Springer, New York, 2004)
44. R.Y. Rubinstein, *Simulation and the Monte Carlo Method* (Wiley, New York, 1981)
45. S.L. Scott, A.W. Blocker, F.V. Bonassi, H.A. Chipman, E.I. George, R.E. McCulloch, Bayes and big data: the consensus Monte Carlo algorithm. Int. J. Manag. Sci. Eng. Manag. **11**(2):78–88 (2016)
46. S. Srivastava, V. Cevher, Q. Dinh, D. Dunson, WASP: scalable Bayes via barycenters of subset posteriors. Artif. Intell. Stat. **38**, 912–920 (2015)
47. Y. Sun, J. Schmidhuber, F.J. Gomez, Improving the asymptotic performance of Markov chain Monte-carlo by inserting vortices, in *Advances in Neural Information Processing Systems (NIPS-10)* (2010), pp. 2235–2243
48. P. Vanetti, A. Bouchard-Côté, G. Deligiannidis, A. Doucet, Piecewise deterministic Markov chain Monte Carlo. Preprint, arXiv:1707.05296 (2017)
49. X. Wang, D. Dunson, Parallelizing MCMC via Weierstrass sampler. Preprint, arXiv:1312.4605 (2013)
50. X. Wang, F. Guo, K. Heller, D. Dunson, Parallelizing MCMC with random partition trees, in *Advances in Neural Information Processing Systems* (2015), pp. 451–459
51. M. Welling, Y. Teh, Bayesian learning via stochastic gradient Langevin dynamics, in *Proceedings of the 28th International Conference on Machine Learning (ICML-11)* (2011), pp. 681–688
52. C. Wu, Acceleration strategies of Markov chain Monte Carlo algorithms for Bayesian computation. PhD thesis, Université Paris Dauphine PSL, 2018

Chapter 5
Bayesian Variable Selection

Matthew Sutton

Abstract In this chapter we survey Bayesian approaches for variable selection and model choice in regression models. We explore the methodological developments and computational approaches for these methods. In conclusion we note the available software for their implementation.

5.1 Introduction

Bayesian variable selection methodology has been progressing rapidly in recent years. While the seminal work of the Bayesian spike and slab prior [1] remains the main approach, continuous shrinkage priors have received a large amount of attention. There is growing interest in speeding up inference with these sparse priors using modern Bayesian computational approaches. Moreover, the subject of inference for these sparse models has become an increasingly important area of discussion among statisticians. A common theme among Bayesian variable selection methods is that they aim to select variables while also quantifying uncertainty through selection probabilities and variability of the estimates. This chapter gives a survey of relevant methodological and computational approaches in this area, along with some descriptions of available software.

5.2 Preliminaries

5.2.1 The Variable Selection Problem

In the context of variable selection for a regression model we consider the following canonical problem in Bayesian analysis. Suppose we want to model a sample of n

M. Sutton (✉)
Queensland University of Technology, Brisbane, QLD, Australia
e-mail: m.sutton5@lancaster.ac.uk

© The Editor(s) (if applicable) and The Author(s), under exclusive
licence to Springer Nature Switzerland AG 2020
K. L. Mengersen et al. (eds.), *Case Studies in Applied Bayesian Data Science*,
Lecture Notes in Mathematics 2259, https://doi.org/10.1007/978-3-030-42553-1_5

observations of a response variable $Y \in \mathbb{R}^n$ and a set of p potential explanatory variables $X_1, \ldots, X_p$, where $X_j \in \mathbb{R}^n$. The variable selection problem is to find the 'best' model between the response Y and a subset of $X_1, \ldots, X_p$ where there is uncertainty in which subset to use. Throughout this chapter, we index each of the possible 2^p subset choices by the vector

$$\gamma = (\gamma_1, \ldots, \gamma_p)^T,$$

where $\gamma_j = 1$ if variable X_j is included in the model, and $\gamma_j = 0$ otherwise. We let $s_\gamma = \sum_{j=1}^p \gamma_j$ denote the number of selected variables for a model indexed by γ. Given γ, suppose that Y has density $p(Y \mid \beta_\gamma, \gamma)$ where β_γ is a vector of unknown parameters corresponding to the variables indexed by γ. The Bayesian approach assigns a prior probability to the space of models $p(\gamma)$, and a prior to the parameters of each model $p(\beta_\gamma \mid \gamma)$.

The probability for the model with the selected variables γ conditional on having observed Y, is the posterior model probability

$$p(\gamma \mid Y) = \frac{p(Y \mid \gamma) p(\gamma)}{\sum_{\gamma' \in \{0,1\}^p} p(Y \mid \gamma') p(\gamma')},$$

where

$$p(Y \mid \gamma) = \int p(Y \mid \gamma, \beta_\gamma) p(\beta_\gamma \mid \gamma) \mathrm{d}(\beta_\gamma),$$

is the marginal likelihood of Y. The priors $p(\beta_\gamma \mid \gamma)$ and $p(\gamma)$ provide an initial representation of model uncertainty and the posterior adjusts for the information in Y, allowing us to quantify the uncertainty of the variable selection. The actual variable selection in a Bayesian analysis can proceed in several ways. Two common approaches are:

1. Select the variables with the highest estimated posterior probability $p(\gamma \mid Y)$, also known as the highest posterior density model (HPD),
2. Select variables with estimated posterior probability of inclusion $p(\gamma_j = 1 \mid Y)$ greater than 0.5, also known as the median probability model (MPM).

The appropriateness of the HPD and MPM model have been studied in detail [2, 3]. It has been shown that for orthogonal linear regression, the optimal model from a Bayesian predictive objective is the MPM rather than the HPD.

In a Bayesian framework, the accuracy of the variable selection method depends on the specification of the priors for the model space and parameters. In this section, we survey priors which fall into one of four possible categories, priors on the model space, spike and slab priors, shrinkage priors and projection methods.

5.2.2   Model Space Priors

We begin by considering priors on the model space $p(\gamma)$. A common prior on the model space assumes that the γ_j are independent and Bernoulli distributed,

$$p(\gamma) = \prod_{j=1}^{p} w_j^{\gamma_j} (1 - w_j)^{1-\gamma_j}, \tag{5.1}$$

is computationally inexpensive and has been found to give sensible results in practice [4–7]. Under this prior, each variable X_j will enter the model with probability $p(\gamma_j = 1) = w_j$. A common variant of this method is to place a Beta prior on $w \sim Beta(a, b)$ which yields

$$p(\gamma) = \frac{B(a + s_\gamma,\ b + p - s_\gamma)}{B(a,\ b)},$$

where $B(a, b)$ is the beta function with hyper-parameters a and b. The choice of $a = b = 1$ corresponds to an uninformative prior on the model space. This type of prior is also recommended in [8], where the choice of hyper-parameters is considered asymptotically. More generally, one can put a prior $h(s_\gamma)$ on the model dimension and let

$$p(\gamma) = \binom{p}{s_\gamma}^{-1} h(s_\gamma),$$

which allows for the belief that the optimal models are sparse [16]. Priors of this form are considered generally by Scott in [9]. The priors described so far are useful when there is no structural information about the predictors.

Structured priors have also been considered, for example [10] propose a model space prior which incorporates known correlation in the predictors. They assume that the covariates have an underlying graphical structure and use an Ising prior to incorporate the structural information (see [11] for a survey on the Ising model). This structural information is used to capture underlying biological processes in the modelling.

5.2.3   Spike and Slab Priors

We now consider the specification of the prior for the parameters $p(\beta_\gamma \mid \gamma)$. Arguably, one of the simplest and most natural classes of prior distributions is given by the spike and slab type priors. In the original formulation [1, 12] the spike and slab distribution was defined as a mixture of a Dirac measure concentrated at zero and a uniform diffuse component. Similar to [13], we use a more general version of the prior. In this chapter we refer to a spike and slab as any mixture of two distributions where one component is peaked at zero and the other is diffuse. More specifically, we define a spike and slab to have the form,

$$\beta_j \mid \gamma_j \sim (1 - \gamma_j)G_0(\beta_j) + \gamma_j G_1(\beta_j),$$

for $j = 1, \ldots, p$ where G_0 and G_1 are probability measures on $\mathbb{R}$ and $\gamma \sim p(\gamma)$, where $p(\gamma)$ is a prior on the model space. This framework naturally extends the model space prior discussed in the previous section. The original spike and slab (Mitchell et al. [1]) corresponds to a Dirac mass at zero δ_0 for G_0 and a uniform slab distribution for G_1.

For this section, we will assume an independent Bernoulli prior for γ_j, where $\gamma_j \sim Bernoulli(w_j)$, and $w_j \in [0, 1]$ for $j = 1, \ldots, p$. Using this prior on the model space the spike and slab can be written as the mixture

$$\beta_j \mid w_j \sim (1 - w_j)G_0(\beta_j) + w_j G_1(\beta_j),$$

where we have marginalised over the binary term γ_j. There are a number of prior specifications which use this hierarchical setup but differ in the distributions chosen for G_0 and G_1 [14]:

Kuo and Mallick The Bernoulli–Gaussian or Binary Mask model is due to [15]. This prior takes a Dirac for the spike $G_0 = \delta_0$ and a Gaussian for the slab G_1,

$$\beta_j \mid \gamma_j \sim (1 - \gamma_j)\delta_0 + \gamma_j N(0, \sigma_\beta^2),$$

where $N(\mu_\beta, \sigma_\beta^2)$ denotes a Normal distribution with mean μ_β and standard deviation σ_β. The slab distribution is chosen with sufficiently large variance to allow the non-zero coefficients to spread over large values. As noted by O'Hara and Sillanpää [14] this method can suffer poor mixing in an MCMC implementation due to the sharp shrinkage properties of the Dirac measure.

Stochastic Search Variable Selection (SSVS) A related method for variable selection is the stochastic search variable selection (SSVS) or Normal-Normal formulation proposed by George and McCulloch [6]. This prior has the aim of excluding variable β_j from the model whenever $|\beta_j| < \epsilon_j$ given $\epsilon_j > 0$ and where $|\cdot|$ denotes the absolute value. The idea is that ϵ_j is a practical threshold that can aid the identification of variables with effect size larger than some specified value. The prior has the form,

$$\beta_j \mid \gamma_j \sim (1 - \gamma_j)N(0, \tau_j^2) + \gamma_j N(0, c_j \tau_j^2),$$

where the separation between the two components is controlled through the tuning parameters τ_j and $c_j > 0$ which control the variance of the spike τ_j^2 and the variance of the slab $\tau_j^2 c$. To help guide the choice of these tuning parameters, [6] and [16] note that the two Gaussians intersect at the points $\pm\epsilon_j$ where

$$\epsilon_j = \tau_j\sqrt{2\log(c_j)c_j^2/(c_j - 1)}.$$

Thus posterior coefficients within the interval $[-\epsilon_j, \epsilon_j]$ can be considered "practically zero". They suggest using this to aid in the selection of the hyper-parameters τ_j and c_j. A variant of this prior is called the Gibbs variable selection (GVS) method suggested by Dellaportas et al. [17] and Carlin and Chib [18]. This method was motivated to improve convergence in MCMC implementations by reducing the sharp shrinkage of the Dirac. Their method suggests that the distribution G_1 corresponding to $\gamma_j = 0$ should be chosen so that it has no effect on the posterior. When the likelihood is Normal this method follows a similar form as the SSVS method where G_1 is a normal distribution with mean and variance chosen to minimise the effect on the posterior. This method can have good mixing properties but is difficult to tune in practice [14].

A recent extension of the SSVS type of prior was proposed by Narisetty and He [19] who propose a spike and slab priors that are Normal, but where the prior parameters depend explicitly on the sample size to achieve appropriate shrinkage. They establish model selection consistency in a high-dimensional setting, where p can grow nearly exponentially with n.

Normal Mixture of Inverse Gamma (NMIG)   For linear regression, [20] proposed to move the spike and slab to the variance term rather than placing a prior on the parameter itself. The form of their prior parameterised the variance as a product of random variables with inverse gamma distribution (IG) and a Dirac. We state the equivalent parameterisation of this spike and slab model [21]

$$\beta_j \mid \tau_j^2 \sim N(0, \tau_j^2) \tag{5.2}$$

$$\tau_j^2 \mid \gamma_j \sim (1 - \gamma_j)IG(a, \frac{d_0}{b}) + \gamma_j IG(a, \frac{d_1}{b}) \tag{5.3}$$

where d_0 and d_1 now have the role of τ_j^2 and c_j from the SSVS prior. Integrating over the variance terms the prior on β_j can be seen as a mixture of two scaled t-distributions. A similar argument based on the desired "practical effect" can be made for this prior to assist in the choice of hyper-parameters (see [20] and [21]).

Spike and Slab Lasso   More recently priors with thicker tails have been considered for the distributions of the spike and slab. In particular, [22] propose a version of the spike and slab distribution,

$$\beta_j \mid \gamma_j \sim (1 - \gamma_j)Lap(\lambda_0) + \gamma_j Lap(\lambda_1),$$

where $Lap(\lambda) = \frac{\lambda}{2}e^{-\lambda|\beta|}$ denotes a Laplace (double exponential) distribution. Taking λ_1 small and λ_0 large enables the distribution to mimic the original [1] prior with Dirac spike and diffuse slab. Taking instead $\lambda_0 = \lambda_1 = \lambda$, the prior is equivalent to a single Laplace with parameter λ. This method provides a bridge between the weak shrinkage of the Laplace distribution and the harsh shrinkage of the original spike and slab. Additional computational advantages for mode detection are also possible due to the choice of Laplace shrinkage.

Heavy Tailed Spike and Slab Recent work of [13], have considered using distributions with heavier tails than the Laplace distribution. They advocate the use of priors of the form

$$\beta_j \mid \gamma_j \sim (1 - \gamma_j)\delta_0 + \gamma_j Cauchy(1),$$

where $Cauchy(1)$ denotes a standard Cauchy distribution. In particular they find that for the prior $\gamma_j \sim Bernoulli(w)$ for all $j = 1, \ldots, p$, if the hyper parameter w is calibrated via marginal maximum likelihood empirical Bayes, the Laplace slab is shown to lead to a suboptimal rate for the empirical Bayes posterior [13]. Heavier tailed distributions are required in order to make the empirical posterior contract at the optimal rate.

Nonlocal Priors Each of the priors considered so far places local prior densities on regression coefficients in the model. That is, the slab G_1 distributions all have positive prior density at the origin 0, which can make it more difficult to distinguish between models with small coefficients. Johnson and Rossell [23] proposed two new classes of priors which are zero at and around the origin. These priors are motivated from a Bayesian model averaging perspective and assign a lower weight to more complex models [24, 25].

5.2.4 Shrinkage Priors

Due to high computational costs spike and slab methods are often not able to scale to very high dimensional problems. This is due largely to the discrete γ variable and the large model space. Consequently, this has motivated the development of a wealth of priors that aim to provide continuous alternatives to the spike and slab. One of the earliest methods that received attention for this purpose is the Bayesian Lasso (least absolute shrinkage and selection) [26]. This method was motivated largely by the Lasso penalisation approach which has been celebrated in the statistics community for its computational efficiency and variable selection performance. For a detailed survey of the lasso and related Penalised regression methods see [27]. The Bayesian Lasso corresponds to the use of a Laplace prior on the regression coefficient. The resulting posterior mode for the Bayesian lasso is equivalent to the solution for the Lasso regression problem. While the Lasso estimate has been shown to have good variable selection properties, the Bayesian Lasso does not. Castillo et al. [8] show that the Bayesian Lasso does not make the posterior concentrate near the true value in large samples.

In recent years, continuous Bayesian priors with good shrinkage properties have been introduced to the literature. One broad class of priors is referred to as global-local shrinkage priors [28] which have the hierarchical form,

$$\beta_j \mid \eta_j, w \sim \mathcal{N}(0, w\eta_j), \tag{5.4}$$

$$\eta_j \sim \pi(\eta_j), \tag{5.5}$$

$$w \sim \pi(w) \tag{5.6}$$

where η_js are known as the local shrinkage parameters and control the degree of shrinkage for each individual coefficient β_j, while the global parameter w causes an overall shrinkage. If the prior $\pi(\eta_j)$ is appropriately heavy-tailed, then the coefficients of nonzero variables will not incur a strong shrinkage effect. This hierarchical formulation essentially places a scale mixture of Normal distributions using (5.5) and (5.6) and is found frequently in the Bayesian literature. This includes the normal-gamma [29], Horseshoe prior [30], generalised double Pareto [31], Dirichlet-Laplace (DL) prior [32] and the Horseshoe+ prior [33]. These priors all contain a significant amount of mass at zero so that coefficients are shrunk to zero.

Ghosh et al. [34] observed that for a large number of global-local shrinkage priors, the parameter η_j has a distribution that can be written as,

$$\pi(\eta_j) = K\eta_j^{-a-1}L(\eta_j), \tag{5.7}$$

where $K > 0$ and $a > 0$ are positive constants, and L is a positive measureable function. Table 1 from [35] provides a list of the more well known global-local shrinkage priors that fall into this form, their corresponding density for η_j, and the component $L(\eta_j)$. Theoretical properties and uncertainty quantification has also been considered for these types of shrinkage priors [36]. Importantly, point estimates using only shrinkage priors on the regression coefficients are not able to produce exact zeros. Quantification of the selected variables is often achieved using the estimated credible intervals. Additional inference on the regression coefficients may also be achieved using the decoupling shrinkage and selection (DSS) framework developed by Hahn and Carvalho [37].

5.3   Computational Methods

In this section we survey some of the standard methods used in computational Bayesian statistics to compute posterior inference in the Bayesian variable selection methods. For each method we outline the general implementation details. For illustrative purposes, we show how these methods may be used for a linear regression analysis with the following hierarchical framework:

$$Y \mid \beta_\gamma, \gamma, \sigma \sim N_n(X_\gamma \beta_\gamma, \sigma^2 I) \tag{5.8}$$

$$\beta_\gamma \mid \sigma, \gamma \sim N_{s_\gamma}(\mu_\beta, \sigma^2 \Sigma_\gamma), \tag{5.9}$$

$$\sigma^2 \sim IG(d/2, d\lambda/2), \tag{5.10}$$

$$\gamma_j \overset{iid}{\sim} Bern(w) \text{ for } j = 1, \ldots, p, \tag{5.11}$$

where X_γ and β_γ denote subvectors of the covariates and regression parameters corresponding to the selected indices in γ and $\Sigma_\gamma \in \mathbb{R}^{s_\gamma \times s_\gamma}$ is the $s_\gamma \times s_\gamma$ prior covariance matrix for the selected regressors. Since $\gamma_j \overset{iid}{\sim} Bern(w)$ with w fixed, this prior on the model space favours models with wp selected variables. This prior specification for $\beta \mid \gamma$ corresponds to the Normal-Binomial or Kuo and Mallick spike and slab.

5.3.1 Markov Chain Monte Carlo Methods

The most widely used tool for fitting Bayesian models are sampling techniques based on Markov chain Monte Carlo (MCMC), in which a Markov chains is designed with stationary distribution that matches the desired posterior. In Bayesian variable selection, MCMC procedures are used to generate a sequence

$$\gamma^{(1)}, \gamma^{(2)}, \ldots \tag{5.12}$$

from a Markov chain with stationary distribution $p(\gamma \mid Y)$. In situations where there is no closed form expression for $p(\gamma \mid Y)$ we can attain a sequence of the form

$$\gamma^{(1)}, \beta^{(1)}, \sigma^{(1)}, \gamma^{(2)}, \beta^{(2)}, \sigma^{(2)} \ldots \tag{5.13}$$

from a Markov chain with distribution $p(\beta, \sigma, \gamma \mid Y)$. In the next two subsections we described various MCMC algorithms which may be used for simulating from (5.12) and (5.13). These algorithms are variants of the Metropolis–Hastings (MH) and Gibbs sampler algorithms, respectively. For more information on these algorithms and other MCMC methods for variable selection see the lecture notes [16].

5.3.2 Metropolis–Hastings

Algorithm 1 gives a generic description of an iteration of a Hastings–Metropolis algorithm that samples from $p(\gamma \mid Y)$. The MH algorithm works by sampling from an arbitrary probability transition kernel $q(\gamma^* \mid \gamma)$ (the distribution of the proposal γ^*) and imposing a random rejection step.

Input: γ
Output: γ'
1. Sample $\gamma^* \sim q(\gamma^* | \gamma)$
2. With Probability

$$\alpha = \min\left(1, \frac{q(\gamma \mid \gamma^*)p(\gamma^* \mid Y)}{q(\gamma^* \mid \gamma)p(\gamma \mid Y)}\right)$$

Set $\gamma' \leftarrow \gamma^*$, otherwise $\gamma' \leftarrow \gamma$.
Algorithm 1: Metropolis–Hastings (MH) algorithm

The simplest transition kernel would be to take $q(\gamma^* \mid \gamma) = 1/p$ if a single component of γ is changed. This yields a Metropolis algorithm which simulates a new proposal by randomly changing one component of γ. This algorithm was originally proposed for graphical model selection by Madigan et al. [38] and is named MC^3 (Markov chain Monte Carlo model composition). Alternative transition kernels could be constructed to propose changes in d components of γ, or more generally to change a random number of components in γ. We note that the MH approach for variable selection has inspired a number of methods that are able to effectively explore a large model space. The stochastic search methods developed by Hans et al. [39] explores multiple candidate models in parallel at each iteration and moves more aggressively toward regions of higher probability. Parallel tempering together with genetic algorithms have also been adapted to help assist the exploration of the large feature space in a method called Evolutionary MCMC (EMC) [40]. This was later adapted to Bayesian variable selection by Bottolo and Richardson [41]. For variable selection problems where $p(\gamma \mid Y)$ is not easily attained, MH methods will need to sample both β_γ and γ, so care must be taken in choosing the appropriate transition kernel.

Example Details A valuable feature of the prior in (5.8) is that, due to conjugacy of the priors [16], the parameters β_γ and σ can be eliminated from $p(Y, \beta_\gamma, \sigma \mid \gamma)$ to yield,

$$p(Y \mid \gamma) \propto |X_\gamma^T X_\gamma + \Sigma_\gamma^{-1}|^{-1/2}|\Sigma_\gamma|^{-1/2}(d\lambda + S_\gamma^2)^{(-(n+d)/2)}$$

where,

$$S_\gamma^2 = Y^T Y - Y^T X_\gamma (X_\gamma^T X_\gamma + \Sigma_\gamma^{-1})^{-1} X_\gamma^T Y.$$

Thus, for the model prior $p(\gamma) = w^{s_\gamma}(1 - w)^{p-s_\gamma}$ the posterior is proportional to

$$p(\gamma \mid Y) \propto p(Y \mid \gamma)p(\gamma) = g(\gamma).$$

Taking the previously defined transition kernel $q(\gamma^* \mid \gamma)$ and making use of the fact that $g(\gamma)/g(\gamma') = p(\gamma \mid Y)/p(\gamma' \mid Y)$, the MH algorithm follows the steps in Algorithm 1.

5.3.3 Gibbs Sampling

A well known MCMC approach to variable selection when the conditional distributions of the parameters are known is to apply Gibbs sampling. Unfortunately a drawback of Gibbs sampling is that it is not very generic and implementation depends strongly on the prior and model. When the prior is analytically tractable and a function $g(\gamma) \propto p(\gamma \mid Y)$ is available, the standard way to draw samples from the posterior $p(\gamma \mid Y)$ is by sampling the p components $(\gamma_1, \ldots, \gamma_p)$ as,

$$\gamma_j \sim p(\gamma_j \mid Y, \gamma_{(-j)}), \quad j = 1, \ldots, p,$$

where $\gamma_{(-j)} = (\gamma_1, \ldots, \gamma_{j-1}, \gamma_{j+1}, \ldots, \gamma_p)$ and where components γ_j may be drawn in fixed or random order. By computing the ratios

$$\frac{p(\gamma_j = 1, \gamma_{(-j)} \mid Y)}{p(\gamma_j = 0, \gamma_{(-j)} \mid Y)} = \frac{g(\gamma_j = 1, \gamma_{(-j)})}{g(\gamma_j = 0, \gamma_{(-j)})},$$

we can make use of the following [16]

$$p(\gamma_j = 1 \mid Y, \gamma_{(-j)}) = \frac{p(\gamma_j = 1, \gamma_{(-j)} \mid Y)}{p(\gamma_j = 0, \gamma_{(-j)} \mid Y)} \left(1 + \frac{p(\gamma_j = 1, \gamma_{(-j)} \mid Y)}{p(\gamma_j = 0, \gamma_{(-j)})}\right)^{-1}.$$

It is worth noting the recent work of Zanella and Roberts [42] which proposes an importance sampling version of the Gibbs sampling method with application to Bayesian variable selection. Additional computational advantages may be possible by drawing the components of γ in groups rather than one at a time. In this case the potential advantage of group updates would perform best if correlated variables are jointly updated.

Example Details As before, we have the function $g(\gamma)$

$$p(Y \mid \gamma) \propto g(\gamma) = |X_\gamma^T X_\gamma + \Sigma_\gamma^{-1}|^{-1/2} |\Sigma_\gamma|^{-1/2} (d\lambda + S_\gamma^2)^{(-(n+d)/2)}$$

where,

$$S_\gamma^2 = Y^T Y - Y^T X_\gamma (X_\gamma^T X_\gamma + \Sigma_\gamma^{-1})^{-1} X_\gamma^T Y.$$

The Bayesian update for $\gamma_j \mid Y, \gamma_{(-j)}$ is a Bernoulli draw with probability

$$p(\gamma_j = 1 \mid Y, \gamma_{(-j)}) = \frac{g(\gamma_j = 1, \gamma_{(-j)})}{g(\gamma_j = 0, \gamma_{(-j)})} \left(1 + \frac{g(\gamma_j = 1, \gamma_{(-j)})}{g(\gamma_j = 0, \gamma_{(-j)})} \right)^{-1}.$$

5.4   Software Implementations

There is a vast supply of software available to perform Bayesian variable selection. For this survey we restrict the scope to packages built for the R programming language [43]. These packages are free and available on the comprehensive R archive network CRAN (cran.r-project.org).

We start by noting that computational implementation of the priors and models described can be easily implemented in a number of generic Bayesian software. Ntzoufras [44] provide interesting examples of variable selection for the programs WinBUGS [45] and JAGS [46]. Code has also been made available for JAGS implementations of variable selection priors in the tutorial [14]. General purpose Bayesian software such as STAN [47] is not able to model discrete parameters so the spike and slab priors cannot be implemented. However, a large range of shrinkage priors such as the Horseshoe and Horseshoe+ are available. Practical examples for the analysis of variable selection has been proposed using STAN [48] (Table 5.1).

In addition to the general probabilistic programming languages, there are a large number of specific variable selection R packages. A survey of available R packages for variable selection has compared and contrasted popular software available as recent as February 17, 2017 [49]. In this chapter, we note some recent packages which were found using the PKGSEARCH R package [50]. The key words searched were *Bayesian variable selection, Bayesian model averaging* and *Bayesian feature selection*. From this search we note the following packages: *EMVS, basad, varbvs, BAS, spikeSlabGAM, BVSNLP, BayesS5, mombf, BoomSpikeSlab, R2GUESS, BMA, SSLASSO*.

BoomSpikeslab [51] implements a fast Gibbs sampling procedure for Bayesian modelling using a variant of the SSVS spike and slab prior. BMA implements a Metropolis Hastings (MC^3) algorithm for linear and some nonlinear sparse Bayesian models. BAS is similar to BMA in that it provides Bayesian model averaging methods. However, the sampler in BAS makes use of adaptive MCMC methods to give more efficient estimates. The mombf package provides a Gibbs sampler for the non-local and local priors (see Sect. 5.2.3). spikeSlabGAM implements a Gibbs sampler using a variant of the SSVS prior for generalised additive mixed models. Varbvs [52] implements a variational Bayesian variable selection method. As an alternative to MCMC, this package returns approximate estimates of posterior probabilities. These methods can scale much better with the dimension of the data than MCMC methods but suffer an approximation bias. R2GUESS provides an evolutionary stochastic search algorithm for both single and multiple response linear

Table 5.1 Recent packages for variable selection found using the R package PKGSEARCH

Package	Last release	Downloads	Description
BoomSpikeSlab	2019	214, 663	MCMC for Spike and Slab regression
BMA	2018	159, 652	Bayesian model averaging
BAS	2018	80, 286	Bayesian variable selection and model averaging using Bayesian adaptive sampling
mombf	2019	39, 764	Bayesian model selection and averaging for non-local and local priors
spikeSlabGAM	2018	21, 332	Bayesian variable selection and model choice for generalized additive mixed models
Varbvs	2019	14, 781	Large-scale Bayesian variable selection using variational methods
R2GUESS	2018	14, 595	A graphics processing unit-based R package for Bayesian variable selection regression of multivariate responses
BayesS5	2018	11, 295	Bayesian variable selection using simplified shotgun stochastic search with screening (S5)
BVSNLP	2019	10, 985	Bayesian variable selection in high dimensional settings using nonlocal priors
basad	2017	6187	Bayesian variable selection with Shrinking and diffusing priors
SSLASSO	2018	4407	The Spike and Slab LASSO
EMVS	2018	3816	The expectation-maximization approach to Bayesian variable selection

Year of the last release of the package, number of package downloads (calculated using CRANLOGS as of 28th July 2019)

models. BayesS5 is an efficient algorithm based on a variation of the stochastic search method and screening steps to improve computation time in high dimensions. The package BVSNLP implements considers local and nonlocal priors (similar to mombf) for binary and survival data [53]. The package basad implements variable selection with shrinking and diffusing spike and slab priors [19]. SSLASSO provides an implementation of the spike and slab lasso [22] for fast variable selection with Laplacian distributions for both the spike and slab. Finally, EMVS provides an expectation maximisation approach for Bayesian variable selection. The method provides a deterministic alternative to the stochastic search methods in order to find posterior modes.

Acknowledgement The author would like to acknowledge the Australian Research Council Centre of Excellence in Mathematical and Statistical Frontiers for funding.

References

1. T.J. Mitchell, J.J. Beauchamp, Bayesian variable selection in linear regression. J. Am. Stat. Assoc. **83**(404), 1023–1032 (1988)
2. M. Barbieri, J.O. Berger, E.I. George, V. Rockova, The median probability model and correlated variables. arXiv:1807.08336 (2020)
3. M.M. Barbieri, J.O. Berger, Optimal predictive model selection. Ann. Stat. **32**(3), 870–897 (2004)
4. F. Liang, Q. Song, K. Yu, Bayesian subset modeling for high-dimensional generalized linear models. J. Am. Stat. Assoc. **108**(502), 589–606 (2013)
5. W. Jiang, Bayesian variable selection for high dimensional generalized linear models: convergence rates of the fitted densities. Ann. Stat. **35**(4), 1487–1511 (2007)
6. E.I. George, R.E. McCulloch, Variable selection via Gibbs sampling. J. Am. Stat. Assoc. **88**(423), 881–889 (1993)
7. M. Smith, R. Kohn, A Bayesian approach to nonparametric bivariate regression. J. Am. Stat. Assoc. **92**(440), 1522–1535 (1997)
8. I. Castillo, J. Schmidt-Hieber, A. van der Vaart, Bayesian linear regression with sparse priors. Ann. Stat. **43**(5), 1986–2018 (2015)
9. J.G. Scott, J.O. Berger, Bayes and empirical-bayes multiplicity adjustment in the variable-selection problem. Ann. Stat. **38**(5), 2587–2619 (2010)
10. F. Li, N.R. Zhang, Bayesian variable selection in structured high-dimensional covariate spaces with applications in genomics. J. Am. Stat. Assoc. **105**(491), 1202–1214 (2010)
11. M.E.J. Newman, G.T. Barkema, *Monte Carlo Methods in Statistical Physics* (Clarendon Press, Oxford, 1999)
12. E.E. Leamer, *Specification Searches: Ad hoc Inference with Nonexperimental Data*, vol. 53 (Wiley, Hoboken, 1978)
13. I. Castillo, R. Mismer, Empirical bayes analysis of spike and slab posterior distributions. Electron. J. Stat. **12**, 3953–4001 (2018)
14. R.B. O'Hara, M.J. Sillanpää, A review of Bayesian variable selection methods: what, how and which. Bayesian Anal. **4**(1), 85–117 (2009)
15. L. Kuo, B. Mallick, Variable selection for regression models. Sankhyā Indian J. Stat. Ser. B (1960–2002) **60**(1), 65–81 (1998)
16. H. Chipman, E.I. George, R.E. McCulloch, *The Practical Implementation of Bayesian Model Selection*. Lecture Notes–Monograph Series, vol. 38 (Institute of Mathematical Statistics, Beachwood, 2001), pp. 65–116. https://doi.org/10.1214/lnms/1215540964
17. P. Dellaportas, J.J. Forster, I. Ntzoufras, Bayesian variable selection using the Gibbs sampler. BIOSTATISTICS-BASEL- **5**, 273–286 (2000)
18. B.P. Carlin, S. Chib, Bayesian model choice via Markov chain Monte Carlo methods. J. R. Stat. Soc. Ser. B Stat. Methodol. **57**(3), 473–484 (1995)
19. N.N. Narisetty, X. He, Bayesian variable selection with shrinking and diffusing priors. Ann. Stat. **42**(2), 789–817 (2014)
20. H. Ishwaran, J.S. Rao, Detecting differentially expressed genes in microarrays using Bayesian model selection. J. Am. Stat. Assoc. **98**(462), 438–455 (2003)
21. L. Fahrmeir, T. Kneib, S. Konrath, Bayesian regularisation in structured additive regression: a unifying perspective on shrinkage, smoothing and predictor selection. Stat. Comput. **20**(2), 203–219 (2010)
22. V. Ročková, E.I. George, The spike-and-slab lasso. J. Am. Stat. Assoc. **113**(521), 431–444 (2018)
23. V.E. Johnson, D. Rossell, Bayesian model selection in high-dimensional settings. J. Am. Stat. Assoc. **107**(498), 649–660 (2012)
24. D. Rossell, D. Telesca, Non-local priors for high-dimensional estimation. J. Am. Stat. Assoc. **112**(517), 254–265 (2017)

25. A. Nikooienejad, W. Wang, V.E. Johnson, Bayesian variable selection for binary outcomes in high-dimensional genomic studies using non-local priors. Bioinformatics **32**(9), 1338–1345 (2016)
26. R. Tibshirani, Regression shrinkage and selection via the lasso. J. R. Stat. Soc. Ser. B Stat. Methodol. **58**(1), 267–288 (1996)
27. J. Fan, J. Lv, A selective overview of variable selection in high dimensional feature space. Stat. Sin. **20**(1), 101–148 (2010)
28. N.G. Polson, J.G. Scott, Local shrinkage rules, lévy processes and regularized regression. J. R. Stat. Soc. Ser. B Stat. Methodol. **74**(2), 287–311 (2012)
29. J.E. Griffin, P.J. Brown, Inference with normal-gamma prior distributions in regression problems. Bayesian Anal. **5**(1), 171–188 (2010)
30. C.M. Carvalho, N.G. Polson, J.G. Scott, The horseshoe estimator for sparse signals. Biometrika **97**(2), 465–480 (2010)
31. A. Armagan, D.B. Dunson, J. Lee, Generalized double pareto shrinkage. Stat. Sin. **23**(1), 119–143 (2013)
32. A. Bhattacharya, D. Pati, N.S. Pillai, D.B. Dunson, Dirichlet–laplace priors for optimal shrinkage. J. Am. Stat. Assoc. **110**(512), 1479–1490 (2015)
33. A. Bhadra, J. Datta, N.G. Polson, B. Willard, The horseshoe+ estimator of ultra-sparse signals. Bayesian Anal. **12**(4), 1105–1131 (2017)
34. P. Ghosh, X. Tang, M. Ghosh, A. Chakrabarti, Asymptotic properties of bayes risk of a general class of shrinkage priors in multiple hypothesis testing under sparsity. Bayesian Anal. **11**(3), 753–796 (2016)
35. R. Bai, M. Ghosh, High-dimensional multivariate posterior consistency under global–local shrinkage priors. J. Multivar. Anal. **167**, 157–170 (2018)
36. S. van der Pas, B. Szabó, A. van der Vaart, Uncertainty quantification for the horseshoe (with discussion). Bayesian Anal. **12**(4), 1221–1274 (2017)
37. P.R. Hahn, C.M. Carvalho, Decoupling shrinkage and selection in Bayesian linear models: a posterior summary perspective. J. Am. Stat. Assoc. **110**(509), 435–448 (2015)
38. D. Madigan, J. York, D. Allard, Bayesian graphical models for discrete data. Int. Stat. Rev./Rev. Int. de Stat. **63**(2), 215–232 (1995)
39. C. Hans, A. Dobra, M. West, Shotgun stochastic search for "large p" regression. J. Am. Stat. Assoc. **102**(478), 507–516 (2007)
40. F. Liang, W.H. Wong, Evolutionary monte carlo: applications to C p model sampling and change point problem. Stat. Sin. **10**(2), 317–342 (2000)
41. L. Bottolo, S. Richardson, Evolutionary stochastic search for Bayesian model exploration. Bayesian Anal. **5**(3), 583–618 (2010)
42. G. Zanella, G. Roberts, Scalable importance tempering and Bayesian variable selection. J. R. Statist. Soc. B **81**, 489–517 (2019)
43. R Core Team, *R: A Language and Environment for Statistical Computing* (R Foundation for Statistical Computing, Vienna, 2013)
44. I. Ntzoufras, Gibbs variable selection usingbugs. J. Stat. Softw. **7**(7), 1–19 (2002)
45. D.J. Lunn, A. Thomas, N. Best, D. Spiegelhalter, Winbugs-a Bayesian modelling framework: concepts, structure, and extensibility. Stat. Comput. **10**(4), 325–337 (2000)
46. M. Plummer, et al., JAGS: A program for analysis of Bayesian graphical models using gibbs sampling, in *Proceedings of the 3rd International Workshop on Distributed Statistical Computing*, vol. 124 (2003)
47. B. Carpenter, A. Gelman, M.D. Hoffman, D. Lee, B. Goodrich, M. Betancourt, M. Brubaker, J. Guo, P. Li, A. Riddell, Stan: A probabilistic programming language. J. Stat. Softw. **76**(1), 1–32 (2017)
48. J. Piironen, A. Vehtari, Projection predictive model selection for gaussian processes, in *2016 IEEE 26th International Workshop on Machine Learning for Signal Processing (MLSP)* (2016), pp. 1–6

49. A. Forte, G. Garcia-Donato, M. Steel, Methods and tools for Bayesian variable selection and model averaging in normal linear regression. Int. Stat. Rev./Rev. Int. de Stat. **86**(2), 237–258 (2018)
50. G. Csárdi, *pkgsearch: Search CRAN R Packages*. R package version 2.0.1. (2018). https://CRAN.R-project.org/package=pkgsearch
51. H. Ishwaran, U.B. Kogalur, J.S. Rao, spikeslab: prediction and variable selection using spike and slab regression. R J. **2**, 68–73 (2010)
52. P. Carbonetto, M. Stephens, Scalable variational inference for Bayesian variable selection in regression, and its accuracy in genetic association studies. Bayesian Anal. **7**, 73–108 (2012)
53. D. Rossell, J.D. Cook, D. Telesca, P. Roebuck, mombf: moment and inverse moment bayes factors. R Package Version 1. 0, vol. 3 (2008)

Chapter 6
Bayesian Computation with Intractable Likelihoods

Matthew T. Moores, Anthony N. Pettitt, and Kerrie L. Mengersen

Abstract This chapter surveys computational methods for posterior inference with intractable likelihoods, that is where the likelihood function is unavailable in closed form, or where evaluation of the likelihood is infeasible. We survey recent developments in pseudo-marginal methods, approximate Bayesian computation (ABC), the exchange algorithm, thermodynamic integration, and composite likelihood, paying particular attention to advancements in scalability for large datasets. We also mention R and MATLAB source code for implementations of these algorithms, where they are available.

Keywords Composite likelihood · Likelihood-free inference · Markov random fields · Pseudo-marginal methods

The likelihood function plays an important role in Bayesian inference, since it connects the observed data with the statistical model. Both simulation-based (e.g. MCMC) and optimisation-based (e.g. variational Bayes) algorithms require the likelihood to be evaluated pointwise, up to an unknown normalising constant. However, there are some situations where this evaluation is analytically and computationally intractable. For example, when the complexity of the likelihood grows at a combinatorial rate in terms of the number of observations, then likelihood-based inference quickly becomes infeasible for the scale of data that is regularly encountered in applications.

M. T. Moores (✉)
National Institute for Applied Statistics Research Australia, School of Mathematics & Applied Statistics, University of Wollongong, Keiraville, NSW, Australia
e-mail: mmoores@uow.edu.au

A. N. Pettitt · K. L. Mengersen
School of Mathematical Sciences, Queensland University of Technology, Brisbane, QLD, Australia
e-mail: a.pettitt@qut.edu.au; k.mengersen@qut.edu.au

© The Editor(s) (if applicable) and The Author(s), under exclusive licence to Springer Nature Switzerland AG 2020
K. L. Mengersen et al. (eds.), *Case Studies in Applied Bayesian Data Science*, Lecture Notes in Mathematics 2259, https://doi.org/10.1007/978-3-030-42553-1_6

Intractable likelihoods arise in a variety of contexts, including models for DNA mutation in population genetics [43, 64], models for the spread of disease in epidemiology [46, 60], models for the formation of galaxies in astronomy [12], and estimation of the model evidence in Bayesian model choice [29]. This chapter will mainly focus on Markov random field (MRF) models with discrete state spaces, such as the Ising, Potts, and exponential random graph models (ERGM). These models are used for image segmentation or analysis of social network data, two areas where millions of observations are commonplace. There is therefore a need for scalable inference algorithms that can handle these large volumes of data.

The Ising, Potts, or ERGM likelihood functions can be expressed in the form of an exponential family:

$$p(\mathbf{y} \mid \boldsymbol{\theta}) = \frac{\exp\left\{\boldsymbol{\theta}^T \mathbf{s}(\mathbf{y})\right\}}{C(\boldsymbol{\theta})}, \tag{6.1}$$

where the observed data $\mathbf{y} = y_1, \ldots, y_n$ is in the form of an undirected graph, $\boldsymbol{\theta}$ is a vector of unknown parameters, $\mathbf{s}(\mathbf{y})$ is a corresponding vector of jointly-sufficient statistics for these parameters, and $C(\boldsymbol{\theta})$ is an intractable normalising constant, also known as a partition function:

$$C(\boldsymbol{\theta}) = \sum_{\mathbf{y} \in \mathcal{Y}} \exp\left\{\boldsymbol{\theta}^T \mathbf{s}(\mathbf{y})\right\}, \tag{6.2}$$

where the sum is over all possible configurations of states, $\mathbf{y} \in \mathcal{Y}$.

In the case of an Ising model, a single node can take one of two possible values, $y_i \in \{0, 1\}$. For example, in image analysis the value 1 might represent a foreground pixel, while 0 represents the background. The q-state Potts model generalises this construction to more than two states, so $y_i \in \{1, \ldots, q\}$. The cardinality of the configuration space, $\#\mathcal{Y}$, is then q^n. Even with only 2 states and $n = 100$ pixels, computation of (6.2) requires more than 10^{30} floating point operations. It would take a supercomputer with 100 PetaFLOPS over 400,000 years to find an answer.

Both the Ising and Potts models possess a single parameter, β, known as the inverse temperature. The corresponding sufficient statistic is then:

$$s(\mathbf{y}) = \sum_{i \sim \ell \in \mathcal{E}} \delta(y_i, y_\ell), \tag{6.3}$$

where $\mathcal{E}$ is the set of all unique pairs of neighbours $i \sim \ell$ in the observed graph, and $\delta(x, y)$ is the Kronecker delta function, which equals 1 when $x = y$ and 0 otherwise. We assume a first-order neighbourhood structure, so a given pixel y_i would have up to 4 neighbours in a regular 2D lattice, or 6 neighbours in 3D. Pixels on the boundary of the image domain have less than 4 (or 6) neighbours, so $\#\mathcal{E} = 2(n - \sqrt{n})$ for a square 2D lattice, or $3(n - n^{2/3})$ for a cube.

The observed data for an ERGM can be represented as a binary adjacency matrix Y, encoding the presence or absence of a neighbourhood relationship between nodes i and j: $[Y]_{i,j} = 1$ if $i \sim j$; $[Y]_{i,j} = 0$ otherwise. $\#\mathcal{Y}$ for an ERGM is equal to 2^M, where $M = n(n - 1)/2$ is the maximum number of ties in an undirected graph with n nodes. As with the Ising or Potts models, computing the normalising constant (6.2) is therefore intractable for non-trivial graphs. Various kinds of ERGM can be defined by the choice of sufficient statistics. The simplest example is the Bernoulli random graph [23], which has a single statistic $s_1(Y) = m$, the number of connected neighbours in the graph. In an undirected graph, this is half the number of nonzero entries in the adjacency matrix. An important class of graph statistics are the numbers of k-stars [26], which can be defined in terms of the degree distribution [59]:

$$n_k = \sum_{i=1}^{n} \binom{d_i}{k}, \tag{6.4}$$

where the degree d_i is the number of neighbours of node i:

$$d_i = \sum_{j=1}^{n} [Y]_{ij}. \tag{6.5}$$

Note that under this definition $n_1 = 2m$, since each tie is counted twice. An alternative definition, which avoids double-counting, is given by:

$$n_1 = \sum_{i<j} [Y]_{ij} \qquad \text{number of edges}$$
$$n_2 = \sum_{i<j<k} [Y]_{ik}[Y]_{jk} \qquad \text{number of 2-stars}$$
$$n_3 = \sum_{i<j<l<k} [Y]_{ik}[Y]_{jk}[Y]_{lk} \quad \text{number of 3-stars.}$$

The remainder of this chapter will describe various MCMC methods that target the posterior distribution $\pi(\theta \mid \mathbf{y})$, or some approximation thereof. This will be in the context of a random walk Metropolis (RWM) algorithm that proposes a new value of θ' at iteration t using a (multivariate) Gaussian proposal distribution, $q(\cdot \mid \theta^{t-1}) \sim \mathcal{N}(\theta^{t-1}, \Sigma_t)$. Methods for tuning the proposal bandwidth Σ_t have been described by Andrieu and Thoms [3] and Roberts and Rosenthal [67]. Normally, the proposed parameter value would be accepted with probability $\min\{1, \rho_t\}$, or else rejected, where ρ_t is the Radon–Nikodým derivative:

$$\rho_t = \frac{q\left(\theta^{(t-1)} \mid \theta'\right) p\left(\mathbf{y} \mid \theta'\right) \pi_0\left(\theta'\right)}{q\left(\theta' \mid \theta^{(t-1)}\right) p\left(\mathbf{y} \mid \theta^{(t-1)}\right) \pi_0\left(\theta^{(t-1)}\right)}, \tag{6.6}$$

$\pi_0(\theta)$ is the prior density for the parameter/s, and $p(\mathbf{y} \mid \theta)$ is the likelihood (6.1). If we use a symmetric proposal distribution q and a uniform prior π_0, then these terms will cancel, leaving:

$$\rho_t = \frac{\psi\left(\mathbf{y} \mid \boldsymbol{\theta}'\right)}{\psi(\mathbf{y} \mid \boldsymbol{\theta}^{(t-1)})} \frac{C(\boldsymbol{\theta}^{(t-1)})}{C\left(\boldsymbol{\theta}'\right)}, \tag{6.7}$$

which is the ratio of unnormalised likelihoods $\psi = \exp\left\{\boldsymbol{\theta}^T \mathbf{s}(\mathbf{y})\right\}$, multiplied by the ratio of intractable normalising constants (6.2). It is clearly infeasible to evaluate (6.7) directly, so alternative algorithms are required. One option is to estimate ρ_t by simulation, which we categorise as auxiliary variable methods: pseudo-marginal algorithms, the exchange algorithm, and approximate Bayesian computation (ABC). Other methods include path sampling, also known as thermodynamic integration (TI), pseudolikelihood, and composite likelihood.

6.1 Auxiliary Variable Methods

6.1.1 Pseudo-Marginal Algorithms

Pseudo-marginal algorithms [2, 6] are computational methods for fitting latent variable models, that is where the observed data $\mathbf{y}$ can be considered as noisy observations of some unobserved or hidden states, $\mathbf{x}$. For example, hidden Markov models (HMMs) are commonly used in time series analysis and signal processing. Models of this form can also arise as the result of data augmentation approaches, such as for mixture models [17, 73]. The marginal likelihood is of the following form:

$$p(\mathbf{y} \mid \boldsymbol{\theta}) = \int_{\mathcal{X}} p(\mathbf{y} \mid \mathbf{x}) \, p(\mathbf{x} \mid \boldsymbol{\theta}) \, d\mathbf{x}, \tag{6.8}$$

which can be intractable if the state space is very high-dimensional and non-Gaussian. In this case, we can substitute an unbiased, non-negative estimate of the likelihood.

O'Neill et al. [60] introduced the Monte Carlo within Metropolis (MCWM) algorithm, which replaces both $p\left(\mathbf{y} \mid \boldsymbol{\theta}'\right)$ and $p\left(\mathbf{y} \mid \boldsymbol{\theta}^{(t-1)}\right)$ in the Metropolis–Hastings ratio ρ_t (6.6) with importance-sampling estimates:

$$\tilde{p}_{IS}(\mathbf{y} \mid \boldsymbol{\theta}) \approx \frac{1}{M} \sum_{m=1}^{M} p(\mathbf{y} \mid X_m) \frac{p(X_m \mid \boldsymbol{\theta})}{q(X_m \mid \boldsymbol{\theta})}, \tag{6.9}$$

where the samples $X_1, \ldots, X_M$ are drawn from a proposal distribution $q(X_m \mid \boldsymbol{\theta})$ for $\boldsymbol{\theta}'$ and $\boldsymbol{\theta}^{(t-1)}$. MCWM is generally considered as an approximate algorithm, since it does not target the exact posterior distribution for $\boldsymbol{\theta}$. However,

Medina-Aguayo et al. [47] have established some conditions under which MCWM converges to the correct target distribution as $M \rightarrow \infty$. See also [55] and [1] for further theoretical analysis of approximate pseudo-marginal methods.

Beaumont [6] introduced the grouped independence Metropolis–Hastings (GIMH) algorithm, which does target the exact posterior. The key difference is that $\tilde{p}_{IS}\left(\mathbf{y} \mid \boldsymbol{\theta}^{(t-1)}\right)$ is reused from the previous iteration, rather than being recalculated every time. The theoretical properties of this algorithm have been an active area of research, with notable contributions by Andrieu and Roberts [2], Maire et al. [41], Andrieu and Vihola [4], and Sherlock et al. [69]. Andrieu et al. [5] introduced the particle MCMC algorithm, which is a pseudo-marginal method that uses sequential Monte Carlo (SMC) in place of importance sampling. This is particularly useful for HMMs, where SMC methods such as the bootstrap particle filter provide an unbiased estimate of the marginal likelihood [62]. Although importance sampling and SMC are both unbiased estimators, it is necessary to use a large enough value of M so that the variance is kept at a reasonable level. Otherwise, the pseudo-marginal algorithm can fail to be variance-bounding or geometrically ergodic [39]. Doucet et al. [18] recommend choosing M so that the standard deviation of the log-likelihood estimator is between 1 and 1.7.

Pseudo-marginal algorithms can be computationally intensive, particularly for large values of M. One strategy to reduce this computational burden, known as the Russian Roulette algorithm [40], is to replace $\tilde{p}_{IS}(\mathbf{y} \mid \boldsymbol{\theta})$ (6.9) with a truncated infinite series:

$$\tilde{p}_{RR}(\mathbf{y} \mid \boldsymbol{\theta}) = \sum_{j=0}^{\tau} V_{\boldsymbol{\theta}}^{(j)}, \qquad (6.10)$$

where τ is a random stopping time and $V_{\boldsymbol{\theta}}^{(j)}$ are random variables such that (6.10) is almost surely finite and $\mathbb{E}[\tilde{p}_{RR}(\mathbf{y} \mid \boldsymbol{\theta})] = p(\mathbf{y} \mid \boldsymbol{\theta})$. There is a difficulty with this method, however, in that the likelihood estimates are not guaranteed to be non-negative. Jacob and Thiery [37] have established that there is no general solution to this sign problem, although successful strategies have been proposed for some specific models.

Another important class of algorithms for accelerating pseudo-marginal methods involve approximating the intractable likelihood function using a surrogate model. For example, the delayed-acceptance (DA) algorithm of [14] first evaluates the Metropolis–Hastings ratio (6.6) using a fast, approximate likelihood $\tilde{p}_{DA}(\mathbf{y} \mid \boldsymbol{\theta})$. The proposal $\boldsymbol{\theta}'$ is rejected at this screening stage with probability $1 - \min\{1, \rho_t\}$. Otherwise, a second ratio $\rho_{DA}^{(2)}$ is calculated using a full evaluation of the likelihood function (6.9). The acceptance probability $\min\{1, \rho_{DA}^{(2)}\}$ is modified at the second stage according to:

$$\rho_{DA}^{(2)} = \frac{\tilde{p}_{IS}(\mathbf{y} \mid \boldsymbol{\theta}') \pi_0(\boldsymbol{\theta}')}{\tilde{p}_{IS}(\mathbf{y} \mid \boldsymbol{\theta}^{(t-1)}) \pi_0(\boldsymbol{\theta}^{(t-1)})} \frac{\tilde{p}_{DA}(\mathbf{y} \mid \boldsymbol{\theta}^{(t-1)}) \pi_0(\boldsymbol{\theta}^{(t-1)})}{\tilde{p}_{DA}(\mathbf{y} \mid \boldsymbol{\theta}') \pi_0(\boldsymbol{\theta}')}, \qquad (6.11)$$

which corrects for the conditional dependence on acceptance at the first stage and therefore preserves the exact target distribution. DA has been used for PMCMC by Golightly et al. [33], where the linear noise approximation [25] was used for $\tilde{p}_{DA}$. Sherlock et al. [70] instead used k-nearest-neighbours for $\tilde{p}_{DA}$ in a pseudo-marginal algorithm.

Drovandi et al. [22] proposed an approximate pseudo-marginal algorithm, using a Gaussian process (GP) as a surrogate log-likelihood. The GP is trained using a pilot run of MCWM, then at each iteration $\log \tilde{p}(\mathbf{y} \mid \boldsymbol{\theta}')$ is either approximated using the GP or else using SMC or importance sampling, depending on the level of uncertainty in the surrogate model for $\boldsymbol{\theta}'$. MATLAB source code is available from http://www.runmycode.org/companion/view/2663. Stuart and Teckentrup [71] have shown that, under certain assumptions, a GP provides a consistent estimator of the negative log-likelihood, and they provide error bounds on the approximation.

6.1.2 Exchange Algorithm

Møller et al. [50] introduced a MCMC algorithm for the Ising model that targets the exact posterior distribution for β. An auxiliary variable $\mathbf{x}$ is defined on the same state space as $\mathbf{y}$, so that $\mathbf{x}, \mathbf{y} \in \mathcal{Y}$. This is a data augmentation approach, where we simulate from the joint posterior $\pi(\beta, \mathbf{x} \mid \mathbf{y})$, which admits the posterior for β as its marginal. Given a proposed parameter value β', a proposal $\mathbf{x}'$ is simulated from the model to obtain an unbiased sample from (6.1). This requires perfect simulation methods, such as coupling from the past [65], perfect slice sampling [49], or bounding chains [8, 35]. Refer to [36] for further explanation of perfect simulation. Instead of (6.7), the joint ratio for β' and $\mathbf{x}'$ becomes:

$$\rho_t = \frac{\psi\left(\mathbf{y} \mid \beta'\right)}{\psi\left(\mathbf{y} \mid \beta^{(t-1)}\right)} \frac{\psi\left(\mathbf{x}' \mid \tilde{\beta}\right)}{\psi\left(\mathbf{x}^{(t-1)} \mid \tilde{\beta}\right)} \frac{\psi(\mathbf{x}^{(t-1)} \mid \beta^{(t-1)})}{\psi\left(\mathbf{x}' \mid \beta'\right)}, \tag{6.12}$$

where the normalising constants $C(\beta')$ and $C(\beta^{(t-1)})$ cancel out with each other. This is analogous to an importance-sampling estimate of the normalising constant with $M = 1$ samples, since:

$$\mathbb{E}_{\mathbf{x}}\left[\frac{\psi\left(\mathbf{x} \mid \beta\right)}{q\left(\mathbf{x} \mid \beta\right)}\right] = C(\beta), \tag{6.13}$$

where the proposal distribution $q(\mathbf{x} \mid \beta)$ is (6.1). This algorithm is therefore closely-related with pseudo-marginal methods such as GIMH.

Murray et al. [54] found that (6.12) could be simplified even further, removing the need for a fixed value of $\tilde{\beta}$. The exchange algorithm replaces (6.7) with the ratio:

$$\rho_t = \frac{\psi\left(\mathbf{y} \mid \beta'\right)}{\psi\left(\mathbf{y} \mid \beta^{(t-1)}\right)} \frac{\psi\left(\mathbf{x}' \mid \beta^{(t-1)}\right)}{\psi\left(\mathbf{x}' \mid \beta'\right)}. \tag{6.14}$$

However, perfect sampling is still required to simulate $\mathbf{x}'$ at each iteration, which can be infeasible when the state space is very large. Cucala et al. [15] proposed an approximate exchange algorithm (AEA) by replacing the perfect sampling step with 500 iterations of Gibbs sampling. Caimo and Friel [9] were the first to employ AEA for fully-Bayesian inference on the parameters of an ERGM. AEA for the hidden Potts model is implemented in the R package 'bayesImageS' [51] and AEA for ERGM is implemented in 'Bergm' [10].

6.1.3 Approximate Bayesian Computation

Like the exchange algorithm, ABC uses an auxiliary variable $\mathbf{x}$ to decide whether to accept or reject the proposed value of θ'. In the terminology of ABC, $\mathbf{x}$ is referred to as "pseudo-data." Instead of a Metropolis–Hastings ratio such as (6.7), the summary statistics of the pseudo-data and the observed data are directly compared. The proposal is accepted if the distance between these summary statistics is within the ABC tolerance, ϵ. This produces the following approximation:

$$p\left(\theta \mid \mathbf{y}\right) \approx \pi_\epsilon\left(\theta \mid \|\mathbf{s}(\mathbf{x}) - \mathbf{s}(\mathbf{y})\| < \epsilon\right), \tag{6.15}$$

where $\|\cdot\|$ is a suitable norm, such as Euclidean distance. Since $\mathbf{s}(\mathbf{y})$ are jointly-sufficient statistics for Ising, Potts, or ERGM, the ABC approximation (6.15) approaches the true posterior as $n \to \infty$ and $\epsilon \to 0$. In practice there is a tradeoff between the number of parameter values that are accepted and the size of the ABC tolerance.

Grelaud et al. [34] were the first to use ABC to obtain an approximate posterior for β in the Ising/Potts model. Everitt [24] used ABC within sequential Monte Carlo (ABC-SMC) for Ising and ERGM. ABC-SMC uses a sequence of target distributions $\pi_{\epsilon_t}\left(\theta \mid \|\mathbf{s}(\mathbf{x}) - \mathbf{s}(\mathbf{y})\| < \epsilon_t\right)$ such that $\epsilon_1 > \epsilon_2 > \cdots > \epsilon_T$, where the number of SMC iterations T can be determined dynamically using a stopping rule. The ABC-SMC algorithm of [19] uses multiple MCMC steps for each SMC iteration, while the algorithm of [16] uses multiple replicates of the summary statistics for each particle. Everitt [24] has provided a MATLAB implementation of ABC-SMC with the online supplementary material accompanying his paper.

The computational efficiency of ABC is dominated by the cost of drawing updates to the auxiliary variable, as reported by Everitt [24]. Thus, we would expect that the execution time for ABC would be similar to AEA or pseudo-marginal methods. Various approaches to improving this runtime have recently been proposed. "Lazy ABC" [63] involves early termination of the simulation step at a random stopping time, hence it bears some similarities with Russian Roulette. Surrogate models have also been applied in ABC, using a method known as Bayesian indirect likelihood [BIL; 20, 21]. Gaussian processes (GPs) have been used as surrogate models by Wilkinson [75] and Meeds and Welling [48]. Järvenpää et al. [38] used a heteroskedastic GP model and demonstrated how the output of the precomputation step could be used for Bayesian model choice. Moores et al. [52] introduced a piecewise linear approximation for ABC-SMC with Ising/Potts models. Boland et al. [7] derived a theoretical upper bound on the bias introduced by this and similar piecewise approximations. They also developed a piecewise linear approximation for ERGM. Moores et al. [53] introduced a parametric functional approximate Bayesian (PFAB) algorithm for the Potts model, which is a form of BIL where $\tilde{p}_{BIL}(\mathbf{y} \mid \boldsymbol{\theta})$ is derived from an integral curve.

6.2 Other Methods

6.2.1 Thermodynamic Integration

Since the Ising, Potts, and ERGM are all exponential families of distributions, the expectation of their sufficient statistic/s can be expressed in terms of the normalising constant:

$$\mathbb{E}_{\mathbf{y}|\boldsymbol{\theta}}[\mathbf{s}(\mathbf{y})] = \frac{\mathrm{d}}{\mathrm{d}\boldsymbol{\theta}} \log\{C(\boldsymbol{\theta})\}. \tag{6.16}$$

Gelman and Meng [31] derived an approximation to the log-ratio of normalising constants for the Ising/Potts model, using the path sampling identity:

$$\log\left\{\frac{C(\beta_{t-1})}{C(\beta')}\right\} = \int_{\beta'}^{\beta_{t-1}} \mathbb{E}_{\mathbf{y}|\beta}[s(\mathbf{y})] \, \mathrm{d}\beta, \tag{6.17}$$

which follows from (6.16). The value of the expectation can be estimated by simulating from the Gibbs distribution (6.1) for fixed values of β. At each iteration, $\log\{\rho_t\}$ (6.7) can then be approximated by numerical integration methods, such as Gaussian quadrature or the trapezoidal rule. Figure 6.1 illustrates linear interpolation of $\mathbb{E}_{\mathbf{y}|\beta}[s(\mathbf{y})]$ on a 2D lattice for $q = 6$ labels and β ranging from 0 to 2 in increments of 0.05. This approximation was precomputed using the algorithm of [72].

TI is explained in further detail by Chen et al. [13, chap. 5]. A reference implementation in R is available from the website accompanying [42]. Friel and Pettitt [29] introduced the method of power posteriors to estimate the marginal likelihood or model evidence using TI. Calderhead and Girolami [11] provide bounds on the discretisation error and derive an optimal temperature schedule by

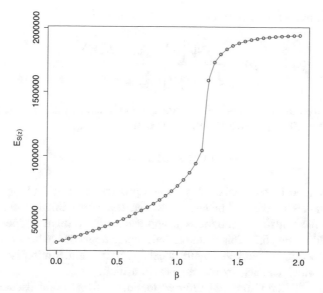

Fig. 6.1 Approximation of $\mathbb{E}_{\mathbf{y}|\beta}[s(\mathbf{y})]$ by simulation for fixed values of β, with linear interpolation

minimising the variance of the Monte Carlo estimate. Oates et al. [56] introduced control variates for further reducing the variance of TI.

The TI algorithm has an advantage over auxiliary variable methods because the additional simulations are performed prior to fitting the model, rather than at each iteration. This is particularly the case when analysing multiple images that all have approximately the same dimensions. Since these simulations are independent, they can make use of massively parallel hardware. However, the computational cost is still slightly higher than pseudolikelihood, which does not require a pre-computation step.

6.2.2 Composite Likelihood

Pseudolikelihood is the simplest of the methods that we have considered and also the fastest. Rydén and Titterington [68] showed that the intractable distribution (6.1) could be approximated using the product of the conditional densities:

$$\tilde{p}_{PL}(\mathbf{y} \mid \boldsymbol{\theta}) \approx \prod_{i=1}^{n} p(y_i \mid y_{\backslash i}, \boldsymbol{\theta}). \tag{6.18}$$

This enables the Metropolis–Hastings ratio ρ_t (6.6) to be evaluated using (6.18) to approximate both $p\left(\mathbf{y} \mid \boldsymbol{\theta}'\right)$ and $p\left(\mathbf{y} \mid \boldsymbol{\theta}^{(t-1)}\right)$ at each iteration. The conditional density function for the Ising/Potts model is given by:

$$p(y_i \mid y_{\backslash i}, \beta) = \frac{\exp\left\{\beta \sum_{\ell \in \partial(i)} \delta(z_i, z_\ell)\right\}}{\sum_{j=1}^{q} \exp\left\{\beta \sum_{\ell \in \partial(i)} \delta(j, z_\ell)\right\}}, \tag{6.19}$$

where $\ell \in \partial(i)$ are the first-order (nearest) neighbours of pixel i. The conditional density for an ERGM is given by the logistic function:

$$p([Y]_{ij} = 1 \mid [Y]_{\backslash ij}, \boldsymbol{\theta}) = \text{logit}^{-1}\left\{\boldsymbol{\theta}^T \mathbf{s}(Y)\right\}. \tag{6.20}$$

Pseudolikelihood is exact when $\boldsymbol{\theta} = 0$ and provides a reasonable approximation for small values of the parameters. However, the approximation error increases rapidly for the Potts/Ising model as β approaches the critical temperature, β_{crit}, as illustrated by Fig. 6.2. This is due to long-range dependence between the labels, which is inadequately modelled by the local approximation. Similar issues can arise for ERGM, which can also exhibit a phase transition.

Rydén and Titterington [68] referred to Eq. (6.18) as point pseudolikelihood, since the conditional distributions are computed for each pixel individually. They suggested that the accuracy could be improved using block pseudolikelihood. This is where the likelihood is calculated exactly for small blocks of pixels, then (6.18) is modified to be the product of the blocks:

$$\tilde{p}_{BL}(\mathbf{y} \mid \boldsymbol{\theta}) \approx \prod_{i=1}^{N_B} p(\mathbf{y}_{B_i} \mid \mathbf{y}_{\backslash B_i}, \boldsymbol{\theta}) \tag{6.21}$$

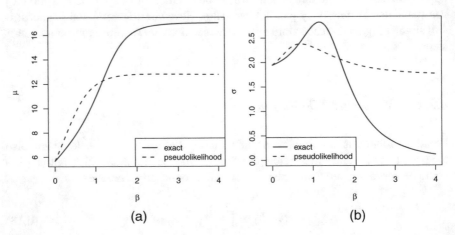

Fig. 6.2 Approximation error of pseudolikelihood for $n = 12$, $q = 3$ in comparison to the exact likelihood calculated using a brute force method: (**a**) $\sum_{\mathbf{y} \in \mathcal{Y}} s(\mathbf{y}) p(\mathbf{y} \mid \beta)$ using either Eq. (6.1) or (6.18); (**b**) $\sqrt{\sum_{\mathbf{y} \in \mathcal{Y}} \left(s(\mathbf{y}) - \mathbb{E}_{\mathbf{y} \mid \beta}[s(\mathbf{y})]\right)^2 p(\mathbf{y} \mid \beta)}$. (**a**) Expectation. (**b**) Standard deviation

where N_B is the number of blocks, $\mathbf{y}_{B_i}$ are the labels of the pixels in block B_i, and $\mathbf{y}_{\backslash B_i}$ are all of the labels except for $\mathbf{y}_{B_i}$. This is a form of composite likelihood, where the likelihood function is approximated as a product of simplified factors [74]. Friel [27] compared point pseudolikelihood to composite likelihood with blocks of 3×3, 4×4, 5×5, and 6×6 pixels. Friel showed that (6.21) outperformed (6.18) for the Ising ($q = 2$) model with $\beta < \beta_{crit}$. Okabayashi et al. [58] discuss composite likelihood for the Potts model with $q > 2$ and have provided an open source implementation in the R package 'potts' [32].

Evaluating the conditional likelihood in (6.21) involves the normalising constant for $\mathbf{y}_{B_i}$, which is a sum over all of the possible configurations $\mathcal{Y}_{B_i}$. This is a limiting factor on the size of blocks that can be used. The brute force method that was used to compute Fig. 6.2 is too computationally intensive for this purpose. Pettitt et al. [61] showed that the normalising constant can be calculated exactly for a cylindrical lattice by computing eigenvalues of a $k^r \times k^r$ matrix, where r is the smaller of the number of rows or columns. The value of (6.2) for a free-boundary lattice can then be approximated using path sampling. Friel and Pettitt [28] extended this method to larger lattices using a composite likelihood approach.

The reduced dependence approximation (RDA) is another form of composite likelihood. Reeves and Pettitt [66] introduced a recursive algorithm to calculate the normalising constant using a lag-r representation. Friel et al. [30] divided the image lattice into sub-lattices of size $r_1 < r$, then approximated the normalising constant of the full lattice using RDA:

$$C(\beta) \approx \frac{C_{r_1 \times n}(\beta)^{r-r_1+1}}{C_{r_1-1 \times n}(\beta)^{r-r_1}} \tag{6.22}$$

McGrory et al. [44] compared RDA to pseudolikelihood and the exact method of [50], reporting similar computational cost to pseudolikelihood but with improved accuracy in estimating β. Ogden [57] showed that if r is chosen proportional to n, then RDA gives asymptotically valid inference when $\beta < \beta_{crit}$. However, the error increases exponentially as β approaches the phase transition. This is similar to the behaviour of pseudolikelihood in Fig. 6.2. Source code for RDA is available in the online supplementary material for McGrory et al. [45].

6.3 Conclusion

This chapter has surveyed a variety of computational methods for Bayesian inference with intractable likelihoods. Auxiliary variable methods, such as the exchange algorithm and pseudo-marginal algorithms, target the exact posterior distribution. However, their computational cost can be prohibitive for large datasets. Algorithms such as delayed acceptance, Russian Roulette, and "lazy ABC" can accelerate inference by reducing the number of auxiliary variables that need to be simulated, without modifying the target distribution. Bayesian indirect likelihood (BIL) algo-

148

M. T. Moores et al.

rithms approximate the intractable likelihood using a surrogate model, such as a Gaussian process or piecewise function. As with thermodynamic integration, BIL can take advantage of a precomputation step to train the surrogate model in parallel. This enables these methods to be applied to much larger datasets by managing the tradeoff between approximation error and computational cost.

Acknowledgments This research was conducted by the Australian Research Council Centre of Excellence for Mathematical and Statistical Frontiers (project number CE140100049) and funded by the Australian Government.

References

1. P. Alquier, N. Friel, R. Everitt, A. Boland, Noisy Monte Carlo: convergence of Markov chains with approximate transition kernels. Stat. Comput. **26**(1–2), 29–47 (2016). https://doi.org/10.1007/s11222-014-9521-x
2. C. Andrieu, G.O. Roberts, The pseudo-marginal approach for efficient Monte Carlo computations. Ann. Statist. **37**(2), 697–725 (2009). https://doi.org/10.1214/07-AOS574
3. C. Andrieu, J. Thoms, A tutorial on adaptive MCMC. Stat. Comput. **18**(4), 343–373 (2008). https://doi.org/10.1007/s11222-008-9110-y
4. C. Andrieu, M. Vihola, Convergence properties of pseudo-marginal Markov chain Monte Carlo algorithms. Ann. Appl. Prob. **25**(2), 1030–1077, 04 (2015). https://doi.org/10.1214/14-AAP1022
5. C. Andrieu, A. Doucet, R. Holenstein, Particle Markov chain Monte Carlo methods. J. R. Stat. Soc. Ser. B **72**(3), 269–342 (2010). https://doi.org/10.1111/j.1467-9868.2009.00736.x
6. M.A. Beaumont, Estimation of population growth or decline in genetically monitored populations. Genetics **164**(3), 1139–1160 (2003)
7. A. Boland, N. Friel, F. Maire, Efficient MCMC for Gibbs random fields using pre-computation. Electron. J. Statist. **12**(2), 4138–4179 (2018). https://doi.org/10.1214/18-EJS1504.
8. C.T. Butts, A perfect sampling method for exponential family random graph models. J. Math. Soc. **42**(1), 17–36 (2018). https://doi.org/10.1080/0022250X.2017.1396985
9. A. Caimo, N. Friel, Bayesian inference for exponential random graph models. Soc. Networks **33**(1), 41–55 (2011). https://doi.org/10.1016/j.socnet.2010.09.004
10. A. Caimo, N. Friel, Bergm: Bayesian exponential random graphs in R. J. Stat. Soft. **61**(2), 1–25 (2014). https://doi.org/10.18637/jss.v061.i02
11. B. Calderhead, M. Girolami, Estimating Bayes factors via thermodynamic integration and population MCMC. Comput. Stat. Data Anal. **53**(12), 4028–4045 (2009). https://doi.org/10.1016/j.csda.2009.07.025
12. E. Cameron, A.N. Pettitt, Approximate Bayesian computation for astronomical model analysis: a case study in galaxy demographics and morphological transformation at high redshift. Mon. Not. R. Astron. Soc. **425**(1), 44–65 (2012). https://doi.org/10.1111/j.1365-2966.2012.21371.x
13. M.-H. Chen, Q.-M. Shao, J.G. Ibrahim, *Monte Carlo Methods in Bayesian Computation*. Springer Series in Statistics (Springer, New York, 2000)
14. J.A. Christen, C. Fox, Markov chain Monte Carlo using an approximation. J. Comput. Graph. Stat. **14**(4), 795–810 (2005). https://doi.org/10.1198/106186005X76983
15. L. Cucala, J.-M. Marin, C.P. Robert, D.M. Titterington, A Bayesian reassessment of nearest-neighbor classification. J. Am. Stat. Assoc. **104**(485), 263–273 (2009). https://doi.org/10.1198/jasa.2009.0125

16. P. Del Moral, A. Doucet, A. Jasra, An adaptive sequential Monte Carlo method for approximate Bayesian computation. Stat. Comput. **22**(5), 1009–20 (2012). https://doi.org/10.1007/s11222-011-9271-y
17. A.P. Dempster, N.M. Laird, D.B. Rubin, Maximum likelihood from incomplete data via the EM algorithm. J. R. Stat. Soc. Ser. B **39**(1), 1–38 (1977).
18. A. Doucet, M. Pitt, G. Deligiannidis, R. Kohn, Efficient implementation of Markov chain Monte Carlo when using an unbiased likelihood estimator. Biometrika **102**(2), 295–313 (2015). https://doi.org/10.1093/biomet/asu075
19. C.C. Drovandi, A.N. Pettitt, Estimation of parameters for macroparasite population evolution using approximate Bayesian computation. Biometrics **67**(1), 225–233 (2011). https://doi.org/10.1111/j.1541-0420.2010.01410.x
20. C.C. Drovandi, A.N. Pettitt, M.J. Faddy, Approximate Bayesian computation using indirect inference. J. R. Stat. Soc. Ser. C **60**(3), 317–337 (2011). https://doi.org/10.1111/j.1467-9876.2010.00747.x
21. C.C. Drovandi, A.N. Pettitt, A. Lee, Bayesian indirect inference using a parametric auxiliary model. Stat. Sci. **30**(1), 72–95 (2015). https://doi.org/10.1214/14-STS498
22. C.C. Drovandi, M.T. Moores, R.J. Boys, Accelerating pseudo-marginal MCMC using Gaussian processes. Comput. Stat. Data Anal. **118**, 1–17 (2018). https://doi.org/10.1016/j.csda.2017.09.002
23. P. Erdős, A. Rényi, On random graphs. Publ. Math. Debr. **6**, 290–297 (1959)
24. R.G. Everitt, Bayesian parameter estimation for latent Markov random fields and social networks. J. Comput. Graph. Stat. **21**(4), 940–960 (2012). https://doi.org/10.1080/10618600.2012.687493
25. P. Fearnhead, V. Giagos, C. Sherlock, Inference for reaction networks using the linear noise approximation. Biometrics **70**(2), 457–466 (2014). https://doi.org/10.1111/biom.12152
26. O. Frank, D. Strauss, Markov graphs. J. Amer. Stat. Assoc. **81**(395), 832–842 (1986)
27. N. Friel, Bayesian inference for Gibbs random fields using composite likelihoods, in ed. by C. Laroque, J. Himmelspach, R. Pasupathy, O. Rose, A.M. Uhrmacher, *Proceedings of the 2012 Winter Simulation Conference (WSC)* (2012), pp. 1–8. https://doi.org/10.1109/WSC.2012.6465236
28. N. Friel, A.N. Pettitt, Likelihood estimation and inference for the autologistic model. J. Comp. Graph. Stat. **13**(1), 232–246 (2004). https://doi.org/10.1198/1061860043029
29. N. Friel, A.N. Pettitt, Marginal likelihood estimation via power posteriors. J. R. Stat. Soc. Ser. B **70**(3), 589–607 (2008). https://doi.org/10.1111/j.1467-9868.2007.00650.x
30. N. Friel, A.N. Pettitt, R. Reeves, E. Wit, Bayesian inference in hidden Markov random fields for binary data defined on large lattices. J. Comp. Graph. Stat. **18**(2), 243–261 (2009). https://doi.org/10.1198/jcgs.2009.06148
31. A. Gelman, X.-L. Meng, Simulating normalizing constants: from importance sampling to bridge sampling to path sampling. Statist. Sci. **13**(2), 163–185 (1998). https://doi.org/10.1214/ss/1028905934
32. C.J. Geyer, L. Johnson, *potts: Markov Chain Monte Carlo for Potts Models*. R package version 0.5-2 (2014). http://CRAN.R-project.org/package=potts
33. A. Golightly, D.A. Henderson, C. Sherlock, Delayed acceptance particle MCMC for exact inference in stochastic kinetic models. Stat. Comput. **25**(5), 1039–1055 (2015). https://doi.org/10.1007/s11222-014-9469-x
34. A. Grelaud, C.P. Robert, J.-M. Marin, F. Rodolphe, J.-F. Taly, ABC likelihood free methods for model choice in Gibbs random fields. Bayesian Anal. **4**(2), 317–336 (2009). https://doi.org/10.1214/09-BA412
35. M.L. Huber, A bounding chain for Swendsen-Wang. Random Struct. Algor. **22**(1), 43–59 (2003). https://doi.org/10.1002/rsa.10071
36. M.L. Huber, *Perfect Simulation* (Chapman & Hall/CRC Press, London/Boca Raton, 2016)
37. P.E. Jacob, A.H. Thiery, On nonnegative unbiased estimators. Ann. Statist. **43**(2), 769–784 (2015). https://doi.org/10.1214/15-AOS1311

38. M. Järvenpää, M. Gutmann, A. Vehtari, P. Marttinen, Gaussian process modeling in approximate Bayesian computation to estimate horizontal gene transfer in bacteria. Ann. Appl. Stat. **12**(4), 2228–2251 (2018). https://doi.org/10.1214/18-AOAS1150
39. A. Lee, K. Łatuszyński, Variance bounding and geometric ergodicity of Markov chain Monte Carlo kernels for approximate Bayesian computation. Biometrika **101**(3), 655–671 (2014). https://doi.org/10.1093/biomet/asu027
40. A.-M. Lyne, M. Girolami, Y. Atchadé, H. Strathmann, D. Simpson, On Russian roulette estimates for Bayesian inference with doubly-intractable likelihoods. Statist. Sci. **30**(4), 443–467 (2015). https://doi.org/10.1214/15-STS523
41. F. Maire, R. Douc, J. Olsson, Comparison of asymptotic variances of inhomogeneous Markov chains with application to Markov chain Monte Carlo methods. Ann. Statist. **42**(4), 1483–1510, 08 (2014). https://doi.org/10.1214/14-AOS1209
42. J.-M. Marin, C.P. Robert, *Bayesian Core: A Practical Approach to Computational Bayesian Statistics*. Springer Texts in Statistics (Springer, New York, 2007)
43. P. Marjoram, J. Molitor, V. Plagnol, S. Tavaré, Markov chain Monte Carlo without likelihoods. Proc. Natl Acad. Sci. **100**(26), 15324–15328 (2003). https://doi.org/10.1073/pnas.0306899100
44. C.A. McGrory, D.M. Titterington, R. Reeves, A.N. Pettitt, Variational Bayes for estimating the parameters of a hidden Potts model. Stat. Comput. **19**(3), 329–340 (2009). https://doi.org/10.1007/s11222-008-9095-6
45. C.A. McGrory, A.N. Pettitt, R. Reeves, M. Griffin, M. Dwyer, Variational Bayes and the reduced dependence approximation for the autologistic model on an irregular grid with applications. J. Comput. Graph. Stat. **21**(3), 781–796 (2012). https://doi.org/10.1080/10618600.2012.632232
46. T.J. McKinley, I. Vernon, I. Andrianakis, N. McCreesh, J.E. Oakley, R.N. Nsubuga, M. Goldstein, R.G. White, et al., Approximate Bayesian computation and simulation-based inference for complex stochastic epidemic models. Statist. Sci. **33**(1), 4–18 (2018). https://doi.org/10.1214/17-STS618
47. F.J. Medina-Aguayo, A. Lee, G.O. Roberts, Stability of noisy Metropolis-Hastings. Stat. Comput. **26**(6), 1187–1211 (2016). https://doi.org/10.1007/s11222-015-9604-3
48. E. Meeds, M. Welling, GPS-ABC: Gaussian process surrogate approximate Bayesian computation, in *Proceedings of the 30th Conference on Uncertainty in Artificial Intelligence*, Quebec City, Canada (2014)
49. A. Mira, J. Møller, G.O. Roberts, Perfect slice samplers. J. R. Stat. Soc. Ser. B **63**(3), 593–606 (2001). https://doi.org/10.1111/1467-9868.00301
50. J. Møller, A.N. Pettitt, R. Reeves, K.K. Berthelsen, An efficient Markov chain Monte Carlo method for distributions with intractable normalising constants. Biometrika **93**(2), 451–458 (2006). https://doi.org/10.1093/biomet/93.2.451
51. M.T. Moores, D. Feng, K. Mengersen, *bayesImageS: Bayesian Methods for Image Segmentation Using a Potts Model*. R package version 0.5-3 (2014). URL http://CRAN.R-project.org/package=bayesImageS
52. M.T. Moores, C.C. Drovandi, K. Mengersen, C.P. Robert, Pre-processing for approximate Bayesian computation in image analysis. Stat. Comput. **25**(1), 23–33 (2015). https://doi.org/10.1007/s11222-014-9525-6
53. M.T. Moores, G.K. Nicholls, A.N. Pettitt, K. Mengersen, Scalable Bayesian inference for the inverse temperature of a hidden Potts model. Bayesian Anal. **15**, 1–27 (2020). https://doi.org/10.1214/18-BA1130.
54. I. Murray, Z. Ghahramani, D.J.C. MacKay, MCMC for doubly-intractable distributions, in *Proceedings of the 22nd Conference on Uncertainty in Artificial Intelligence*, Arlington (AUAI Press, Tel Aviv-Yafo, 2006), pp. 359–366
55. G.K. Nicholls, C. Fox, A. Muir Watt, Coupled MCMC with a randomized acceptance probability (2012).Preprint arXiv:1205.6857 [stat.CO]. https://arxiv.org/abs/1205.6857
56. C.J. Oates, T. Papamarkou, M. Girolami, The controlled thermodynamic integral for Bayesian model evidence evaluation. J. Am. Stat. Assoc. **111**(514), 634–645 (2016). https://doi.org/10.1080/01621459.2015.1021006

57. H.E. Ogden, On asymptotic validity of naive inference with an approximate likelihood. Biometrika **104**(1), 153–164 (2017). https://doi.org/10.1093/biomet/asx002
58. S. Okabayashi, L. Johnson, C.J. Geyer, Extending pseudo-likelihood for Potts models. Statistica Sinica **21**, 331–347 (2011)
59. E. Olbrich, T. Kahle, N. Bertschinger, N. Ay, J. Jost, Quantifying structure in networks. Eur. Phys. J. B **77**(2), 239–247 (2010). https://doi.org/10.1140/epjb/e2010-00209-0
60. P.D. O'Neill, D.J. Balding, N.G. Becker, M. Eerola, D. Mollison, Analyses of infectious disease data from household outbreaks by Markov chain Monte Carlo methods. J. R. Stat. Soc. Ser. C **49**(4), 517–542 (2000). https://doi.org/10.1111/1467-9876.00210
61. A.N. Pettitt, N. Friel, R. Reeves, Efficient calculation of the normalizing constant of the autologistic and related models on the cylinder and lattice. J. R. Stat. Soc. Ser. B **65**(1), 235–246 (2003). https://doi.org/10.1111/1467-9868.00383
62. M.K. Pitt, R. dos Santos Silva, P. Giordani, R. Kohn, On some properties of Markov chain Monte Carlo simulation methods based on the particle filter. J. Econometr. **171**(2), 134–151 (2012). https://doi.org/10.1016/j.jeconom.2012.06.004
63. D. Prangle, Lazy ABC. Stat. Comput. **26**(1), 171–185 (2016). https://doi.org/10.1007/s11222-014-9544-3
64. J.K. Pritchard, M.T. Seielstad, A. Perez-Lezaun, M.W. Feldman, Population growth of human Y chromosomes: a study of Y chromosome microsatellites. Mol. Biol. Evol. **16**(12), 1791–1798 (1999). https://doi.org/10.1093/oxfordjournals.molbev.a026091
65. J.G. Propp, D.B. Wilson, Exact sampling with coupled Markov chains and applications to statistical mechanics. Random Struct. Algor. **9**(1–2), 223–252 (1996). https://doi.org/10.1002/(SICI)1098-2418(199608/09)9:1/2<223::AID-RSA14>3.0.CO;2-O
66. R. Reeves, A.N. Pettitt, Efficient recursions for general factorisable models. Biometrika **91**(3), 751–757 (2004). https://doi.org/10.1093/biomet/91.3.751
67. G.O. Roberts, J.S. Rosenthal, Examples of adaptive MCMC. J. Comput. Graph. Stat. **18**(2), 349–367 (2009). https://doi.org/10.1198/jcgs.2009.06134
68. T. Rydén, D.M. Titterington, Computational Bayesian analysis of hidden Markov models. J. Comput. Graph. Stat. **7**(2), 194–211 (1998). https://doi.org/10.1080/10618600.1998.10474770
69. C. Sherlock, A.H. Thiery, G.O. Roberts, J.S. Rosenthal, On the efficiency of pseudo-marginal random walk Metropolis algorithms. Ann. Statist. **43**(1), 238–275, 02 (2015). https://doi.org/10.1214/14-AOS1278
70. C. Sherlock, A. Golightly, D.A. Henderson, Adaptive, delayed-acceptance MCMC for targets with expensive likelihoods. J. Comput. Graph. Stat. **26**(2), 434–444 (2017). https://doi.org/10.1080/10618600.2016.1231064
71. A.M. Stuart, A.L. Teckentrup, Posterior consistency for Gaussian process approximations of Bayesian posterior distributions. Math. Comp. **87**, 721–753 (2018). https://doi.org/10.1090/mcom/3244
72. R.H. Swendsen, J.-S. Wang, Nonuniversal critical dynamics in Monte Carlo simulations. Phys. Rev. Lett. **58**, 86–88 (1987). https://doi.org/10.1103/PhysRevLett.58.86
73. M.A. Tanner, W.H. Wong, The calculation of posterior distributions by data augmentation. J. Am. Stat. Assoc. **82**(398), 528–40 (1987)
74. C. Varin, N. Reid, D. Firth, An overview of composite likelihood methods. Statistica Sinica **21**, 5–42 (2011)
75. R.D. Wilkinson, Accelerating ABC methods using Gaussian processes, in ed. by S. Kaski, J. Corander, *Proceedings of the 17th International Conference on Artificial Intelligence and Statistics AISTATS (JMLR: Workshop and Conference Proceedings)* , vol. 33 (2014), pp. 1015–1023

Part II
Real World Case Studies in Health

Chapter 7
A Bayesian Hierarchical Approach to Jointly Model Cortical Thickness and Covariance Networks

Marcela I. Cespedes, James M. McGree, Christopher C. Drovandi, Kerrie L. Mengersen, Lee B. Reid, James D. Doecke, and Jurgen Fripp

Abstract Estimation of structural biomarkers and covariance networks from MRI have provided valuable insight into the morphological processes and organisation of the human brain. State-of-the-art analyses such as linear mixed effects (LME) models and pairwise descriptive correlation networks are usually performed independently, providing an incomplete picture of the relationships between the biomarkers and network organisation. Furthermore, descriptive network analyses do not generalise to the population level. In this work, we develop a Bayesian generative model based on wombling that allows joint statistical inference on biomarkers and connectivity covariance structure. The parameters of the wombling model were estimated via Markov chain Monte Carlo methods, which allow for simultaneous inference of the brain connectivity matrix and the association of participants' biomarker covariates. To demonstrate the utility of wombling on real data, the method was used to characterise intrahemispheric cortical thickness and networks in a study cohort of subjects with Alzheimer's disease (AD), mild-cognitive impairment and healthy ageing. The method was also compared with state-of-the-art alternatives. Our Bayesian modelling approach provided posterior probabilities for the connectivity matrix of the wombling model, accounting for the uncertainty for each connection. This provided superior inference in comparison with descriptive networks. On the study cohort, there was a loss of connectivity across diagnosis levels from healthy to Alzheimer's disease for all network connec-

M. I. Cespedes (✉) · L. B. Reid · J. D. Doecke · J. Fripp
CSIRO Health and Biosecurity/Australian E-Health Research Centre, Level 5, Royal Brisbane and Women's Hospital, Herston, QLD, Australia
e-mail: Marcela.Cespedes@csiro.au; Lee.Reid@csiro.au; James.Doecke@csiro.au; Jurgen.Fripp@csiro.au

J. M. McGree · C. C. Drovandi · K. L. Mengersen
ARC Centre of Excellence for Mathematical and Statistical Frontiers, Brisbane, QLD, Australia

School of Mathematical Sciences, Queensland University of Technology, Brisbane, QLD, Australia
e-mail: James.McGree@qut.edu.au; C.Drovandi@qut.edu.au; K.Mengersen@qut.edu.au

© The Editor(s) (if applicable) and The Author(s), under exclusive licence to Springer Nature Switzerland AG 2020
K. L. Mengersen et al. (eds.), *Case Studies in Applied Bayesian Data Science*, Lecture Notes in Mathematics 2259, https://doi.org/10.1007/978-3-030-42553-1_7

155

tions (posterior probability ≥ 0.7). In addition, we found that wombling and LME model approaches estimated that cortical thickness progressively decreased along the dementia pathway. The major advantage of the wombling approach was that spatial covariance among the regions and global cortical thickness estimates could be estimated. Joint modelling of biomarkers and covariance networks using our novel wombling approach allowed accurate identification of probabilistic networks and estimated biomarker changes that took into account spatial covariance. The wombling model provides a novel tool to address multiple brain features, such as morphological and connectivity changes facilitating a better understanding of disease pathology.

Keywords Conditional autoregressive model · Markov chain Monte Carlo · Spatial statistics · Wombling · Cortical thickness · Alzheimer's disease · Structural MRI

7.1 Introduction

Alzheimer's disease (AD) is the most common form of dementia [13, 67]. While clinical diagnosis of AD is often derived from psychological assessments, neuroimaging studies have found that the structural and functional changes in the brain that align with AD pathology can be identified prior to the detection of cognitive symptoms [2, 65].

Structural neuroimaging studies typically use two approaches: region of interest (ROI) analyses to estimate morphological biomarkers for each region, such as thickness, volume and the rate of tissue loss; and cortical networks to investigate associations between multiple ROIs. This two pronged approach is important as biomarkers in one region are likely to influence the morphological properties of connected regions. For example, highly correlated ROIs (often quantified through cortical networks) are often a part of a system that is known to be associated with particular behavioural or cognitive functions [3, 47]. Nonetheless these approaches are often performed independently, providing valuable insight into the differences in brain organisation and degeneration patterns for multiple regions between healthy and pathological groups [9, 15, 34, 55, 58, 62, 72]. For example, Bernal-Rusiel et al. [9] found that their models for ROIs were able to characterise changes in individuals' measurements at multiple time points while handling up to 45.5% patient drop out. Furthermore, analyses on cortical thickness networks have demonstrated a reduction in connectivity efficiency between healthy groups and groups with neurological disorders such as schizophrenia and AD [8, 15, 37, 46, 59, 71].

An advantage of analyses conducted on a single region is the direct biological interpretation on the estimation of tissue features, such as thickness and estimated annual rate of tissue loss [9]. However, it is difficult to ascertain a brain-wide picture of all ROIs under such analyses, as this requires multiple comparison corrections in order to account for the high number of hypothesis tests [21, 34, 60, 72]. Alternatively, cortical networks provide a summary measure on the topological brain network organisation which conveniently encompasses the complex information across all ROIs [15, 37, 46, 58]. However, direct biological interpretations of such networks are difficult as the relationship between the ROI node and corresponding links represent a covariance measure among ROIs, and not physical connections [3]. Furthermore, generalising to a population cortical network is difficult to achieve from descriptive analyses as such methods are not generative models and do not take into account the variability of each connection [46, 59–61, 71].

In practice both approaches complement each other with participants who are healthy, in general, tend to have thicker cortical tissue and highly organised networks compared to pathological groups such as AD [9, 17, 18, 37, 55, 71]. The aforementioned shortcomings of these methods could be resolved by combining both approaches into a unified framework. Such a framework could avoid multiple comparisons and provide a cortical network whose links reflect the uncertainty of the data.

In this work, we propose a Bayesian hierarchical (generative) model that jointly performs network-based inference in conjunction with neuroimaging biomarker estimates. This approach enforces consistency between any spatial interactions and biomarker estimates (for network and cortical thickness) at the population and participant level, while handling correlated measures from within and between individuals in conjunction with covariates in a statistically principled manner.

7.1.1 Technical Survey of Previous Work of Bayesian Hierarchical Models

Bayesian hierarchical models have been extensively applied to unify independent analyses, for example, combining the joint estimation of voxel and ROI analyses [12, 19, 69], and combining diffusion and functional MRI into a single model [70]. Accommodating both within and between participant variation from longitudinal observations, as well as high patient drop out (unbalanced design) has been previously achieved through a related method called the mass univariate Bayesian hierarchical analysis [72]. Previously, Bayesian linear mixed effect (LME) models, which are a type of hierarchical model, have been applied independently to key ROIs associated with AD progression [17]. An advantage of Bayesian inference is that it can detect significant differences among groups of interest through the direct comparison of the marginal posterior distributions, without the need for hypothesis tests or multiple comparisons corrections. However, LME models (Bayesian and non-Bayesian) applied to neuroimaging data usually analyse each ROI independently and do not account for the covarying measurements between several brain regions.

Recent probabilistic brain networks in the Bayesian framework have shown great potential to estimate a population network for clinical groups. Bayesian brain networks are probabilistic rather than binary, and so are easy to interpret [38, 41, 61]. However, these probabilistic networks are not easily extended to include additional neuroimaging biomarker estimation, such as cortical thickness, volumes, or fluorodeoxyglucose uptake as measured by positron emission topography. To account for the correlation between measurements on ROIs, several neuroimaging studies [33, 36, 48, 53] have used spatial dependence modelling via a Gaussian Markov random field (GMRF, Gössl et al. [32] and Woolrich et al. [68]). However, an underlying and potentially invalid assumption is that the adjacency structure of the correlations are known and fixed, and most are constrained to nearest neighbour configurations. This was highlighted in the Bayesian hierarchical spatial models by Bowman [11] and Bowman et al. [12] suggesting that the underlying physical and biological processes may not always be contiguous, and relationships among ROIs are not restricted to regions which are immediate anatomical neighbours.

7.1.1.1 Previous Work on Wombling

Wombling refers to the estimation of a neighbourhood matrix through the covariance structure of a GMRF that is estimated under a Bayesian framework [42, 49, 50]. This neighbourhood structure can be incorporated as an additional parameter in the Bayesian hierarchical model, and can be estimated in addition to participant specific covariates such as gender and other biomarkers associated with AD factors [22, 24, 31].

7.1.2 Overview of Our Work

In this work, we propose Bayesian hierarchical wombling models that jointly performs network based inference in conjunction with regional biomarker estimates. This approach estimates the complex covariance associations among several regions without assuming contiguous relationships via estimation of a connectivity structure. Furthermore, biomarker estimates at the population and participant level handle correlated measures from within and between individuals in conjunction with covariates. This enables full statistical inference of biomarker estimates and produces a probabilistic network.

To this end, this chapter is organised as follows: Sect. 7.2.2 outlines the proposed Bayesian hierarchical wombling model. The wombling model is validated via a simulation study described in Sect. 7.2.3 and results are reported in Sect. 7.3.1. Sections 7.2.1 and 7.3.2 present the application of brain wombling on the Australian Imaging, Biomarkers and Lifestyle (AIBL) study of ageing data on healthy controls (HC), mild cognitive impaired (MCI) and AD diagnosed groups as well as in age ranges discretised into three groups. A comparison of the results from the wombling approach with comparable independent analyses are presented in Sects. 7.3.2.2 and 7.3.2.4. A discussion of our work appears in Sect. 7.4.

7.2 Materials and Methods

The overarching objective of this work is to develop and validate a joint analysis of biomarker and covariance networks facilitated by the proposed wombling approach. The flowchart in Fig. 7.1 provides an overview of the experiments presented in this work, showing the inputs and outputs for each analysis. Case study data will be based on cortical thickness estimates into a study of into a study of healthy ageing, MCI and AD participants. The wombling method will be compared to Pearson pairwise correlation networks and Bayesian LME models. In addition, as this work was the first to investigate the wombling approach for joint analysis of cortical networks and biomarker estimates, a simulation study was used to evaluate the performance of the wombling algorithm to recover the true connectivity structure and simulated biomarker values.

7.2.1 AIBL Study of Ageing

In this work, we applied our proposed method to data from the Australian Imaging, Biomarkers and Lifestyle (AIBL) longitudinal study of ageing. AIBL

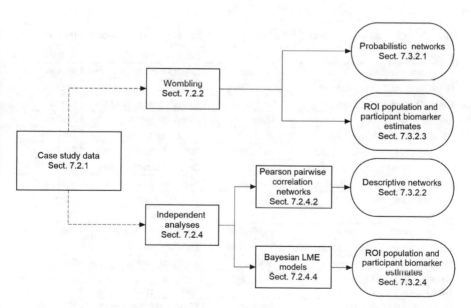

Fig. 7.1 Overview of the analysis workflow. Arrows show the relationships between the rectangle methods sections. The results from both methods are denoted by the circular plots on the far right and are compared to each other to assess the performance of joint analyses facilitated by the wombling approach (Sect. 7.2.2) in comparison with the state-of-the-art independent analyses (Sect. 7.2.4)

is an ongoing study which aims to discover which biomarkers such as cognitive assessment, neuroimaging, lifestyle and demographic factors potentially influence the development of AD. The AIBL study was approved by the institutional ethics committees of Austin Health, St Vincent's Health, Hollywood Private Hospital and Edith Cowan University. All study volunteers gave written informed consent prior to participating in the study. MRI data were collected at baseline and at several ~18 month follow-up intervals (replicates) from a subset of 167 participants. This resulted in a total of 597 sets of ROI observations. Only those observations from participants with two or more replicates were retained; these included 120 HC, 21 and 26 clinically diagnosed MCI individuals and AD participants respectively. Of the 167 participants, 77 were male (46%) and 90 were female (54%). Mean baseline ages was numerically higher in those diagnosed with MCI (HC: 73.1 ± 6.7, MCI: 77.0 ± 6.4 and AD: 73.8 ± 7.5, $p = 0.055$).

The structural T1W MRI images were first segmented into grey/white matter and cerebral spinal fluid using an in-house implementation of the expectation maximisation algorithm applied to a Gaussian mixture model [64]. Cortical thickness was computed along the grey matter based on a combined Lagrangian-Eulerian partial differential equation approach [1]. The grey matter was parcellated following the Automated Anatomical Labelling (AAL) atlas [63] using a multi-atlas registration approach [10]. For this work, we used 35 ROI cortical thickness regions from the left hemisphere of the brain, as listed in Table 7.1.

Table 7.1 Cortical regions from the left hemisphere of the brain, as parcellated via the AAL

Region name	Abbrev.	Region name	Abbrev.
Precentral gyrus	PreCent	Superior occipital gyrus	SupOcc
Superior frontal gyrus dorsolateral	SupFrDorso	Middle occipital gyrus	MidOcc
Superior frontal gyrus orbital	SupFrOrb	Inferior occipital gyrus	InfOcc
Middle frontal gyrus	MidFr	Fusiform gyrus	Fusifrm
Middle frontal gyrus-orbital	MidFrOpen	Postcentral gyrus	Post
Inferior frontal gyrus-opercular	InfFrOpec	Superior parietal gyrus	SupPar
Inferior frontal gyrus-triangular	InFrTri	Inferior parietal gyrus	InfPar
Inferior frontal gyrus-orbital	InFrOrb	Supramarginal gyrus	SupMar
Supplementary motor area	SuppMtr	Angular gyrus	Angular
Olfactory cortex	Olfac	Precuneus	Precun
Superior frontal gyrus-medial	SupFrMed	Paracentral Lobule	ParacenLob
Superior frontal gyrus-medial orbital	SupFrMedOrb	Heschl gyrus	Heschl
Gyrus rectus	GrRcts	Superior temporal gyrus	SupTemp
Anterior cingulate and paracingulate gyri	AntCingPara	Temporal pole:superior temporal gyrus	TempPolSup
Posterior cingulate gyrus	PostCing	Middle temporal gyrus	MidTemp
Calcarine fissure and surrounding cortex	CalFiss	Temporal pole:middle temporal gyrus	TempPolMid
Cuneus	Cuneus	Inferior temporal gyrus	InfTemp
Lingual gyrus	Ling		

7.2.2 *Wombling Model Formulation and Parameter Estimation*

In this section, we present the wombling generative model used to jointly estimate cortical brain connectivity and thickness in a Bayesian hierarchical framework. Wombling is a type of a LME model that accounts for correlations between regions, after accounting for fixed effects. In this work, wombling does not provide age related estimates as it is not a longitudinal model, such an extension is beyond the scope of this work and motivates future work. The wombling model comprises of two parts; a mixed effect model and connectivity estimation with their respective set of assumptions. LME model assumptions include: a linear relationship exists between the response and the exploratory variables; the response is normally distributed about a mean, although for non-normal responses we may extend this assumption to the exponential family and apply generalised linear mixed models [51]; the variances across fixed and random effects are unknown but constant, and observations for a region can be correlated with its neighbours, but observations between non-neighbouring regions are assumed to be conditionally independent. Connectivity matrix assumptions are twofold. Firstly, the underlying connectivity structure quantified by matrix W is the same across all individuals in a specified group. Secondly, relationships between regions are equally weighted, as our framework estimates the probability of each pairwise connection and not the connection strength. This implies that if region j is a neighbour of region k, then region k is also neighbour of region j, and regions are not neighbours with themselves, $w_{ii} = 0 \ \forall i$.

The hierarchical structure of the model separates the variation of the data into two levels; fixed effects (A) and random effects (B) shown in Fig. 7.2. At level A, the linear predictor for person i, at repeated measure r on region k comprises of participant i's covariate vector $\mathbf{x}_i$ (covariate matrix for all participants is denoted by X), parameter vector $\boldsymbol{\beta}$, spatial random effects $\mathbf{b}_i$ and residual variance σ^2. Level B consists of the spatial random effects $\mathbf{b}_i$ which follow a multivariate normal distribution with a mean of $\mathbf{0}$ and a covariance matrix $\sigma_s^2 Q$. The product, $\sigma_s^2 Q$, comprises of the spatial scale variance term, σ_s^2, which controls the variation of the random effects and a function of the connectivity structure matrix Q.

The cortical thickness of region $k = 1, 2, \ldots, K$ within participant $i = 1, 2, \ldots, I$ who has $r = 1, \ldots, R_i$ replicates is y_{irk} measured in millimetres. The brain wombling model is of the following form:

$$y_{irk} | b_{ik}, \boldsymbol{\beta}, \sigma^2 \sim N(\mathbf{x}_i \boldsymbol{\beta} + b_{ik}, \sigma^2)$$

$$\mathbf{b}_i | \sigma_s^2, W \sim MVN(\mathbf{0}, \sigma_s^2 Q)$$

$$Q^{-1} = \rho(D_w - W) + (1 - \rho)\mathbb{I}. \tag{7.1}$$

Details of the formulation for the connectivity structure are as follows: matrix D_w is a diagonal matrix with elements given by the row sums (or number of neighbours) $\sum_{j=1}^{K} w_{jk}$ for $k = 1, 2, \ldots, K$. The matrix W is a zero-diagonal, binary symmetric

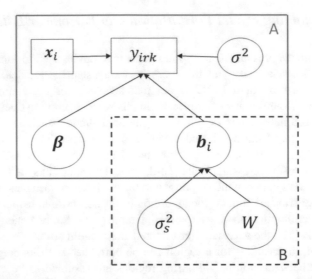

Fig. 7.2 Visualisation of brain wombling model (7.1) via a directed acyclic graph. Nodes in circles and rectangle denote parameters and observed variables respectively. Direction of arrows indicates direction of influence or dependence. Rectangle B denotes the second layer of model, which accounts for spatial dependence among ROI conditional and nested in rectangle A, which is the upper-most layer with fixed effects β parameter. The spatial random effects vector for participant i, $\mathbf{b}_i$, is modelled as a multivariate normal, whose covariance structure is a function of the binary symmetric adjacency matrix W, of dimension $K \times K$ and $w_{jk} = 1$ or $w_{jk} = 0$ implies that region j and k are connected or disconnected respectively

matrix, with elements $w_{jk} = 1$ if regions j and k are neighbours or zero otherwise, and identity matrix $\mathbb{I}$ has dimension $K \times K$. The value of ρ determines the global level of the spatial correlation [43] where values of ρ close to zero correspond to (near) independence in the spatial random effect, and ρ close to one denotes high spatial correlation. While ρ can be an additional parameter in our wombling model, in this application we fix $\rho = 0.9$, to enforce high spatial correlation and avoid the difficult and computationally intensive task of estimating ρ, as described in Lu et al. [50] and Lee [42]. For completeness we investigated the effect of ρ at various values to assess the recovery of W; refer to Sect. 7.2.3 for further details.

The parametrisation of Q^{-1} defined in (7.1) was chosen due to its superior ability to handle a range of spatial strengths [42, 45]. This parametrisation has also been favoured in other wombling and spatial clustering applications [4, 43]. Visualisation of model parameters conditional on the observed regional biomarker response, such as cortical thickness, and participant specific covariates are shown in Fig. 7.2.

In a Bayesian framework the likelihood corresponding to the model in (7.1) is $p(\mathbf{y}|\mathbf{b}, \sigma^2, \boldsymbol{\beta}, X)$, which is conditional on the spatial random effects and the model parameters. Note the data is conditionally independent of the network structure W and spatial scale variance, σ_s^2. The resultant joint posterior distribution for the model parameters and the random effects given the data is

$$p(W, \sigma^2, \sigma_s^2, \boldsymbol{\beta} | \mathbf{y}, X) \propto \left[\prod_{i=1}^{I} \prod_{r=1}^{R_i} \prod_{k=1}^{K} p(y_{irk} | b_{ik}, \sigma^2, \boldsymbol{\beta}, \mathbf{x}_i) \right] \left[\prod_{i=1}^{I} p(\mathbf{b}_i | \sigma_s^2, W) \right]$$

$$p(\boldsymbol{\beta}) p(\sigma^2) p(\sigma_s^2) p(W). \tag{7.2}$$

In the Bayesian paradigm the population parameters are random variables, and priors $p(\boldsymbol{\beta})$, $p(\sigma^2)$, $p(\sigma_s^2)$ and $p(W)$ are assigned to each parameter. Details on prior specification are described in Sects. 7.2.2.1 and 7.2.2.3. Markov chain Monte Carlo (MCMC) methods were used to sample from the joint posterior probability distribution of the parameters [57], which samples from the marginal posterior distributions as a by-product [29]. At each MCMC step, samples are iteratively drawn from the full conditionals of the parameters with a Metropolis-Hastings (M-H, Chib and Greenberg [20], Metropolis et al. [52]) update for W. Following a burn-in period, samples will eventually be drawn from the joint posterior distribution of the parameters [12].

Full conditional distributions in closed form were derived for parameters $\boldsymbol{\beta}, \sigma_s^2$ and σ^2 which were sampled via a Gibbs sampler as described in Sect. 7.2.2.1. As the matrix W is symmetric, the off-diagonal, upper triangular elements were updated one at a time via a M-H sampler as described in Sect. 7.2.2.3.

7.2.2.1 Prior and Conditional Distributions for σ_s^2, σ^2 and β

A semi-conjugate prior in the form of an inverse gamma distribution, $IG(c, d)$, was chosen for the spatial scale variance σ_s^2, with shape and rate values c and d, respectively. Likewise, the prior for the residual variance σ^2 was an $IG(e, f)$ distribution. Hyperparameters were chosen to provide support over a wide range of possible values for σ_s^2 and σ^2. The full conditional distributions for σ_s^2 and σ^2 are as follows

$$p(\sigma_s^2 | W, \mathbf{b}_i) \sim IG \left(\frac{IK + 2c}{2}, \frac{1}{2} \sum_{i=1}^{I} \mathbf{b}_i Q^{-1} \mathbf{b}_i + d \right), \tag{7.3}$$

and

$$p(\sigma^2 | \mathbf{b}, \mathbf{y}, \boldsymbol{\beta}, X) \sim IG \left(\frac{N + 2e}{2}, \frac{1}{2} \left(\sum_{i=1}^{I} \sum_{r=1}^{R_i} \sum_{k=1}^{K} (y_{irk} - \mathbf{x}_i \boldsymbol{\beta} - b_{ik})^2 \right) + f \right), \tag{7.4}$$

where N is the total number of observations.

The prior for the fixed effect parameter $\boldsymbol{\beta}$ is a multivariate normal distribution $MVN(\boldsymbol{\mu}_0, \Sigma_0)$, and in keeping with wombling literature [50], $\boldsymbol{\mu}_0$ and Σ_0 were chosen so that the prior on $\boldsymbol{\beta}$ is vague. It can be shown that the full conditional distribution for $\boldsymbol{\beta}$ is

$$p(\boldsymbol{\beta}|\mathbf{b}, \mathbf{y}, X, \sigma^2) \sim MVN\left(\left[\frac{1}{\sigma^2}\overset{L}{X}^T\overset{L}{X} + \Sigma_0^{-1}\right]^{-1}\left[\frac{1}{\sigma^2}\overset{L}{X}^T(\mathbf{y} - \mathbf{b}) + \Sigma_0^{-1}\boldsymbol{\mu}_0\right],\right.$$

$$\left.\left[\frac{1}{\sigma^2}\overset{L}{X}^T\overset{L}{X} + \Sigma_0^{-1}\right]^{-1}\right). \tag{7.5}$$

The response in long vector form is $\mathbf{y} = [y_{111}, y_{112}, \ldots, y_{11K}, y_{121}, \ldots, y_{IR_IK}]$ and the covariate matrix X with superscript L is $\overset{L}{X} = [\mathbf{x}_1, \mathbf{x}_1, \ldots, \mathbf{x}_2, \mathbf{x}_2, \ldots, \mathbf{x}_I]$, hence the individual specific covariate vector $\mathbf{x}_i$ is repeated R_i times, where $\overset{L}{X}$ is an N by p matrix; where p is the total number of covariates for the model, including the intercept. Similarly, the long vector form for the spatial random effects is $\mathbf{b} = [b_{111}, b_{121}, \ldots, b_{1KR_1}, b_{211}, \ldots, b_{IR_IK}]$.

7.2.2.2 Full Conditional Distribution for Spatial Random Effects $\mathbf{b}_i$

From Model (7.1) we can derive meaningful participant specific estimates of cortical thickness for each of the ROIs analysed, and investigate how this deviates from the population average (β_0). The individual-specific estimates of cortical thickness for each ROI in our analysis are derived from the full conditional distribution of $\mathbf{b}_i$ given by

$$p(\mathbf{b}_i|\boldsymbol{\beta}, \mathbf{y}, \sigma^2, \sigma_s^2, W) \sim MVN\left(\left[\frac{R_i}{\sigma^2}\mathbb{I} + \frac{1}{\sigma_s^2}Q^{-1}\right]^{-1}\left[\sum_{r=1}^{R_i}\frac{\mathbf{y}_{ir}}{\sigma^2} - \frac{R_i}{\sigma^2}(\mathbf{x}_i\boldsymbol{\beta})\mathbf{e}\right],\right.$$

$$\left.\left[\frac{R_i}{\sigma^2}\mathbb{I} + \frac{1}{\sigma_s^2}Q^{-1}\right]^{-1}\right), \tag{7.6}$$

where the unit vector $\mathbf{e} = [1, 1, \ldots, 1]$ is of length K.

7.2.2.3 Prior and Posterior Sampling for Brain Connectivity Matrix W

According to the posterior distribution in (7.2), the full conditional for matrix W is of the form

$$p(W|\sigma_s^2, \mathbf{b}) \propto \left[\prod_{i=1}^{I} p(\mathbf{b}_i|\sigma_s^2, W)\right] p(W).$$

Elements of the matrix W are updated one at a time. As W is symmetric, we only require estimation of the off-diagonal, upper triangular elements. To facilitate a data driven method to estimate the brain connectivity matrix, our prior knowledge of the probability of a link between any pair of ROIs is 0.5, that is, $p(w_{ij} = 1) = p(w_{ij} = 0) = 0.5$ for all values of i and j.

We use the M-H algorithm within a Gibbs sampler to draw posterior simulations for W. We update W element-wise by drawing independent proposals, w_{kj}^*, from the prior of W and accepting a proposal with probability

$$
\alpha = \min \left\{ 1, \frac{\prod_{i=1}^{I} p(\mathbf{b}_i | \sigma_s^2, Q_{w_{kj}^*}^{-1})}{\prod_{i=1}^{I} p(\mathbf{b}_i | \sigma_s^2, Q_{w_{kj}}^{-1})} \right\}, \tag{7.7}
$$

where the covariance precision evaluated at the proposed value is $Q_{w_{kj}^*}^{-1}$.

7.2.3 Simulation Studies

The proposed Bayesian brain wombling approach accommodates for both network based inference and biomarker estimates. For this reason the aims of the simulation study are twofold. Firstly, we aim to evaluate the performance of this model at 'recovering' two underlying connectivity matrices W (structured and contiguous configurations). In the context of this manuscript, by recovery we refer to whether the credible intervals of the estimator contain the true solution. The assumed true matrices for W are shown in Fig. 7.3A and F respectively. Our second aim is to illustrate that our model recovers the simulated biomarker estimates via fixed effect vector $\boldsymbol{\beta}$ in addition to simulated participant specific estimates through their spatial random effects ($\mathbf{b}_i$).

In order to relate the simulation study to the real data application, the values used to generate the simulated study data were chosen to reflect features of the AIBL study, such as the number of simulated ROIs, number of repeated measures (replicates) in the unbalanced design, the number of participants and range of biomarker values.

7.2.3.1 Wombling Simulated Analyses

For both configurations of W, the vector $\boldsymbol{\beta} = [\beta_0, \beta_1] = [3, 0.5]$ was assumed as the intercept and gender effect, and $\mathbf{x}_{sim}$ was specified as a binary vector with male participant as baseline (i.e. $x_{i,sim} = 1$ to simulate a female participant and $x_{i,sim} = 0$ a male participant). The average global human cortical thickness can range from approximately 1 to 4.5 mm [27], hence the prior for $\boldsymbol{\beta}$ was chosen to remain physiologically feasible around this value. For this reason the hyperparameters for the precision matrix Σ_0 had zero off-diagonals and diagonal elements of value 1/10, and the hyperparameter for μ_0 was chosen to be **0**. We note that these are the same priors used for the real data application described in Sect. 7.2.4. Variance terms for both W configurations were set to $\sigma_s^2 = 1$ and $\sigma^2 = 0.5$, with relatively uninformative inverse gamma priors specified as $IG(1, 1)$ and $IG(1, 0.5)$ respectively. Priors for both configurations of W matrices are described in Sect. 7.2.2.3.

Our simulation studies were undertaken by generating 50 independent data sets from Model (7.1). We fitted our model to each data set to obtain (50) posterior distributions for our parameters. Here, we considered a balanced design whereby each simulated participant had the same number of repeated measures, and the more realistic unbalanced alternative, where the number of replicates for each participant varied.

Data for $I = 100$ participants were simulated from Model (7.1), where each participant had $K = 35$ simulated ROI as listed on Table 7.1, and each participant had $R_i = 7$ replicates as a balanced design. The unbalanced simulation design comprised of participants with 4–7 replicates (mean 5.8). Parameter values and prior information as described above were set for balanced and unbalanced designs, whereby each design was explored as structured and contiguous W configurations, for a total of four scenarios.

Each scenario resulted in a mean of posterior means for W, representing the probabilistic network. These scenarios were binarised for ease of comparison to assess the recovery of W. Values $w_{jk} = 1$ if the average posterior probability of a connection between regions j and k was equal to or greater than $\tau = 0.6$, and $w_{jk} = 0$ otherwise. We note that binary W is dependent on the choice of τ, and that $\tau = 0.6$ is sufficiently far away from the prior ($p(w_{ij} = 1) = p(w_{ij} = 0) = 0.5$).

Further details of the simulation analyses including percentage of recovery of the assumed true values and MCMC convergence checks are provided in section "Simulation Study" of Appendix.

7.2.4 Application to Study Cohort

We hypothesised that each population group has an underlying cortical brain network, denoted as matrix W, while expecting differences in W between groups, as each group represented progressive levels of neurodegeneration in both cortical thickness estimates and structural brain networks.

The Bayesian brain wombling Model (7.1) was applied independently to data from three diagnosis groups (HC, MCI and AD) as well as three age groups (A: 59–69y; B: 69–79y; C: 79–93y). In order to compare the wombling model with current state-of-the-art methods that provide cortical networks, population and participant specific estimates, we derived Pearson pairwise correlation networks and Bayesian LME models to the aforementioned case study groups. The subsections below describe how the marginal posterior draws were processed after the wombling model was applied to the AIBL case study, as well as details of the independent analyses methods applied to produce comparable biomarker estimates as described in literature [9, 16, 17, 34, 39].

7.2.4.1 Probabilistic Connectivity Matrices via Wombling

Inference on the brain wombling models were estimated by the MCMC scheme described in Sect. 7.2.2, which was applied to each group and was run using four chains. Each chain ran for $M = 500,000$ iterations. The first 50,000 runs (burn-in) were discarded and every 50th iteration retained (thinning).

The resultant elements of the posterior mean of W matrices are $\bar{w}_{kj}$, and represent the probability that region k is connected to region j in a cortical structural network. These networks represent the underlying average network of a group estimated from our sample. Binary matrices were derived for a given probability threshold ($0 < \tau < 1$) for each element of W. This threshold determines the level of confidence in our brain network, and allows for straightforward comparisons across groups. However as noted in He et al. [37] and Yao et al. [71], a high threshold on brain networks may lead to disconnected networks and this may make topological network metrics difficult to analyse. In this work, we set $\tau = 0.7$ as this value is substantially higher than the prior value of 0.5, and is greater than our 0.6 value from our simulated study in Sect. 7.2.3, thus providing a more stringent level on the certainty of the resultant networks, resulting in a high level of confidence regarding the potential connections between nodes.

7.2.4.2 Descriptive Pearson Cortical Networks

Following the methods of Bassett et al. [8], we applied Pearson pairwise correlation networks at both baseline (which consisted of all observations being independent and identically distributed (IID)) as well as on the whole data with repeated measures treated as IID.

7.2.4.3 Wombled Population and Participant ROI Biomarker Estimates

In the Bayesian paradigm, the posterior distributions of parameters can be compared directly to make probabilistic statements about each other, or in regards to other biologically relevant quantities. The probability that parameter $\beta_{0,A}$ from group A is within the lower 2.5% and upper 97.5% quantiles of the posteriori $\beta_{0,B}$ from group B (denoted by Y_L and Y_H), is estimated by

$$P(Y_L < X < Y_H) = \frac{1}{M} \sum_{m=1}^{M} \mathbb{1}(Y_L < \beta_{0,A}^m < Y_H), \qquad (7.8)$$

Table 7.2 Comparisons of estimated total cortical thickness among groups

Group comparison	Probability
$P(AD_L < HC < AD_H)$	0.02
$P(MCI_L < HC < MCI_H)$	0.49
$P(AD_L < MCI < AD_H)$	0.90
$P(B_L < A < B_H)$	0.79
$P(C_L < B < C_H)$	0.97
$P(C_L < A < C_H)$	0.86

Probabilities of parameter X with respect to the posterior distribution is within the lower (L) 2.5% and upper (H) 97.5% quantiles of the posteriori of Y is expressed by $P(Y_L < X < Y_H)$. A high probability denote posterior distributions overlap among groups and low probability suggest substantial differences in posterior estimates; quantiles of distributions are the box plot whiskers in Fig. 7.6

where the indicator function $\mathbb{1}$ is equal to one if $Y_L < \beta_{0,A}^m < Y_H$ and zero otherwise. The length of the MCMC chain for $\beta_{0,A}$ is M. Comparison of all groups are computed in a similar manner, whose results are listed in Table 7.2 of Sect. 7.3.2.3.

While our algorithm provides cortical thickness estimates on all participants in the analysis for each ROI, we focused on the nine key regions often used to describe the cortical signature of AD [21, 23]: the inferior, medial and superior temporal lobes; supramarginal, angular, posterior cingulate and the precuneus gyrus. Results for all 35 ROI can be found in Figs. 7.17 and 7.18.

Low cortical thickness estimates are often indicative of neurodegeneration. For this reason, at the participant level analyses in Sect. 7.3.2.3, we expected an increasing atrophy pattern to be associated with diagnosis, from AD to MCI to HC, as well as among age groups, from old to young. Participants which differ from this pattern may be showing early signs of AD pathology, thus this analysis could be also be used to flag sub-groups of participants to follow up.

7.2.4.4 Bayesian LME ROI Analyses

Bayesian LME models were applied independently on each ROI in a similar manner as Bernal-Rusiel et al. [9], Guillaume et al. [34], Caselli et al. [16], Holland et al. [39] and Cespedes et al. [17] who applied LME models at the ROI level. Similarly, others who applied these models at the voxel scale in AD and in other neurological applications [35, 72]. Refer to section "Bayesian Linear Mixed Effect Models on Each ROI" in Appendix for model specifications. The wombling and combined Bayesian LME models were compared by the Watanabe-Akaike information criterion (WAIC, Watanabe [66]). The survey by Gelman et al. [30]

describes how the WAIC has been shown to be the preferred approach for model comparison in the Bayesian community. For this reason, it is applied to the models this work.

7.2.5   Statistical Analysis

All analyses were undertaken using the open-source software R [56]. Source code and data used in the simulation study are available at https://github.com/ MarcelaCespedes/Brain_wombling. Simulation experiments were performed using a high performance computer cluster. We note that a single MCMC instance of the Bayesian brain wombling model ran on a single central processing unit (CPU) and took approximately 24 h to run on a standard computer (four core 3.40 GHz Intel i7-4770 processor).

7.3   Results

7.3.1   Simulation Studies

7.3.1.1   Wombling Simulated Analyses

Figure 7.3A and F show the comparison between the W we should recover, and the average estimated W for the structured configuration (Fig. 7.3B and D), and contiguous configuration (Fig. 7.3G and I). Section 7.2.3 describes how the mean of the posterior mean matrices in Fig. 7.3B, D, G and I were binarised. The resultant binarised matrices for the structured balanced and unbalanced designs recovered 83% and 82% of the networks' solution (Fig. 7.3C, E). The binarised matrices for the contiguous balanced and unbalanced designs recovered 70% and 65% of the contiguous configuration (Fig. 7.3H and J). The parameter dimension in the 35 ROI simulation study consisted of the off-diagonals of W, ($K(K - 1)/2 = 595$) in addition to β, σ^2 and σ_s^2, which was a total of 599 parameters. As can be seen by Fig. 7.3, the wombling model showed the desired recovery of the connectivity matrices in both configurations and in the balanced and unbalanced designs, despite the high parameter dimension.

To assess whether the random effects were recovered appropriately, we evaluated their 95% credible intervals. These results showed that the true values of the random effects were recovered on average 95% of the time indicating that the variation of the posterior distribution is appropriate. See Table 7.3 and Fig. 7.9 for details of these results. Likewise, the recovery of the solution vector β was within the 95% of the credible intervals approximately 95% of the time in all simulation configurations and scenarios, demonstrating that in addition to recovery of connectivity networks,

Fig. 7.3 Data generated from binary W matrices for structured (**a**) and contiguous (**f**) configurations. A single wombling simulation run results in a posterior distribution for W, whose mean represents an average connectivity matrix. The mean of the posterior means over all 50 simulations in each scenario are the connectivity matrices shown in **b**, **d**, **g** and **i**, which show the average posterior probability of region j being a neighbour of region k. Top row: Mean of posterior means for structured balanced (**b**) and unbalanced (**d**) simulation designs, with corresponding binary matrices (**c** and **e** respectively) whose elements are equated to one if their value is greater than threshold $\tau = 0.6$ and zero otherwise. Bottom row: Similarly for the contiguous simulation study, mean of posterior means for balanced (**g**) and unbalanced (**i**) matrices with corresponding binarised matrices (**h** and **i**) at probability threshold of 0.6 and above

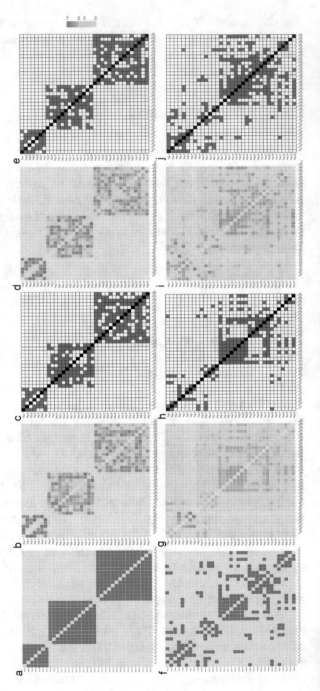

the wombling model was able to recover the biomarker and participant level estimates.

A sensitivity analysis with respect to the prior information on σ^2, σ_s^2, and $\boldsymbol{\beta}$, was conducted on the structured W configuration. This entailed re-running the analysis using various specifications of the prior information. The subsequent posterior summaries did not vary considerably based on different prior information. Hence we postulate that estimation of σ^2, σ_s^2, and $\boldsymbol{\beta}$ are relatively robust to the priors specified in this work.

The results described above relate to two fixed W configurations with the same values on $\boldsymbol{\beta}$, σ^2, and σ_s^2 for each scenario. We investigated the effect of different values for variance terms (σ_s^2 and σ^2) on the recovery of W and fixed and random effects. We found the results to be very similar to those reported here (model results for different variance terms not shown). Furthermore, we investigated the effect of the value of ρ on the recovery of W with $\rho \in \{0.85, 0.9, 0.95, 0.99\}$ using the balanced structured simulation scenario. There is some wombling literature which suggests that the choice of ρ can affect the recovery of σ_s^2 and W [42, 44, 50]; we found a choice of $\rho = 0.9$ provided appropriate recovery of parameters of interest. Refer to Table 7.4 for results on a range of ρ values.

In summary, our simulation study showed the recovery of W proved to be appropriate, which implies that the estimation of Q^{-1} is reliable. However the spatial scale variance (σ_s^2) was typically overestimated, a finding that is not uncommon in wombling literature [50]. Despite this, the simulation study also showed adequate recovery on biomarker and participant estimates, as such our estimates for $\boldsymbol{\beta}$, σ^2 and $\mathbf{b}_i$ are reliable.

7.3.2 Application to Real Data

In this section, we present the results of the joint analysis derived by the wombling model, and compared them with the results from the independent analyses (overview in Fig. 7.1).

The MCMC algorithm was utilised to draw posterior samples from the wombling model applied on diagnosis and age groups of the AIBL case study. As described in Sect. 7.2.5, informal diagnostic measures were assessed, such as trace, density and autocorrelation plots, as well as formal measures to investigate between and within chain variation with the Gelman-Rubin convergence measure [14]. All plots suggested convergence to a stationarity distribution according to the Gelman-Rubin convergence checks. Furthermore, posterior predictive checks on all models in these analyses showed the models fit the data well; there were no systematic departures from the model predictions and 95–99% of all response values were within the 95% credible intervals of the posterior predictive distributions; refer to Table 7.5 and Figs. 7.10, 7.11 for results.

7.3.2.1 Probabilistic Connectivity Matrices via Wombling

The networks corresponding to the probabilistic matrices in Fig. 7.4 show the results for diagnosis levels HC (top: A and B), MCI (middle: C and D) and AD (bottom: E and F). The varying level of uncertainty between matrices is indicated by elements with probability values close to 0.5, in contrast with connections which have high or low probabilities. This is partly due to a sample size effect, as there were 120 participants who were diagnosed as HC compared to MCI (21) and AD (26) participants.

The networks on the right of Fig. 7.4 show those connections with a probability equal to or greater than 0.7. The network configurations reflect the underlying estimated population networks. The total number of edges in these networks show HC participants have a more complex cortical network structure with 156 connections, in comparison with the MCI network which had 124 connections. Furthermore, the AD network has a lower degree (112 connections) in contrast with the MCI and HC networks, suggesting a higher loss of network communication among the ROIs. The middle temporal lobe is one of the earliest regions known to be affected by the onset of AD [40]; with a probability greater than 0.7, our results indicate the number of connections of the HC, MCI and AD networks for this region are 7, 6 and 4 respectively, suggesting a loss of connections along the AD pathway. A similar reduction in node degree, in general, can be observed on the entire cortical mantle, across the frontal, temporal, parietal and occipital lobes.

Baseline age differences are observed in cortical networks in Fig. 7.5. The networks on the right of Fig. 7.5 show a re-organisation of connections, rather than a direct loss of total network degree with an increase of age. The older age Group C (79–93y) consists of 62 participants of which 41 were diagnosed as HC at baseline and 8 were diagnosed as AD. Hence the analysis in this group is dominated by HC, and the resultant network better aligns to healthy ageing rather than onset of AD. The diagnosis ratio of participants in the younger and middle age Groups A and B (59–69y and 69–79y respectively) have higher ratio of AD and MCI participants in contrast to HC. Hence the averaged networks across these groups include participants with a broader spectrum across healthy ageing, and progression to AD or other dementias, in contrast with age Group C.

7.3.2.2 Descriptive Pearson Cortical Networks

The Pearson pairwise network analyses on diagnosis and age groups were sensitive to data with repeated measures, as connections varied across both groups between networks derived from IID and data with replicates. This finding is interesting as the studies by Li et al. [46] and Fan et al. [25] used repeated measures in their pairwise correlation network analyses. However, in our analyses, only the IID Pearson cortical networks were used for comparison with the wombling model. Once the IID correlation networks were binarised by placing a link between ROIs whose absolute correlation values greater than $\tau = 0.7$, the diagnosis group did

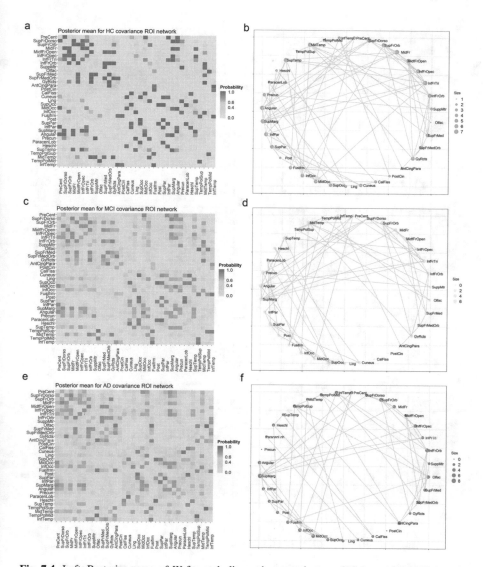

Fig. 7.4 Left: Posterior mean of W for each diagnosis, top to bottom; HC (**a** and **b**), MCI (**c** and **d**) and AD (**e** and **f**). Right: Cortical networks from binarised posterior matrix W with threshold $\tau = 0.7$ for the respective diagnosis groups. Node size reflects the number of edges on each vertice. Total number of edges for each network (top to bottom) are 156, 124 and 112 for HC, MCI and AD networks respectively

not support biologically meaningful networks: the Pearson pairwise correlations were considerably higher in the MCI group, followed by AD and HC with fewer connections. The age correlation matrices were binarised in the same manner, and the resulting sparse networks had a loss of connections from young to older age groups, of A (47) to B (38) to C (19). Refer to Figs. 7.26, 7.27, 7.28, 7.29, 7.30, and 7.31 for full Pearson pairwise correlation network results.

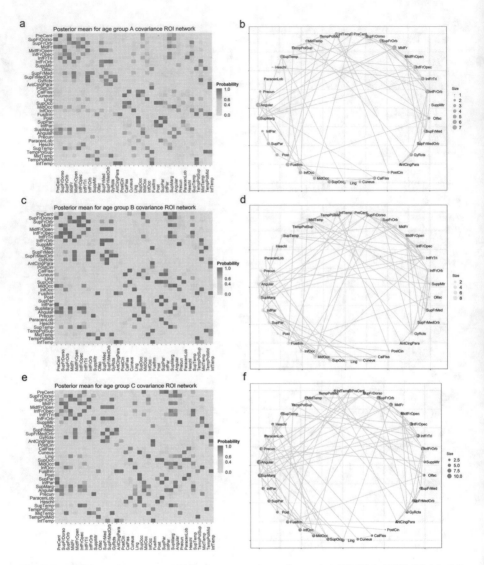

Fig. 7.5 Left: Posterior means of W for age groups (top to bottom) A (59–69y), B (69–79y) and C (79–93y) shown in plots A, C and E respectively. Right: Cortical network from binarised posterior matrix W with threshold $\tau = 0.7$ for the respective age groups for the respective age groups A, B and C shown in plots B, D and F. Node size reflects the number of edges on each vertice

7.3.2.3 Wombled Population and Participant ROI Biomarker Estimates

In our application of brain wombling with Model (7.1), the vector $\boldsymbol{\beta} = [\beta_0, \beta_1]$ contains fixed effect parameters, where the intercept β_0 represents the mean thickness of the left cortex hemisphere of the brain, for a particular group and β_1 is the gender effect, with females as baseline and covariate $x_i = 1$ for male. In all groups analysed, the gender effect was not substantive (95% credible intervals for

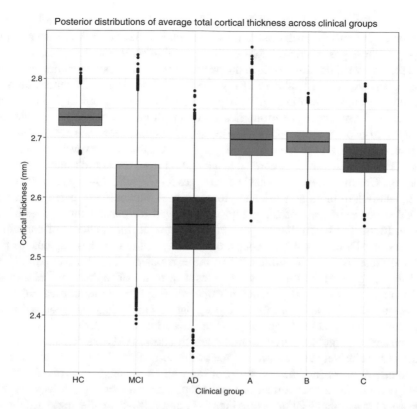

Fig. 7.6 Posterior marginal distributions of total cortical thickness (β_0) across groups. Median of each distribution shown in each box plot, whiskers indicate 95% credible interval for each parameter. Diagnosis groups; healthy control (HC), mild cognitive impaired (MCI), Alzheimer's disease (AD) and age Groups A, B and C correspond to age ranges 59–69y, 69–79y and 79–93y respectively

β_1 included zero), thus we conclude there are no significant gender differences in global cortical thickness between the groups analysed.

The median cortical thickness mantle in HC groups ($\beta_{0,HC}$) is significantly higher than AD clinical diagnosis, as the 95% credible interval of $\beta_{0,HC}$ lies outside of the 95% credible interval of the AD distribution ($\beta_{0,AD}$). While the posterior median of the MCI group was not significantly different from the medians of the HC or AD groups, from Fig. 7.6, the cascading order of degeneration on the cortex can be seen in the disease progression from HC to MCI to AD.

As described in Sect. 7.2.4.3, we can make probabilistic comparisons among the median cortical thickness between groups. The probability that, a posteriori $\beta_{0,HC}$ is within the AD_L and AD_H quantiles of the posterior distribution of the AD is 0.02, which implies there is a significant difference between the cortex of HC and AD groups. The probability that $\beta_{0,HC}$ is within the MCI_L and MCI_H quantiles of the MCI diagnosis ($\beta_{0,MCI}$) is 0.49. This high probability is reflected in the third quartile of the MCI box plot overlapping the HC box plot in Fig. 7.6.

The comparison of the MCI and AD box plots in Fig. 7.6 reflect the overlapping of the upper and lower distribution tail ends, which is reflected in the distribution for $\beta_{0,MCI}$, whose posteriori probability of being within AD_L and AD_H is 0.9.

There were subtle differences in the posterior cortical thickness estimates among age Groups A, B and C shown in Fig. 7.6. Unlike the large differences between diagnosis groups shown in Fig. 7.6 and probability comparisons in Table 7.2, the posteriori of a younger age group lies inside the credible interval of an older age group with a probability ≥ 0.79. These results suggest there were no significant differences between the median cortex of the age groups. However, as expected, there is a cascading order of cortical degeneration from thicker to thinner estimates from age Groups (A, B) to C, that is age ranges 59–79y and 79–93y respectively.

In addition to brain network and global cortical thickness estimates, the hierarchical structure of the wombling approach allowed for participant level estimates for all ROIs. The caterpillar plots in Fig. 7.7 show distinct patterns of participant clusters of AD, MCI and HC groups, particularly in the nine key regions, as they are the most likely to be influenced in the early stages of AD. AD participants had the lowest cortical thickness estimates as a result of higher cortical atrophy. MCI are midway in the degeneration scope with slightly higher cortical thickness estimates than AD, but lower than HC. Regions in which diagnosis groups differed particularly included the temporal poles of the middle and superior temporal gyrus and posterior cingulate gyrus. Regions which showed AD participants were not clustered exclusively at the lowest range of the cortical thickness estimates include the temporal poles (middle and superior) as well as the angular gyrus. Excluding these regions, for the remainder of the diagnosis clusters among participants were consistent in all other ROI plots (see Figs. 7.17 and 7.18). Note that these differences in diagnosis levels are consistent with the loss of network connectivity in Fig. 7.4 and total average cortical thickness estimates in Fig. 7.6.

The results of ranked participants were analysed with respect to age groups and selected regions are shown in Fig. 7.8; refer to Figs. 7.17 and 7.18 for the remaining ROI plots. The regions in Fig. 7.8 showed pronounced age group specific clusters. Key regions which age Group A had consistently higher cortical thickness estimates in contrast with age Groups B and C were the calcarine fissure, fusiform, heschl, middle temporal and precentral gyrus.

7.3.2.4 Bayesian LME ROI Analyses

Participant specific estimates via the Bayesian LME models were assessed and the results align with those from the wombling model in both key ROI which support strong distinctions among groups (particularly in the diagnosis groups) as well as in instances which all ROI showed little difference among groups, such as as those in the age groups; refer to Figs. 7.20, 7.21, 7.22, 7.23 for all Bayesian LME results. For example, the superior middle and inferior temporal regions had distinct HC, MCI and AD participant clusters, as well as the supramarginal and the posterior cingulate. The WAIC values of the wombled model in all groups were found to

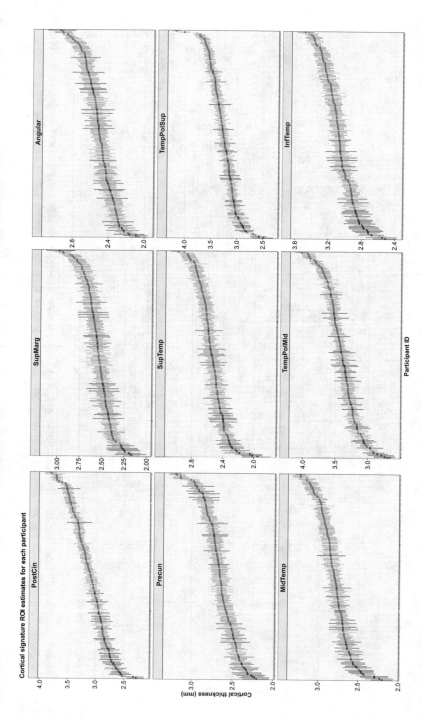

Fig. 7.7 Ordered ROI cortical thickness posterior means and 95% credible intervals over participants random effects $\mathbf{b}_i$, for nine key regions associated with AD cortical signature. Colour coded as follows; AD (red), MCI (orange) and HC (green) diagnosis groups. Refer to Table 7.1 for full name of cortex regions

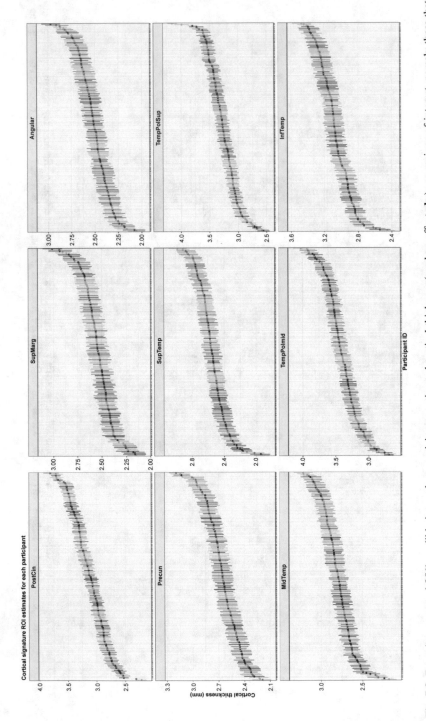

Fig. 7.8 Posterior mean and 95% credible interval on participants estimated cortical thickness (random effects $\mathbf{b}_i$), on regions of interest, namely those that showed the highest differences. Age Groups colour coded as follows; (dark green) A, younger age group; (orange) B, middle age group and (brown) C, oldest age group

be substantially lower than the WAIC values of the Bayesian LME models. These results show that the wombling model is a more parsimonious approach to model biomarker estimates compared to independent LME models on each ROI, and hence a desirable model for this data. Refer to Table 7.6 for WAIC results.

7.4   Discussion

This work demonstrated and validated the Bayesian wombling approach using intrahemispherical cortical thickness observations of the brain in both a simulation study, and applied to an Alzheimer's disease cohort study. Each analysis was applied across HC, MCI and AD diagnosis categories as well as three age groups. Wombling provides a novel way to combine both regression and network analyses into a single unified model. This takes into account the uncertainty of all possible links to estimate a network, but also allows group comparisons from independent measurements (for example, participants' cortical volumes for many ROIs) without the need for multiple comparison correction.

7.4.1   Simulation Study

The ability of the wombling algorithm to successfully recover the underlying connectivity solution while appropriately accounting for the variance was assessed in Sect. 7.3.1. Figure 7.3 shows the overall average performance of the wombling algorithm as probability and thresholded networks. The wombling algorithm consistently and correctly detected the absence of connections in the structured configuration, and recovered 82% and above of the true values of W. On the more difficult contiguous scenario, Fig. 7.3H and I show that 65% and above the contiguous solution was recovered, at a probability threshold greater than 0.6, which as expected was less certain than the structured configuration.

Approximately 95% of the cortical thickness estimates were recovered at the population and participant level. Recovery in the statistical sense refers to whether the intervals of the estimator contain the true solution. Thus, an algorithm that recovers the known solution 100% of the time, could potentially do so by simply overestimating the variance. In our simulation studies, based on 95% credible intervals the wombling parameters were recovered approximately 95% of the time (see Table 7.3). This indicated that the wombled model appropriately estimated the variability in the parameters.

While the simulation study was designed to mimic features typical of longitudinal study data (in this case we matched some of the characteristics of the AIBL study, such as the number of participants, replicates and connectivity configuration), the practical performance of the wombling algorithm is better assessed when it is applied to the real data and directly compared with the alternative state-of-the-art methods.

7.4.2   Application to Study Cohort

7.4.2.1   Cortical Networks

The results from the brain wombling model were compared with those of alternative independent analyses on the AIBL data. Figure 7.4A, C and E shows a decrease in connections from HC (156) to MCI (124) to AD (112), which reflect the biological order of neurodegeneration [15, 18, 58]. As expected from previous work [6], the loss of connections on the wombled networks reflect the strong differences in the diagnosis groups which is also reflected in the wombled cortical thickness estimates shown in Figs. 7.6, 7.7 and Table 7.2, as well as on the Bayesian LME analyses in Figs. 7.20 and 7.22.

At the same threshold as the wombling networks ($\tau = 0.7$), the Pearson pairwise correlation networks on baseline observations did not show a biological decrease of connections. Specifically, both MCI and AD had 34 connections, 12 more than the HC network with 22 connections. These results suggest that in this work, the wombled networks provided superior connectivity information in comparison to the Pearson pairwise correlation method.

Pearson pairwise correlation networks showed a decrease in overall connectivity across baseline age Groups A to B to C with 47, 38 and 19 total connections respectively, suggesting age dependent loss of connections. However, further investigation into these results is required as the Bayesian LME and wombling models did not support participant age clusters; suggesting there were no age differences in the data (see Figs. 7.22 and 7.23). Furthermore, age Group C comprises of predominately HC and MCI participants, as 18 of the 26 AD participants are in age Groups A and B, which suggests age Group C should not reflect high neurodegeneration estimates.

In addition to these improvements, unlike the descriptive networks from the Pearson pairwise correlation approach, the wombling model provided full posterior distributions which quantified the uncertainty in all possible links. As the Pearson pairwise correlation networks do not take into account the uncertainty of each connection, they cannot correctly estimate the group population networks.

One potential question about the modelling approach proposed in this work is whether the inclusion of additional terms in the mean of the model would significantly change the inference about W. Such terms could include fixed effects to estimate ROI means. If we consider the covariance between data for two ROIs, then, in principle, the correlation structure should be unaffected if, for example, the data were standardised such that data for each ROI had a mean of zero and a variance of one. However, such a simplistic scenario may not be directly applicable to the complex model fitted in this work. Thus we investigated this by extending the wombling model to allow for the estimation of ROI means through the inclusion of fixed effect parameters in the mean of the model. The results showed that the inference about W for the HC group was similar to that presented in Sect. 7.3.2.1 (see section "Wombling Cortical Thickness Estimates at the ROI Level" in Appendix). However, with the MCI and AD groups, this model provided large amounts of uncertainty in the posterior distribution for W, limiting our ability to determine whether inference is impacted by the estimation of ROI means. We believe this is due to the additional (35 fixed effect) parameters included into the model, and this appears to have a major impact in the MCI and AD groups as they have much smaller sample sizes (21 and 26, respectively) compared the HC group (120). These smaller sample sizes appear to have led to a loss of information about the network connection for these groups.

The choice of which wombling model to apply, whether it be the model presented in this work or the extended version which includes ROI means depends on the research questions which one wishes to address and the data which are available. If there are only approximately 20 to 30 individuals in a group of interest and intrahemispheric data are available, then the wombling model presented here can provide meaningful inferences about W but not on ROI means. However if there are over 120 individuals in the groups of interest, then the more complex model with additional ROI parameters would provide joint estimates on the ROI means as well as on the participant and covariance networks. We note that in our analyses it was reassuring to find that the estimates for W were similar in both models.

Further, there is potential for the inference about W to change if important covariates are included into the model. That is, perceived covariance may be due to the influence of unobserved covariate information. Our model can easily incorporate covariates, and indeed it also does this in demonstration through the inclusion of sex, and other covariates could be similarly included (and tested for importance).

7.4.2.2 Biomarker Estimates

In all groups analysed, the WAIC values for the wombling model were substantially lower compared to the independent Bayesian LME models combined across all ROIs. In this work, this result shows that the wombling model was the preferred parsimonious approach for modelling biomaker estimates compared to the independent analyses. Refer to Table 7.6 for WAIC results. At the participant level estimates of cortical thickness, both approaches demonstrated comparable differences in diagnosed participant clusters as shown in Figs. 7.7 and 7.8 and the Bayesian LME model estimates in Figs. 7.21 and 7.23. These results further demonstrate the flexibility of the wombling approach to jointly analyse cortical networks in addition to biomarker estimates. Above the third quartile of the MCI posterior distribution had a large degree of overlap with the HC posterior distribution (Fig. 7.6). With a probability of 0.49, the posterior distribution of the HC total cortical thickness average is within the upper and lower 95% quantiles of the MCI distribution (Table 7.2). Such a high probability suggests that this overlap could be due a subset of MCI participants in the study who are not on the AD pathway [26, 54]. Hence further investigation into MCI participants further divided into subgroups, such as participants with documented memory complaints, amnestic and non-amnestic is suggested to identify potential non-AD converters.

7.4.3   Sensitivity Analyses

Two sensitivity analyses were conducted on the application of the Bayesian wombling approach to real data. The first analyses were with respect to the chosen value of ρ, as described in Sect. 7.3.1.1. A number of authors [42, 44, 50] have discussed the limitations of including ρ as an additional parameter to be estimated. Following these recommendations, we fixed $\rho = 0.9$ throughout all our simulations and application studies, and conducted a sensitivity assessment to evaluate the impact of this choice. Table 7.4 showed that at $\rho = 0.9$ and the parameters W, β, σ^2 and $\mathbf{b}_i$ were recovered well. Our results support those of Lu et al. [50], Lee [42] and Lee and Mitchell [44], and we recommend fixing ρ at 0.9 for future wombling model extensions.

The second sensitivity analysis was with respect to the prior specification described in Sect. 7.3.1.1. Since the resulting posterior summaries did not vary considerably based on different prior information, we conclude that our results are relatively robust to the priors specified in this work. The rationale for using vague priors is to ensure that the information in the data primarily governs the results. Alternatively, informative priors may be employed when relevant information is available [12, 70]. In particular, investigating the best use of diffusion or functional network priors (or patient specific networks) would be an interesting future research avenue.

7.4.4   Limitations and Future Work

The intended application of the wombling model in this work is to demonstrate its utility. Due to the limited sample sizes in this study cohort, the biological interpretation and comparison of each group, in this work, is limited to the total number of connections for each network. Additional cortical network metrics which assess the organisational structure, such as small world topology and characteristic path length [7, 15, 58], is beyond the scope of this work. Future work and clinical application of the wombling model will greatly benefit from matched sampled groups which have similar age ranges, number of replicates, gender and other features known to be associated with the pathology of interest.

A primary drawback of wombling models is the computation time. As mentioned by Bowman et al. [12], limitations of a Bayesian hierarchical framework in spatial analysis include extensive and long computational times, often restricting attention to small ROI analysis or localised voxel-wise analysis. For example, the study by Bowman [11] considered only ROI in the cerebellum to limit the computational extent. Although computationally intensive, our brain wombling approach is not prohibitively so: a single MCMC run of the algorithm can also be computed in approximately a day on a standard desktop computer (see Sect. 7.2.5).

As the dimension of W increases, the parameter space increases dramatically, and this is considered a drawback of the wombling model. For example, our 35

ROI model resulted in a 599-dimensional parameter space, which ran for 500, 000 MCMC iterations. This issue motivates future work to investigate inducing sparsity on W based on prior information as suggested in Babacan et al. [5], as this could potentially reduce the computational burden of the wombling model. Nevertheless, in the present study, the added insight and corroboration between networks and cortical thickness estimates were deemed to be worth the additional computational time.

A second limitation of the present study is that our analysis was restricted to participants with four or more repeated measures, as this affected the ability of the wombling model to converge (results not shown). Such repeated measures can be prohibitive in smaller neuroimaging studies, as patient drop out is a common occurrence. For use of this method in neuroimaging studies with a limited number (< 4) of time points, future work detailing the performance of the wombling model is needed. Nonetheless, our algorithm performed remarkably well for small sample sizes (N_{AD} and $N_{MCI} < 21 < 35$ ROI) on data where all participants had repeated measures. We conjecture that the probabilistic networks from the wombling model will better distinguish between a link and the absence of a link (i.e. network probabilities will be closer to zero or one), and result in narrower credible intervals on biomarker estimates as the sample size increases.

The final limitation of the present study was the relative simplicity of the two layered linear random effects model, as shown in Fig. 7.2. This is not a fixed limitation of the approach presented here; the hierarchical Bayesian framework is capable of handling complex models, such as models with two or more nested layers to account for complex data structures [28, 29]. Extensions of this nature will allow the modelling of cerebral morphological features across multiple ROIs over participants' age, and expand our spatial approach into a spatio-temporal domain.

7.4.5 Conclusion

In this work, we have demonstrated the advantages of the Bayesian brain wombling approach applied in the neuroimaging field over state-of-the-art independent analyses. The ability of the wombling model to recover the connectivity and biomarker effect estimates give confidence on our results from the cohort study. Taking into account of the uncertainty of each network, the population wombled networks across diagnosis levels from healthy to Alzheimer's disease showed a loss of connectivity (posterior probability $\geqslant 0.7$). Compared to independent LME models, we found that both approaches estimated cortical thickness progressively along the dementia pathway. Although applied here to cortical thickness, this method can be applied to other types of neuroimaging data, unifying existing previously independent analyses that are aimed at exploring the same underlying biological system. This powerful analysis tool provides the potential to extend our understanding of the human brain functions and effects of brain disorders on both local and network scale.

Acknowledgments We wish to thank the Australian Imaging, Biomarkers and Lifestyle longitudinal study of ageing (www.aibl.csiro.au), including all the clinicians, scientists, participants and their families. MIC was jointly funded by the Research Training Program (RTP), the Commonwealth Scientific and Industrial Research Organisation (CSIRO) Health and Biosecurity division, and supported by the ARC Centre of Excellence for Mathematical & Statistical Frontiers (ACEMS). CCD was supported by an Australian Research Council's Discovery Early Career Researcher Award funding scheme DE160100741. Computational resources and services used in this work were provided by the High Performance Computing (HPC) and Research Support Group, Queensland University of Technology, Brisbane, Australia.

Appendix: Methods and Applications

Additional material to supplement simulation study results, posterior diagnostic checks, wombling ROI cortical thickness estimates at the population and participant levels, independent Bayesian mixed effect model results, WAIC values and Pearson correlation networks can be found in this Appendix. R code to implement the wombling model can be found at the following GitHub repository https://github.com/MarcelaCespedes/Brain_wombling.

Simulation Study

The simulation study described in Sect. 7.2.3 provided a thorough assessment of the Bayesian brain wombling algorithm. The four scenarios in the simulation study are; contiguous balanced (each person had an equal number of replicates) and unbalanced (the number of replicates varied per person), and a structured balanced and unbalanced designs. The results for fixed effect parameters β and residual variance σ^2 are shown in Table 7.3. While the results for the structured configuration show a slightly lower recovery of fixed effect parameters, they do not represent a potential biological configuration. Hence performance of the wombling algorithm is better assessed on the contiguous configuration, whose performance of the recovery of the parameters is approximately 95%.

As discussed in Sect. 7.2.3.1, spatial scale variance σ_s^2 is a biased estimate and was not recovered in our simulation study.

Table 7.3 Percentage (%) of fixed effect and residual variance parameter recovery for four scenarios, each with 50 simulations

	Contiguous Unbalanced	Contiguous Balanced	Structured Unbalanced	Structured Balanced
β_0	100	98	92	92
β_1	100	100	92	90
σ^2	94	98	94	96

Table 7.4 Parameter values for ρ set to 0.85, 0.9, 0.95, 0.99 values

	0.85	0.9	0.95	0.99
β_0	3.1 (2.9, 3.2)	3 (2.8, 3.2)	3.0 (3.2, 3.5)	2.8 (2.2, 3.4)
β_1	0.3 (0.1, 0.5)	0.3 (−0.01, 0.5)	0.1 (−0.3, 0.5)	0.4 (−0.3, .4)
σ^2	0.5 (0.5, 0.7)	0.5 (0.5, 0.5)	0.5 (0.5, 0.5)	0.6 (0.6, 0.7)

True value for β_0 is 3, and β_1, σ are 0.5

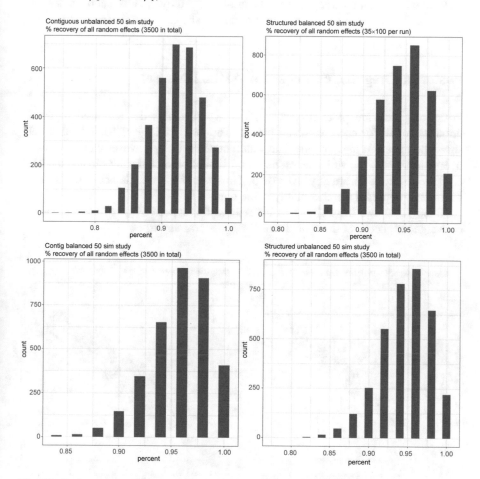

Fig. 7.9 Each scenario (structured balanced and unbalanced, contiguous balanced and unbalanced) had 100 simulated participants and each participant had 35 ROI (3500 random effects in total for each simulation). As each scenario comprised of 50 simulations, there are 50 × 3500 random effects to assess. Each histogram denoted the percentage of the number of random effects recovered (that is random effects whose solution within the 95% credible interval)

Figure 7.9 shows the histograms on the percentage of the recovered random effects for each scenario. The simulation study comprised of 50 independently simulated data sets for each scenario, each data set consisted of $I = 100$ simulated participants, each with $K = 35$ ROI resulting in 3500 random effects per simulated data set to estimate. Overall we can see that there is approximately 95% recovery of the random effects for each scenario.

As ρ in Model (7.1) is a fixed value, we investigated the effect recovering the parameters in the structured scenario for ρ values [0.85, 0.9, 0.95, 0.99]. Table 7.4 summarises the results.

Contact the author for additional simulation study results such as MCMC convergence checks, estimation of credible intervals, and posterior predictive plots.

Posterior Diagnostic Checks for AIBL Data Set

Posterior predictive plots for each AIBL group analysed were used to assess goodness-of-fit for each wombled model. The plots in Figs. 7.10, 7.11 and 7.12 show the expected mean of the data was recovered well, however there is a slight overestimation of the variance, as the proportion of predicted values inside the 95% credible intervals is slightly over 0.95. However these results show our models adequately captured the uncertainty in the data.

Table 7.5 shows the Gelman-Rubin diagnostic, upper 95% credible interval for convergence checks of the four chains for β_0, β_1, σ^2 and σ_s^2.

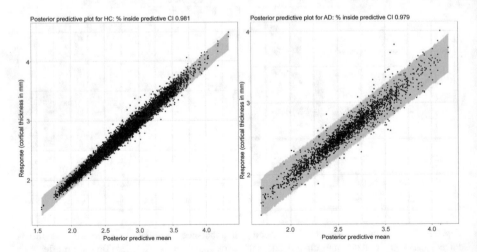

Fig. 7.10 Posterior predictive plots for healthy control (HC, left) and Alzheimer's disease (AD, right) wombling models. The proportion of response values inside the predictive 95% credible intervals (in red) is 0.981 and 0.979 for HC and AD models respectively

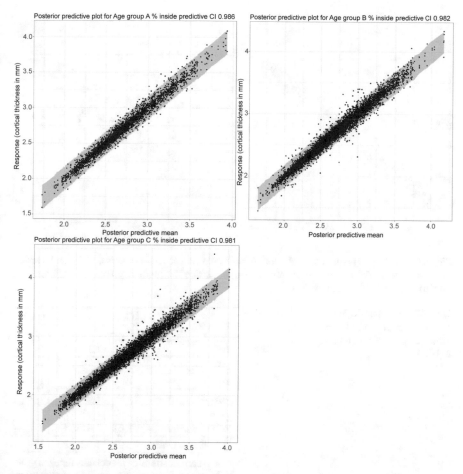

Fig. 7.11 Posterior predictive plots for age groups; A (59–69), B (69–79) and C (79–93). The proportion of response values inside the predictive 95% credible intervals (in red) are 0.986, 0.982 and 0.981 for age groups A, B and C

Wombling Cortical Thickness Estimates at the ROI Level

As discussed in Sect. 7.4.2.1, we investigated an adaptation to the wombling model to account for ROI means via fixed effect parameters. The extended model is of the form

$$y_{irk}|b_{ik}, \boldsymbol{\beta}, \sigma^2 \sim N(\beta_0 + \beta_1 R_2 + \beta_2 R_3 + \ldots + \beta_{34} R_{35} + b_{ik}, \sigma^2)$$

$$\mathbf{b}_i \sim MVN(\mathbf{0}, \sigma_s^2 Q)$$

$$Q^{-1} = \rho(D_w - W) + (1 - \rho)\mathbb{I}. \tag{7.9}$$

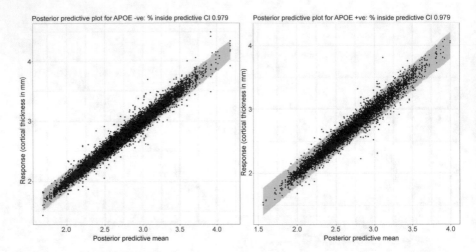

Fig. 7.12 Posterior predictive plots for APOE $\varepsilon4$ non-carriers (negative, left) and carriers (positive, right). The proportion of response values inside the predictive 95% credible intervals (in red) were 0.979 for both models

Table 7.5 Gelman-Rubin diagnostic upper confidence limit values for each group in AIBL study

	Gelman-Rubin diagnostic			
	β_0	β_1	σ^2	σ_s^2
HC	1.15	1.16	1	1
AD	1.09	1.16	1	1
Age A	1.14	1	1	1
Age B	1.04	1.06	1	1.01
Age C	1.07	1	1	1
APOE carrier	1.02	1.01	1	1
APOE non-carrier	1.03	1.02	1	1

As the combinations of four chains for each group had values close to one, we are confident the MCMC algorithm for each group has reached convergence

Where the response (y_{irk}), spatial random effects (b_{ik}), residual (σ^2) and spatial scale variance (σ_s^2) terms are the same as those presented in Sect. 7.2.2. The precentral gyrus is the baseline ROI whose cortical thickness (in mm) is estimated by β_0. The fixed effect parameter β_{k-1} estimates the deviation of ROI k away from β_0 when the binary indicator variable R_k is equal to one. Estimation of β is attained by the same conditional distribution described in Sect. 7.2.2, with minor modifications to account for the design matrix R rather than X. Figures 7.17 and 7.18 show participant specific cortical thickness estimates as caterpillar plots ($\beta_k + b_{ik}$) colour coded for diagnosis and age groups respectively. Figure 7.13 shows the posterior means of W for HC (top), MCI (middle) and AD (bottom) groups. While the posterior mean for the HC group is similar that in Fig. 7.4, with the same 36 links present in both networks and 468 absent connections in common,

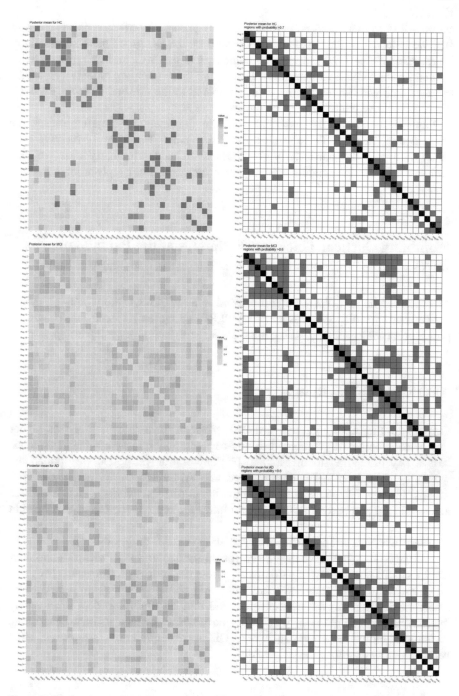

Fig. 7.13 Left column: Posterior means for W for HC (top), MCI (middle) and AD (bottom). Right column: Binarised matrices at posterior probability cut-off values of 0.7 for HC and 0.6 for MCI and AD

the matrices for MCI and AD group show the probability of each link is close to 0.5. We believe that the reason for this is because the HC group has a substantially larger sample size (120 individuals) compared to the MCI and AD groups (with 21 and 26 individuals respectively). Hence, the more complex model in Eq. (7.9) requires data with larger sample sizes, compared to the original wombling model, in order to derive meaningful W estimates.

Figure 7.14 shows the marginal posterior densities for the ROI means for 35 regions. These results resemble the independent Bayesian LME ROI estimates in Fig. 7.20, particularly for ROIs associated with early onset of AD such as the inferior, middle and superior temporal gyrus, posterior cingulate gyrus.

Wombling Cortical Thickness Estimates at the Participant Level

As described in Sect. 7.2.4.4 and discussed in Sect. 7.3.2.4, the wombling model derived participant specific estimates on all ROIs. Figures 7.15 and 7.16 shows the posterior means and 95% credible intervals (as error bars) for each participant.

APOE Wombling Results

Carriers of the Apolipoprotein (APOE) $\varepsilon4$ gene have known to be at higher risk of developing AD compared to non-carriers, hence in neuroimaging studies, it is a key biomarker to investigate. For exploration purposes, we applied the wombling model on AIBL data divided into APOE $\varepsilon4$ carrier and non-carrier groups. Figures 7.17, 7.18 and 7.19 show the cortical networks, global estimates across all ROI and participant specific rankings for key AD regions as described in Sect. 7.2.4.3.

There were no strong differences APOE $\varepsilon4$ carrier and non-carrier groups in any of the ROI. We believe the reason for this is due to APOE ε carrier and non-carrier groups comprising of participants across the entire spectrum (HC, MCI and AD), large variety of ages and many other AD biomarkers, making it difficult to assess the deterioration differences associated with the APOE ε gene. Unfortunately due to our low sample size, we did not have sufficient data to investigate more meaningful biomarker groups such as APOE $\varepsilon4$ carrier and non-carrier groups that were clinically diagnosed as HC or AD.

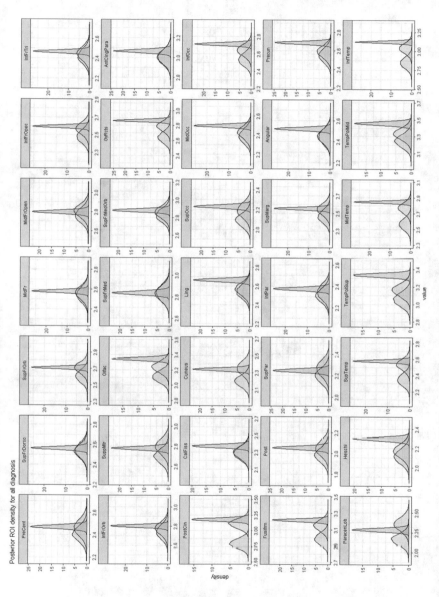

Fig. 7.14 Marginal posterior distributions for 35 ROI means, for HC (green), MCI (blue) and AD (red) groups

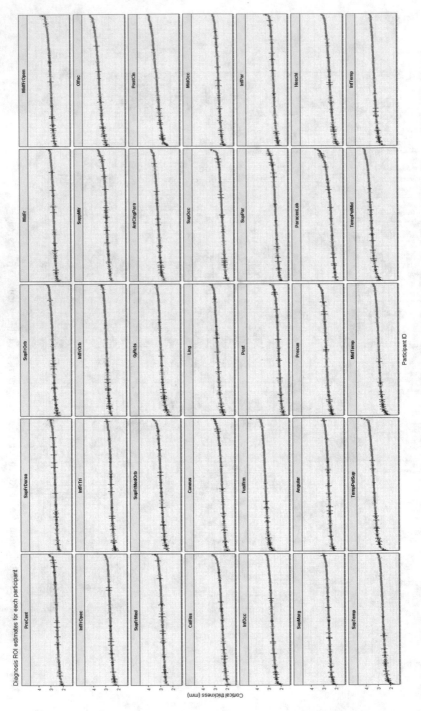

Fig. 7.15 Participant specific cortical thickness estimates for all 35 ROI for HC and AD group comparison, derived by the wombling algorithm

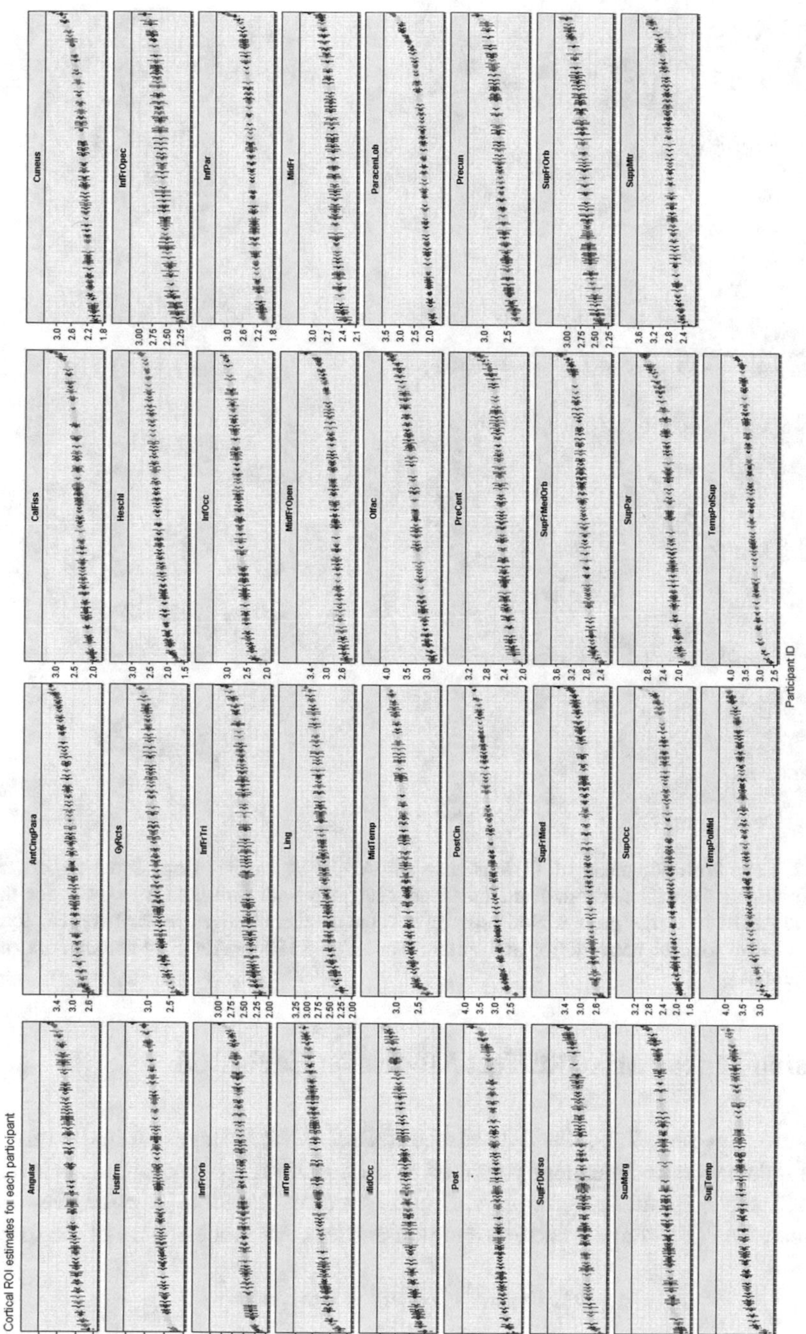

Fig. 7.16 Participant specific cortical thickness estimates for all 35 ROI for age groups A, B and C group comparison, derived by the wombling algorithm

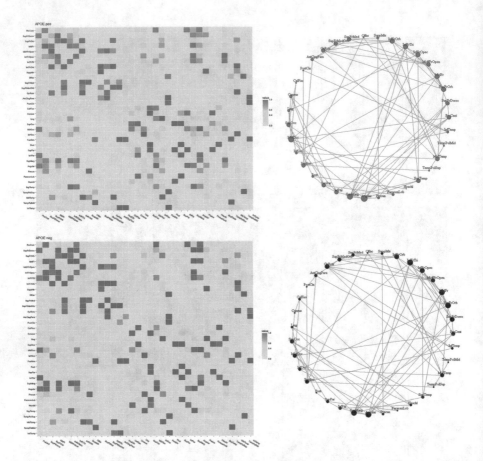

Fig. 7.17 Left: Posterior mean of W heat map for APOE $\varepsilon 4$ carriers (top) and non-carriers (bottom). Right: Cortical network from binarised heat map with threshold $\tau = 0.7$ for the respective APOE $\varepsilon 4$ carrier groups. Node size reflects the number of edges on each vertice. Total number of edges for each network (top and bottom) are 152 and 150 for APOE $\varepsilon 4$ non-carriers and carriers groups

Bayesian Linear Mixed Effect Models on Each ROI

As described in Sect. 7.2.4.4 and discussed in Sect. 7.3.2.4, Bayesian linear mixed effect models were independently applied to each ROI on groups; diagnosis levels HC, MCI and AD and age groups A, B and C. For exploration purposes we also investigated APOE $\varepsilon 4$ allele carriers and non-carriers. All models were of the form

$$y_{ij} | \sigma^2, \beta_1, \mu_{0i} \sim N(\mu_{0i} + \beta_1 x_i, \sigma^2)$$
$$\mu_{0i} | \mu_0, \sigma_0^2 \sim N(\mu_0, \sigma_0^2). \tag{7.10}$$

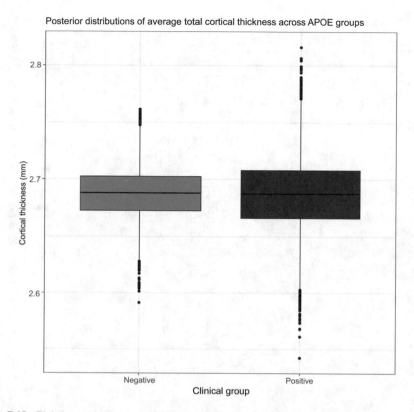

Fig. 7.18 Global posterior cortical thickness means for (red) APOE $\varepsilon4$ carriers and (green) APOE $\varepsilon4$ non-carriers

In order to make the models comparable with the wombling approach, covariate x_i is gender as described in Sect. 7.2.3.1, with $x_i = 1$ for male and 0 otherwise. The residual variance prior for σ^2 and the random effects prior, σ_0^2, is the same as discussed in Sect. 7.2.3.1. Similarly, the prior for the intercept effect μ_0 is also relatively vague with a $N(0, 10)$ distribution.

Figures 7.20, 7.21, 7.22, 7.23, 7.24 and 7.25 show the marginal posterior mean population distributions and participants ranked according to posterior means with 95% credible interval.

WAIC Results

As described in Sect. 7.2.4.4, we applied the WAIC criterion on the wombled and independent Bayesian LME models to assess model choice. Table 7.6 shows the results of the WAIC for the wombling model applied to each group, and the

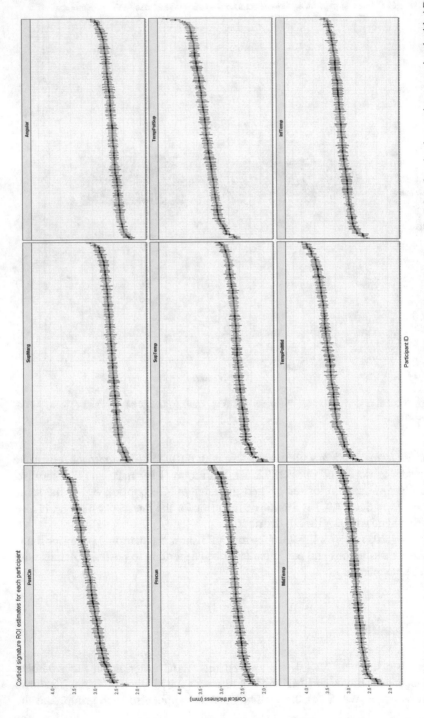

Fig. 7.19 Posterior mean and 95% credible interval on participants estimated cortical thickness (random effects $\mathbf{b}_l$), for nine key regions associated with AD cortical signature. Colour coded as follows; (red) APOE $\varepsilon 4$ carriers and (green) APOE $\varepsilon 4$ non-carriers

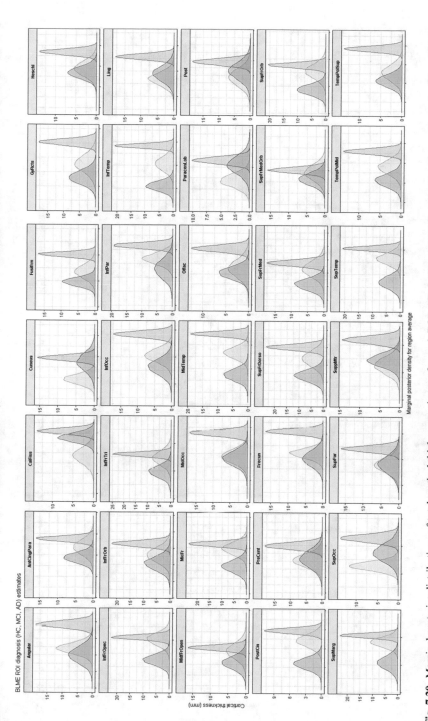

Fig. 7.20 Marginal posterior distributions of total cortical thickness estimates for each ROI. AD, MCI and HC diagnosis are dentoed by red, yellow and green posterior densities

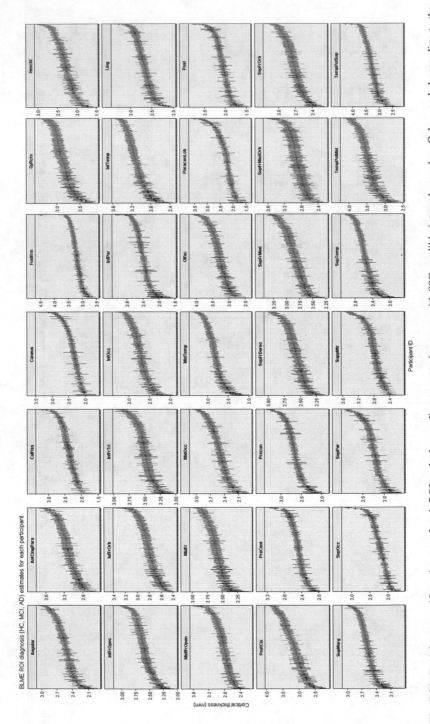

Fig. 7.21 Participant specific estimates of each ROI ranked according to posterior mean with 90% credible interval error bar. Colour coded according to the diagnosis levels HC, MCI and AD (green, yellow and red respectively)

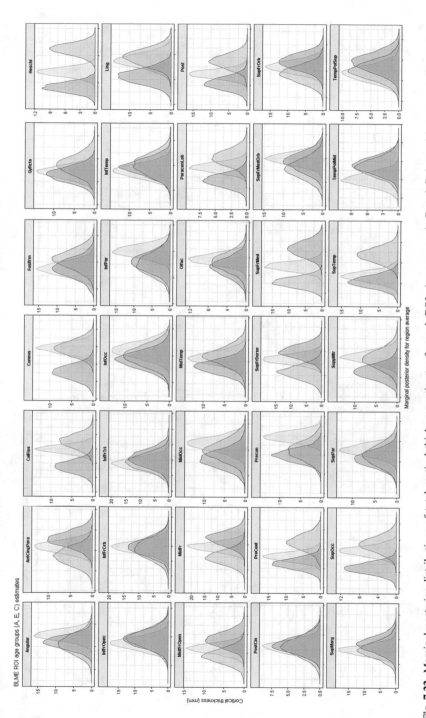

Fig. 7.22 Marginal posterior distributions of total cortical thickness estimates for each ROI. Age groups A, B and C denoted by green, yellow and green posterior densities

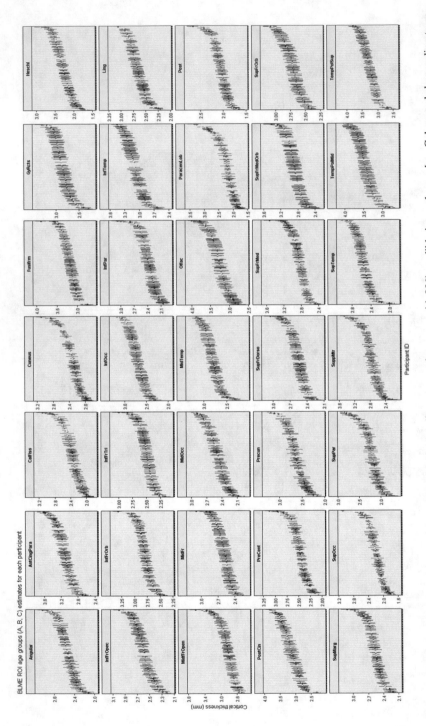

Fig. 7.23 Participant specific estimates of each ROI ranked according to posterior mean with 90% credible interval error bar. Colour coded according to age groups A, B and C (colours: green, yellow and red respectively)

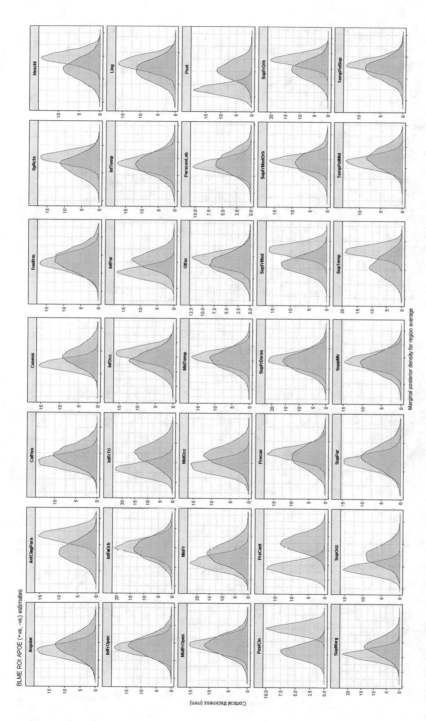

Fig. 7.24 Marginal posterior distributions of total cortical thickness estimates for each ROI. APOE ε4 carrier (red) and non-carrier (green) groups

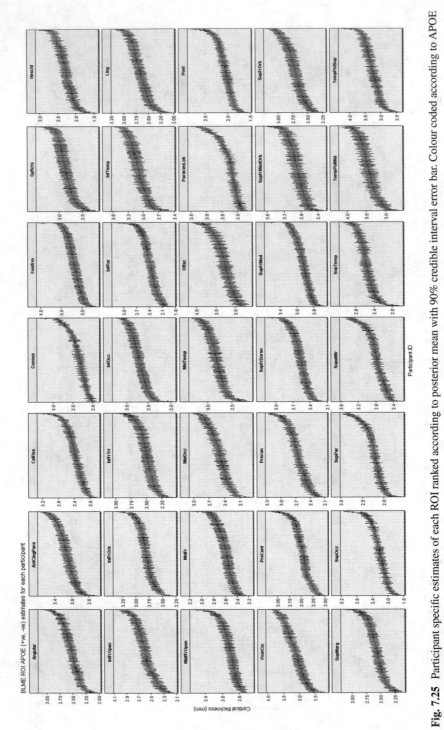

Fig. 7.25 Participant specific estimates of each ROI ranked according to posterior mean with 90% credible interval error bar. Colour coded according to APOE $\varepsilon4$ carrier (red) and non-carrier (green) groups

Table 7.6 WAIC values for diagnosis groups

Group	WAIC wombled model	WAIC LME models
HC	−33,598.26	−12,424.94
MCI	−4355.94	−1589.94
AD	−3045.20	−689.97
Age A	−6029.15	−2870.38
Age B	−13,065.38	−6051.11
Age C	−10,321.63	−4587.98

Smaller WAIC values denotes a more parsimonious model compared to the alternative, here the wombled model is preferred to the independent Bayesian LME models

combined WAIC criterion for the independent Bayesian LME analyses for each region.

Pearson Correlation Networks for Each Group

Cortical networks derived by Pearson's pairwise correlation networks for each group are shown in Figs. 7.26, 7.27, 7.28, 7.29, 7.30, and 7.31. As Pearson's pairwise networks do not accommodate the repeated measure structure of the data, we derived networks at both baseline (independent and identically distributed (IID) observations) as well as on the whole data, with repeated measures treated as IID to investigate any potential differences.

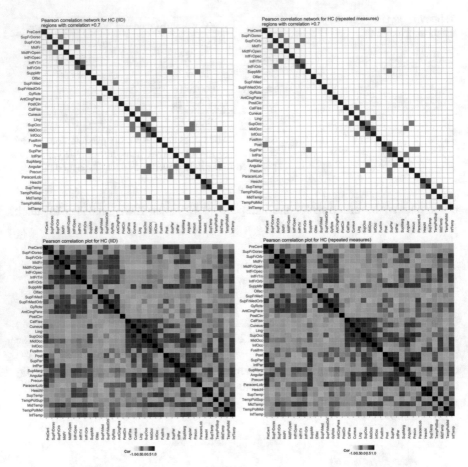

Fig. 7.26 Pearson pairwise correlation plots for baseline (left top and bottom) and repeated measures (right top and bottom) on HC diagnosis. Top: networks binarised according to threshold of $\tau = 0.7$ applied on the absolute value of each element on correlation matrices above

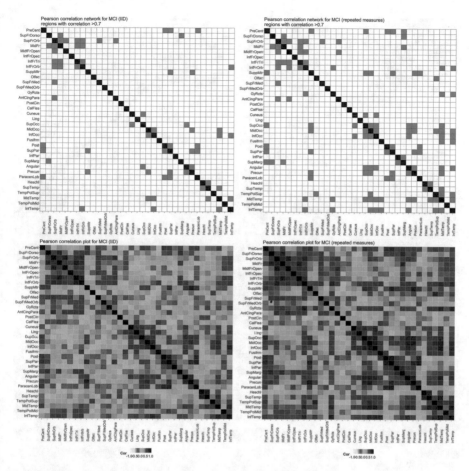

Fig. 7.27 Pearson pairwise correlation plots for baseline (left top and bottom) and repeated measures (right top and bottom) on MCI diagnosis. Top: networks binarised according to threshold of $\tau = 0.7$ applied on the absolute value of each element on correlation matrices above

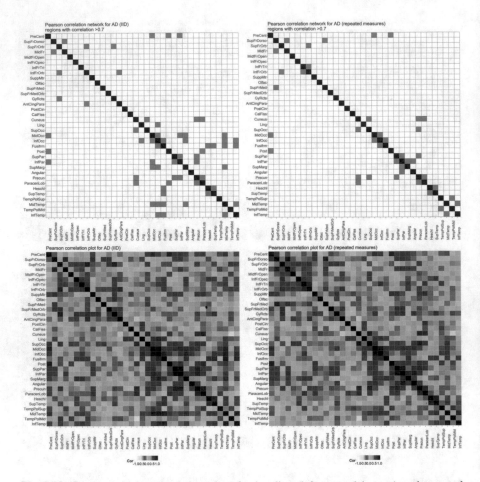

Fig. 7.28 Pearson pairwise correlation plots for baseline (left top and bottom) and repeated measures (right top and bottom) on AD diagnosis. Top: networks binarised according to threshold of $\tau = 0.7$ applied on the absolute value of each element on correlation matrices above

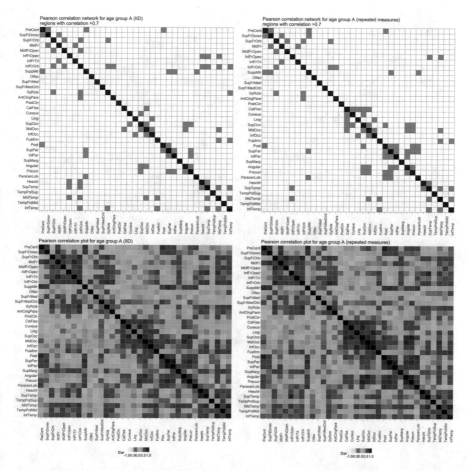

Fig. 7.29 Pearson pairwise correlation plots for baseline (left top and bottom) and repeated measures (right top and bottom) on age group A. Top: networks binarised according to threshold of $\tau = 0.7$ applied on the absolute value of each element on correlation matrices above

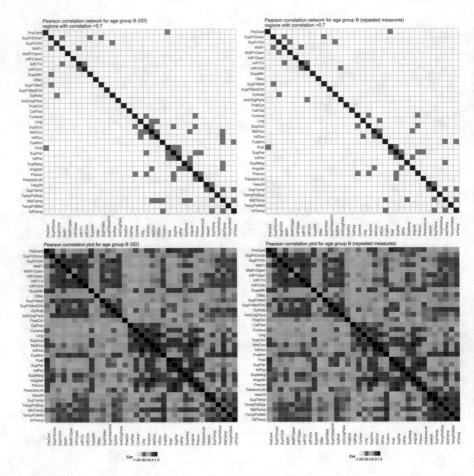

Fig. 7.30 Pearson pairwise correlation plots for baseline (left top and bottom) and repeated measures (right top and bottom) on age group B. Top: networks binarised according to threshold of $\tau = 0.7$ applied on the absolute value of each element on correlation matrices above

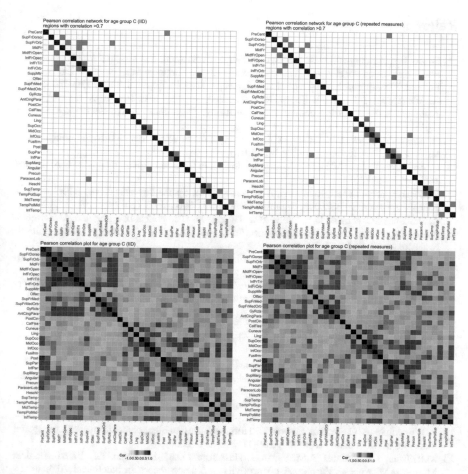

Fig. 7.31 Pearson pairwise correlation plots for baseline (left top and bottom) and repeated measures (right top and bottom) on age group C. Top: networks binarised according to threshold of $\tau = 0.7$ applied on the absolute value of each element on correlation matrices above

References

1. O. Acosta, P. Bourgeat, M.A. Zuluaga, J. Fripp, O. Salvado, S. Ourselin, A.D.N. Initiative, et al., Automated voxel-based 3D cortical thickness measurement in a combined Lagrangian–Eulerian PDE approach using partial volume maps. Med. Image Anal. **13**(5), 730–743 (2009)
2. A. Adaszewski, J. Dukart, F. Kherif, R. Frackowiak, B. Draganski, How early can we predict Alzheimer's disease using computational anatomy. Neurobiol. Aging **34**(12), 2815–2826 (2013)
3. A. Alexander-Bloch, J. N. Giedd, et al., Imaging structural co-variance between human brain regions. Nat. Rev. Neurosci. **14**(5), 322 (2013)
4. C. Anderson, D. Lee, N. Dean, Identifying clusters in Bayesian disease mapping. Biostatistics **15**(3), 457–469 (2014)
5. S.D. Babacan, M. Luessi, R. Molina, A.K. Katsaggelos, Sparse Bayesian methods for low-rank matrix estimation. IEEE Trans. Signal Process. **60**(8), 3964–3977 (2012)
6. A. Bakkour, J.C. Morris, D.A. Wolk, B.C. Dickerson, The effects of aging and Alzheimer's disease on cerebral cortical anatomy: specificity and differential relationships with cognition. NeuroImage **76**, 332–344 (2013)
7. D.S. Bassett, E.T. Bullmore, Small-world brain networks revisited. Neuroscientist (2016). https://doi.org/10.1177/1073858416667720
8. D.S. Bassett, E. Bullmore, B.A. Verchinski, V.S. Mattay, D.R. Weinberger, A. Meyer-Lindenberg, Hierarchical organization of human cortical networks in health and schizophrenia. J. Neurosci. **28**(37), 9239–9248 (2008)
9. J. Bernal-Rusiel, D.N. Greve, M. Reuter, B. Fischl, M.R. Sabuncu, Statistical analysis of longitudinal neuroimage data with Linear Mixed Effects models. NeuroImage **66**, 249–60 (2013)
10. P. Bourgeat, G. Chetelat, V. Villemagne, J. Fripp, P. Raniga, K. Pike, O. Acosta, C. Szoeke, S. Ourselin, D. Ames, et al., β-Amyloid burden in the temporal neocortex is related to hippocampal atrophy in elderly subjects without dementia. Neurology **74**(2), 121–127 (2010)
11. F.D. Bowman, Spatiotemporal models for region of interest analyses of functional neuroimaging data. J. Am. Stat. Assoc. **102**(478), 442–453 (2007)
12. F.D. Bowman, B. Caffo, S.S. Bassett, C. Kilts, A Bayesian hierarchical framework for spatial modeling of fMRI data. NeuroImage **39**(1), 146–156 (2008)
13. M.R. Brier, J.B. Thomas, A.M. Fagan, J. Hassenstab, D.M. Holtzman, T.L. Benzinger, J.C. Morris, B.M. Ances, Functional connectivity and graph theory in preclinical Alzheimer's disease. Neurobiol. Aging **35**(4), 757–768 (2014)
14. S.P. Brooks, A. Gelman, General methods for monitoring convergence of iterative simulations. J. Comput. Graph. Stat. **7**(4), 434–455 (1998)
15. E. Bullmore, O. Sporns, Complex brain networks: graph theoretical analysis of structural and functional systems. Nat. Rev. Neurosci. **10**(3), 186–198 (2009)
16. R.J. Caselli, A.C. Dueck, D. Osborne, M.N. Sabbagh, D.J. Connor, G.L. Ahern, L.C. Baxter, S.Z. Rapcsak, J. Shi, B.K. Woodruff, et al., Longitudinal modeling of age-related memory decline and the APOE ε4 effect. New Engl. J. Med. **361**(3), 255–263 (2009)
17. M.I. Cespedes, J. Fripp, J.M. McGree, C.C. Drovandi, K. Mengersen, J.D. Doecke, Comparisons of neurodegeneration over time between healthy ageing and Alzheimer's disease cohorts via Bayesian inference. BMJ Open, **7**(2), e012174 (2017)
18. Z.J. Chen, Y. He, P. Rosa-Neto, G. Gong, A.C. Evans, Age-related alterations in the modular organization of structural cortical network by using cortical thickness from MRI. NeuroImage **56**(1), 235–245 (2011)
19. S. Chen, F.D. Bowman, H.S. Mayberg, A Bayesian hierarchical framework for modeling brain connectivity for neuroimaging data. Biometrics **72**(2), 596–605 (2016). https://doi.org/10.1111/biom.12433
20. S. Chib, E. Greenberg, Understanding the Metropolis-Hastings algorithm. Am. Stat. **49**(4), 327–335 (1995)

21. B.C. Dickerson, A. Bakkour, D.H. Salat, E. Feczko, J. Pacheco, D.N. Greve, F. Grodstein, C.I. Wright, D. Blacker, H.D. Rosas, et al., The cortical signature of Alzheimer's disease: regionally specific cortical thinning relates to symptom severity in very mild to mild AD dementia and is detectable in asymptomatic amyloid-positive individuals. Cereb. Cortex **19**(3), 497–510 (2009)
22. V. Doré, V.L. Villemagne, P. Bourgeat, J. Fripp, O. Acosta, G. Chetélat, L. Zhou, R. Martins, K.A. Ellis, C.L. Masters, et al., Cross-sectional and longitudinal analysis of the relationship between Aβ deposition, cortical thickness, and memory in cognitively unimpaired individuals and in Alzheimer disease. JAMA Neurol. **70**(7), 903–911 (2013)
23. A.-T. Du, N. Schuff, J.H. Kramer, H.J. Rosen, M.L. Gorno-Tempini, K. Rankin, B.L. Miller, M.W. Weiner, Different regional patterns of cortical thinning in Alzheimer's disease and frontotemporal dementia. Brain **130**(4), 1159–1166 (2007)
24. K.A. Ellis, A.I. Bush, D. Darby, D. De Fazio, J. Foster, P. Hudson, N.T. Lautenschlager, N. Lenzo, R.N. Martins, P. Maruff, et al., The Australian Imaging, Biomarkers and Lifestyle (AIBL) study of aging: methodology and baseline characteristics of 1112 individuals recruited for a longitudinal study of Alzheimer's disease. Int. Psychogeriatr. **21**(04), 672–687 (2009)
25. Y. Fan, F. Shi, J.K. Smith, W. Lin, J.H. Gilmore, D. Shen, Brain anatomical networks in early human brain development. NeuroImage **54**(3), 1862–1871 (2011)
26. F.L. Ferreira, S. Cardoso, D. Silva, M. Guerreiro, A. de Mendonça, S.C. Madeira, Improving prognostic prediction from mild cognitive impairment to Alzheimer's disease using genetic algorithms, in *Alzheimer's Disease: Advances in Etiology, Pathogenesis and Therapeutics*, chapter 14, ed. by K. Iqbal, S.S. Sisodia, B. Winbald (Springer, New York, 2017)
27. B. Fischl, A.M. Dale, Measuring the thickness of the human cerebral cortex from magnetic resonance images. Proc. Natl. Acad. Sci. **97**(20), 11050–11055 (2000)
28. A. Gelman, J. Hill, *Data Analysis Using Regression and Multilevel/Hierarchical Models* (Cambridge University Press, Cambridge, 2006))
29. A. Gelman, J.B. Carlin, H.S. Stern, D.B. Dunson, A. Vehtari, D.B. Rubin, *Bayesian Data Analysis*, 2nd edn. (CRC Press, Boca Raton, 2013)
30. A. Gelman, J. Hwang, A. Vehtari, Understanding predictive information criteria for Bayesian models. Stat. Comput. **24**(6), 997–1016 (2014)
31. A. Goldstone, S.D. Mayhew, I. Przezdzik, R.S. Wilson, J.R. Hale, A.P. Bagshaw, Gender specific re-organization of resting-state networks in older age. Front. Aging Neurosci. **8**, 285 (2016)
32. C. Gössl, D.P. Auer, L. Fahrmeir, Bayesian spatiotemporal inference in functional magnetic resonance imaging. Biometrics **57**(2), 554–562 (2001)
33. A.R. Groves, M.A. Chappell, M.W. Woolrich, Combined spatial and non-spatial prior for inference on MRI time-series. NeuroImage **45**(3), 795–809 (2009)
34. B. Guillaume, X. Hua, P.M. Thompson, L. Waldorp, T.E. Nichols, Fast and accurate modelling of longitudinal and repeated measures neuroimaging data. NeuroImage **94**, 287–302 (2014)
35. Y. Guo, F. DuBois Bowman, C. Kilts, Predicting the brain response to treatment using a Bayesian hierarchical model with application to a study of schizophrenia. Hum. Brain Mapp. **29**(9), 1092–1109 (2008)
36. L.M. Harrison, G.G. Green, A Bayesian spatiotemporal model for very large data sets. NeuroImage **50**(3), 1126–1141 (2010)
37. Y. He, Z. Chen, A. Evans, Structural insights into aberrant topological patterns of large-scale cortical networks in Alzheimer's disease. J. Neurosci. **28**(18), 4756–4766 (2008)
38. M. Hinne, T. Heskes, M.A.J. van Gerven, Bayesian inference of whole-brain networks. 1–10 (2012). arXiv:1202.1696
39. D. Holland, R.S. Desikan, A.M. Dale, L.K. McEvoy, A.D.N. Initiative, et al., Rates of decline in Alzheimer disease decrease with age. PloS One **7**(8), e42325 (2012)
40. C.R. Jack, H.J. Wiste, S.D. Weigand, D.S. Knopman, M.M. Mielke, P. Vemuri, V. Lowe, M.L. Senjem, J.L. Gunter, D. Reyes, et al., Different definitions of neurodegeneration produce similar amyloid/neurodegeneration biomarker group findings. Brain **138**(12), 3747–3759 (2015)

41. R.J. Janssen, M. Hinne, T. Heskes, M.A.J. van Gerven, Quantifying uncertainty in brain network measures using Bayesian connectomics. Front. Comput. Neurosci. **8**, 126 (2014)
42. D. Lee, A comparison of conditional autoregressive models used in Bayesian disease mapping. Spatial Spatio-temporal Epidemiol. **2**(2), 79–89 (2011)
43. D. Lee, R. Mitchell, Boundary detection in disease mapping studies. Biostatistics **13**(3), 415–426 (2012)
44. D. Lee, R. Mitchell, Locally adaptive spatial smoothing using conditional auto-regressive models. J. R. Stat. Soc. Ser. C: Appl. Stat. **62**(4), 593–608 (2013)
45. B.G. Leroux, X. Lei, N. Breslow, Estimation of disease rates in small areas: a new mixed model for spatial dependence, in *Statistical Models in Epidemiology, the Environment, and Clinical Trials*, ed. by H.M. Elizabeth, D. Berry (Springer, New York, 2000), pp. 179–191
46. Y. Li, Y. Wang, G. Wu, F. Shi, L. Zhou, W. Lin, D. Shen, A.D.N. Initiative, et al., Discriminant analysis of longitudinal cortical thickness changes in Alzheimer's disease using dynamic and network features. Neurobiol. Aging **33**(2), 427-e15 (2012)
47. X. Li, F. Pu, Y. Fan, H. Niu, S. Li, D. Li, Age-related changes in brain structural covariance networks. Front. Hum. Neurosci. **7**, 98 (2013)
48. K. Liu, Z.L. Yu, W. Wu, Z. Gu, Y. Li, S. Nagarajan, Bayesian electromagnetic spatio-temporal imaging of extended sources with Markov Random Field and temporal basis expansion. NeuroImage **139**, 385–404 (2016)
49. H. Lu, B.P. Carlin, Bayesian areal wombling for geographical boundary analysis. Geograph. Anal. **37**(3), 265–285 (2005)
50. H. Lu, C.S. Reilly, S. Banerjee, B.P. Carlin, Bayesian areal wombling via adjacency modeling. Environ. Ecol. Stat. **14**(4), 433–452 (2007)
51. P. McCullagh, J.A. Nelder, *Generalized Linear Models*, vol. 37 (CRC Press, Boca Raton, 1989)
52. N. Metropolis, A.W. Rosenbluth, M.N. Rosenbluth, A.H. Teller, E. Teller, Equation of state calculations by fast computing machines. J. Chem. Phys. **21**(6), 1087–1092 (1953)
53. M.F. Miranda, H. Zhu, J.G. Ibrahim, Bayesian spatial transformation models with applications in neuroimaging data. Biometrics **69**(4), 1074–1083 (2013)
54. R.C. Petersen, Mild cognitive impairment as a diagnostic entity. J. Intern. Med. **256**(3), 183–194 (2004)
55. A. Pfefferbaum, T. Rohlfing, M.J. Rosenbloom, W. Chu, I.M. Colrain, E.V. Sullivan, Variation in longitudinal trajectories of regional brain volumes of healthy men and women (ages 10 to 85 years) measured with atlas-based parcellation of MRI. NeuroImage **65**, 176–193 (2013)
56. R Core Team, *R: A Language and Environment for Statistical Computing* (R Foundation for Statistical Computing, Vienna, 2015)
57. C. Robert, G. Casella, *Monte Carlo Statistical Methods*. Springer Texts in Statistics (Springer, New York, 2010)
58. M. Rubinov, O. Sporns, Complex network measures of brain connectivity: uses and interpretations. NeuroImage **52**(3), 1059–1069 (2010)
59. W.W. Seeley, R.K. Crawford, J. Zhou, B.L. Miller, M.D. Greicius, Neurodegenerative diseases target large-scale human brain networks. Neuron **62**(1), 42–52 (2009)
60. S.L. Simpson, F. Bowman, P.J. Laurienti, Analyzing complex functional brain networks: fusing statistics and network science to understand the brain. Stat. Surv. **7**, 1 (2013)
61. M.R. Sinke, R.M. Dijkhuizen, A. Caimo, C.J. Stam, W.M. Otte, Bayesian exponential random graph modeling of whole-brain structural networks across lifespan. NeuroImage **135**, 79–91 (2016)
62. A.B. Storsve, A.M. Fjell, C.K. Tammes, L.T. Westlye, K. Overbye, H.W. Aasland, K.B. Walhovd, Differential longitudinal changes in cortical thickness, surface area and volume across the adult life span: regions of accelerating and decelerating change. J. Neurosci. **34**, 8488–8498 (2014)
63. N. Tzourio-Mazoyer, B. Landeau, D. Papathanassiou, F. Crivello, O. Etard, N. Delcroix, B. Mazoyer, M. Joliot, Automated anatomical labeling of activations in SPM using a macroscopic anatomical parcellation of the MNI MRI single-subject brain. NeuroImage **15**(1), 273–289 (2002)

64. K. Van Leemput, F. Maes, D. Vandermeulen, P. Suetens, Automated model-based tissue classification of MR images of the brain. IEEE Trans. Med. Imaging **18**(10), 897–908 (1999)
65. V. Villemagne, S. Burnham, P. Bourgeat, B. Brown, K. Ellis, O. Salvado, C. Szoeke, S. Macaulay, R. Martins, P. Maruff, D. Ames, C. Rowe, C. Masters, Amyloid β deposition, neurodegeneration and cognitive decline in sporadic Alzheimer's disease. Lancet Neurol. **12**, 357–367 (2013)
66. S. Watanabe, A widely applicable Bayesian information criterion. J. Mach. Learn. Res. **14**, 867–897 (2013)
67. M.W. Weiner, D.P. Veitch, P.S. Aisen, L.A. Beckett, N.J. Cairns, R.C. Green, D. Harvey, C.R. Jack, W. Jagust, E. Liu, et al., The Alzheimer's Disease Neuroimaging Initiative: a review of papers published since its inception. Alzheimer's Dementia **9**(5), e111–e194 (2013)
68. M.W. Woolrich, T.E. Behrens, C.F. Beckmann, M. Jenkinson, S.M. Smith, Multilevel linear modelling for FMRI group analysis using Bayesian inference. NeuroImage **21**(4), 1732–1747 (2004)
69. L. Xu, T.D. Johnson, T.E. Nichols, D.E. Nee, Modeling inter-subject variability in fMRI activation location: a Bayesian hierarchical spatial model. Biometrics **65**(4), 1041–1051 (2009)
70. W. Xue, F.D. Bowman, A.V. Pileggi, A.R. Mayer, A multimodal approach for determining brain networks by jointly modeling functional and structural connectivity. Front. Comput. Neurosci. **9**, 22 (2015)
71. Z. Yao, Y. Zhang, L. Lin, Y. Zhou, C. Xu, T. Jiang, A.D.N. Initiative, et al., Abnormal cortical networks in mild cognitive impairment and Alzheimer's disease. PLoS Comput. Biol. **6**(11), e1001006 (2010)
72. G. Ziegler, W.D. Penny, G.R. Ridgway, S. Ourselin, K.J. Friston, A.D.N. Initiative, et al., Estimating anatomical trajectories with Bayesian mixed-effects modeling. NeuroImage **121**, 51–68 (2015)

Chapter 8
Bayesian Spike Sorting: Parametric and Nonparametric Multivariate Gaussian Mixture Models

Nicole White, Zoé van Havre, Judith Rousseau, and Kerrie L. Mengersen

Abstract The analysis of action potentials is an important task in neuroscience research, which aims to characterise neural activity under different subject conditions. The classification of action potentials, or "spike sorting", can be formulated as an unsupervised clustering problem, and latent variable models such as mixture models are often used. In this chapter, we compare the performance of two mixture-based approaches when applied to spike sorting: the Overfitted Finite Mixture model (OFM) and the Dirichlet Process Mixture model (DPM). Both of these models can be used to cluster multivariate data when the number of clusters is unknown, however differences in model specification and assumptions may affect resulting statistical inference. Using real datasets obtained from extracellular recordings of the brain, model outputs are compared with respect to the number of identified clusters and classification uncertainty, with the intent of providing guidance on their application in practice.

Keywords Mixture model · Dirichlet process · Classification · Spike sorting

8.1 Introduction

Extracellular recordings are a form of electrophysiological data that allows real time monitoring of multiple neurons *in vivo*. Data collection focuses on the measurement of action potentials or "spikes", which characterise local neural activity at a

N. White (✉)
Institute for Health and Biomedical Innovation, Queensland University of Technology, Brisbane, QLD, Australia
e-mail: nm.white@qut.edu.au

Z. van Havre · K. L. Mengersen
School of Mathematical Sciences, Queensland University of Technology, Brisbane, QLD, Australia

J. Rousseau
Department of Statistics, University of Oxford, Oxford, UK

© The Editor(s) (if applicable) and The Author(s), under exclusive
licence to Springer Nature Switzerland AG 2020
K. L. Mengersen et al. (eds.), *Case Studies in Applied Bayesian Data Science*,
Lecture Notes in Mathematics 2259, https://doi.org/10.1007/978-3-030-42553-1_8

215

given point in time. Analysis of these data aims to estimate both the number of active source neurons present and their relative frequency. Comparing the results of analysis across different subject conditions can therefore provide insight into changes in neural activity, for example, in different regions of the brain or in response to various stimuli.

The analysis of extracellular recordings consists of two main stages: (1) spike detection, and (2) the assignment of detected spikes to source neurons. This chapter focuses on the assignment stage, also known as *spike sorting* [1, 2]. A common assumption underpinning spike sorting methods is that different neurons generate action potentials with a characteristic, repeatable shape. Spike sorting can therefore be viewed as an unsupervised clustering problem where spikes with similar features are grouped together, for example, based on summary statistics [3, 4] or low-dimensional transformations of the data, such as wavelet transforms or principal components analysis [1, 5].

Mixture models offer a general solution for unsupervised clustering and are a popular tool for spike sorting, including cases where the number of source neurons (clusters) is unknown. Applications of mixture models to spike sorting have included finite mixtures of Gaussian [2, 6] and t-distributions [7], mixtures of factor analysers [8], Reversible Jump Markov chain Monte Carlo (RJMCMC) [9], and time-dependent mixtures to account for non-stationarity in waveforms [10, 11]. Nonparametric mixture models based on the Dirichlet Process (DP) have also been proposed [12, 13].

Different mixture-based approaches all aim to determine the optimal clustering of a dataset. However, differences in model specification can impact subsequent inferences, for example, the number of clusters identified and/or classification uncertainty for individual observations. This chapter aims to provide insight into this issue by comparing two mixture-based approaches to spike sorting. Both are formulated within the Bayesian framework and represent parametric and nonparametric approaches to mixture modelling. The first model is a finite mixture of multivariate Gaussian distributions, applying methodology proposed by [14]. This model initially overfits the number of clusters expected in the data. The prior distribution for the mixture model weights is then specified in a way that encourages excess clusters in the posterior distribution to have negligible weight [15]. The second model considers a nonparametric approach to mixture estimation which uses the DP as a prior over unknown mixture components. Clustering behaviour induced by properties of the DP is then used to estimate the most likely partition of the data.

Outcomes from each approach are compared with respect to the number of clusters identified, the predicted classification of individuals spikes, and the features of identified clusters.

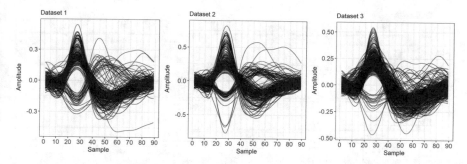

Fig. 8.1 Sampled spikes from three extracellular recordings. Each spike is represented by 89 samples, equivalent to 1 ms of recording. Datasets varied by sample size (L to R): $n = 192, 211, 348$

8.2   Data

Selected approaches were applied to data from three independent extracellular recordings of the brain (Fig. 8.1). Each spike was represented by a waveform consisting of 89 samples, corresponding to 1 millisecond of recording time. The number of detected spikes for analysis was equal to 192, 211 and 348 for Datasets 1, 2 and 3, respectively.

Dimension reduction was performed on sampled waveforms for each dataset in Fig. 8.1 using a robust version of Principal Components Analysis (PCA) [16]. This method was chosen to lessen the influence of outliers on the estimation of principal components. The first four principal components were used as inputs into each mixture model (Fig. 8.2), which explained 83% (Dataset 1), 91% (Dataset 2), and 85% (Dataset 3) of total variation in sampled waveforms.

8.3   Methodology

In this section, key features of each mixture modelling approach are outlined. Common to both approaches is the problem of classifying n spikes into K clusters, where K is *a priori* unknown. Individual spikes in each model are represented by a multivariate vector $\mathbf{y}_i = \{y_{i1}, \ldots, y_{ir}\}$, containing r measurements for spike i.

For the data described in Sect. 8.2, $\mathbf{y}_i$ is assumed to follow a Multivariate Gaussian distribution with mean $\boldsymbol{\mu}_k = [\mu_{1k}, \ldots, \mu_{rk}]$ and variance-covariance matrix Σ_k, $1 \leq k \leq K$. Conditional on assignment to cluster k, the likelihood for $\mathbf{y}_i$ is,

$$p\left(\mathbf{y}_i | z_i = k, \boldsymbol{\theta}_k\right) = N_r\left(\boldsymbol{\mu}_k, \Sigma_k\right), \tag{8.1}$$

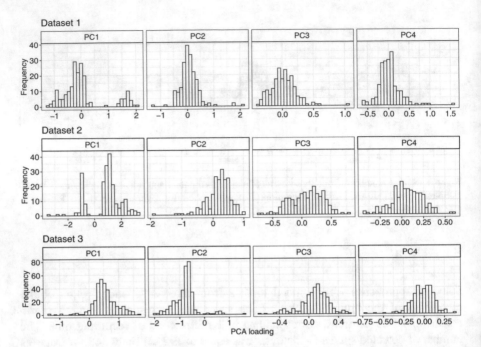

Fig. 8.2 Distribution of the first four principal components of each original dataset. Each row represents a dataset (Dataset 1, 2, 3) and each column represented a principal component (PC1, PC2, PC3, PC4)

with unknown parameters $\theta_k = (\mu_k, \Sigma_k)$. For each cluster, the joint prior distribution for θ_k takes the form:

$$p(\theta_k) = p(\mu_k | \Sigma_k) \, p(\Sigma_k) \qquad (8.2)$$

with

$$p(\mu_k | \Sigma_k) = N_r\left(\mathbf{b}_0, \frac{\Sigma_k}{N_0}\right)$$
$$p(\Sigma_k) = IW(c_0, \mathbf{C}_0). \qquad (8.3)$$

The assignment each spike to available clusters is inferred using a discrete latent variable z_i, where $z_i = k$ if spike i is assigned to cluster k. The inclusion of z_i is a form of data augmentation [17], and is required for sampling from the posterior distribution.

All models were estimated using Markov chain Monte Carlo (MCMC), with details provided Sects. 8.3.1 and 8.3.2. For analyses presented in Sect. 8.4, the following values were chosen for the hyperparameters: $b_0 = \bar{y}$, $N_0 = 0.01$, $c_0 = 5$ and $C_0 = 0.75\text{cov}(y)$. These values were chosen to reflect a plausible range of values for each parameter, whilst remaining relatively non-informative. Other hyperparameter choices for multivariate Gaussian mixture models are discussed in [18].

8.3.1 Overfitted Finite Mixture Model (OFM)

The first approach involves fitting a finite mixture model where the number of clusters is set to be greater than the number of clusters expected in the data. We refer to this approach as the Overfitted Finite Mixture model (OFM) [14]. Assuming $K^* > K$ clusters are fitted to the data, the likelihood of $y = \{y_1, \ldots, y_n\}$ under the OFM is,

$$p(y|\theta, \pi) = \prod_{i=1}^{n} \sum_{k=1}^{K^*} \pi_k N_r\left(\mu_k, \Sigma_k\right), \tag{8.4}$$

where $\pi_k = Pr\,(z_i = k)$, is the prior probability of a randomly selected observation being assigned to cluster k. Collectively, $\pi = \{\pi_1, \ldots, \pi_{K^*}\}$ represent the mixture model weights and are subject to the constraint $\sum_{k=1}^{K^*} \pi_k = 1$.

Under the OFM, the prior distribution for z_i given π is Multinomial,

$$z_i|\pi \sim MN\,(1; \pi_1, \ldots, \pi_{K^*}), \tag{8.5}$$

which allows $z = \{z_1, \ldots, z_n\}$ to be sampled at each MCMC iteration via the posterior probabilities of cluster membership:

$$p(z_i = k|y_i, \theta) = \frac{\pi_k N_r(\mu_k, \Sigma_k)}{\sum_{l=1}^{K} \pi_l N_r(\mu_l, \Sigma_l)} \tag{8.6}$$

$$\propto \pi_k N_r\left(\mu_k, \Sigma_k\right). \tag{8.7}$$

The defining feature of the OFM is the choice of prior distribution for the mixture model weights. As per the specification of a finite mixture model, weights are assumed to follow a Dirichlet distribution,

$$(\pi_1, \ldots, \pi_{K^*}) \sim D(\alpha_1, \ldots, \alpha_{K^*}), \tag{8.8}$$

which is characterised by the hyperparameters $\alpha_1, \ldots, \alpha_{K^*}$. In the absence of prior information, it is common to set these hyperparameters to a common value; i.e. $\alpha_1 = \cdots = \alpha_{K^*} = \gamma$. Building on results by [15], the OFM chooses an appropriate value for γ that results in weights for excess components $\{k = K + 1, \ldots, K^*\}$

being shrunk towards zero. When fitted to the observed data, the number of unique values of $\mathbf{z}$ is an estimate of the true number of clusters, K.

The proposed methodology was recently applied by [14] for the case of univariate Gaussian distributions. A key feature of the methodology was the use of prior tempering on the hyperparameter γ. Briefly, a ladder of T values $\{\gamma^{(1)}, \ldots, \gamma^{(T)}\}$ was created, where each element was chosen *a priori* to promote emptying behaviour, based on the results of [15]. The MCMC algorithm was implemented in parallel in combination with Gibbs sampling steps for the remaining model parameters. Code used to implement the MCMC algorithm for the OFM model presented in this chapter is available online (https://github.com/zoevanhavre/Zmix_devVersion2).

8.3.2   Dirichlet Process Mixture Model (DPM)

The second approach considers a nonparametric alternative to mixture modelling by using the Dirichlet Process (DP) as a prior over unknown mixture components. The DP is a stochastic process which is defined as a distribution over probability measures; i.e. a single draw from the DP is itself a distribution [19]. For a measureable space Θ, the data generating process for $\mathbf{y}_i$ under the DP is,

$$
\begin{aligned}
\mathbf{y}_i | \boldsymbol{\theta}_i &\sim \boldsymbol{\theta}_i \\
\boldsymbol{\theta}_i | G &\sim G \\
G &\sim DP\,(mG_0)\,.
\end{aligned} \tag{8.9}
$$

The random probability measure G follows a DP defined by a base distribution, G_0, and a concentration parameter $m > 0$. G_0 is interpreted as the mean of the DP, and is assigned as suitable distribution according to the form of $\boldsymbol{\theta}_i$.

Under the DP, draws for multiple $\boldsymbol{\theta}_i$ have a non-zero probability of taking the same value. This discreteness property induces clustering of the observed data, which can be seen in different formulations of the DP. Under the stick-breaking construction [20], G is replaced with an infinite weighted sum of point masses:

$$
\begin{aligned}
G &= \sum_{k=1}^{\infty} \pi_k \delta_{\boldsymbol{\theta}_k} \\
\pi_k &= v_k \prod_{l<k} (1 - v_l) \\
v_k &\sim Beta\,(1, m) \\
\theta_k | G_0 &\sim G_0
\end{aligned} \tag{8.10}
$$

where $G_0 = p\,(\boldsymbol{\theta}_k)$ and $\delta_{\boldsymbol{\theta}_k}$ denotes a Dirac mass at $\boldsymbol{\theta}_k$. The term 'stick-breaking' refers to the analogy that the weights $\pi_1, \pi_2, \ldots$ represent portions of a stick with total length equal to 1. Conditional on preceding clusters, each π_k is a randomly drawn proportion of stick length remaining so that $\sum_{k=1}^{\infty} \pi_k = 1$. For this reason, the DPM is often referred to as an infinite mixture model [19].

An alternative construction of the DP is the Polya Urn scheme [21] or Chinese restaurant process. Under this construction, G in integrated out, resulting in the following prior predictive distribution distribution for θ_i,

$$\theta_i | \theta_{i-1}, \ldots, \theta_1, m, G_0 \sim \frac{mG_0}{m+i-1} + \sum_{k=1}^{K-1} \frac{N_k \delta_{\theta_k}}{m+i-1} \tag{8.11}$$

or, in terms of z_i,

$$p(z_i = k | z_1, \ldots, z_{i-1}, m) = \begin{cases} \frac{N_k}{i-1+m} & 1 \le k \le K \\ \frac{m}{i-1+m} & k = K+1. \end{cases} \tag{8.12}$$

where N_k is the number of observations already assigned to cluster k. The DPM therefore assumes that each observation has a probability of being assigned to an existing cluster $(1, \ldots, K)$, or representing a new cluster $(K+1)$.

The DPM includes a additional concentration parameter, m, which influences the level of clustering in the data. For example, under the stick-breaking construction in Eq. (8.10), m influences draws for the stick-breaking weights, $v_1, v_2, \ldots$ which, in turn, are used to compute the mixture weights. This parameter can be treated as an unknown parameter in the DPM; in this chapter, we assume $m \sim \Gamma(1, 1)$.

For results presented in Sect. 8.4, DPM models were estimated using slice sampling [22]. This algorithm is based on the stick-breaking construction (8.10) and involves a modified version of Eq. (8.6) to account for an unspecified number of clusters. Uniform auxiliary variables, $u_i \sim U(0, \pi_{z_i})$, based on current values for the mixture weights are introduced to sample each z_i. Additional clusters are proposed until the condition

$$\sum_{k=1}^{K^*} \pi_i > 1 - \min\{u_1, \ldots, u_n\} \tag{8.13}$$

is met, with K^* being the number of clusters sampled for the current MCMC iteration. R code to implement the DPM slice sampler is available online (https:// github.com/nicolemwhite/spike_sorting_DPM).

8.3.3 Comparing Spike Sorting Solutions

For each dataset in Fig. 8.2, OFM and DPM model outputs were compared to determine the effects of model specification on the estimated number of clusters and classification outcomes.

Number of Clusters The number of non-empty clusters was recorded at the end of each MCMC iteration, as an estimate of the true number of clusters. The resulting distribution of K over all MCMC iteration provided an indication of the most likely number of clusters and associated uncertainty.

Optimal Classification Using MCMC samples for $\mathbf{z}$, pairwise posterior probabilities were calculated to infer the optimal partition of each dataset. For each pair of observations i and i', the posterior pairwise probability $Pr(z_i = z_{i'}|\mathbf{y})$ was calculated as the proportion of MCMC iterations where i and i' were assigned to the same cluster, irrespective of the value of k. A benefit of using these probabilities is that it avoids the need to correct for label switching [23]. The resulting $n \times n$ matrix of probabilities was then used to determine the maximum a posteriori (MAP) estimate of $\mathbf{z}$ [24]. In this chapter, optimal partitions under each DPM were estimated using the Posterior Expected Rand (PEAR) index proposed by [25].

Modelling results were based on 20,000 MCMC iterations, following an initial burn-in phase of 20,000 iterations. OFM estimation assumed an initial estimate of 10 clusters and the proposed tempering algorithm was implemented using $\gamma = 2^{\{-32,-16,-8,-4,-2,0,2,4\}}$. MCMC sampling was further initialised by applying the k-means clustering algorithm to each dataset with $k = 10$.

8.4   Results

Differences in the estimated number of clusters were observed between models, with DPM model outcomes subject to greater posterior uncertainty (Fig. 8.3). Across all datasets, fitted OFM models converged to 4 clusters and showed little to

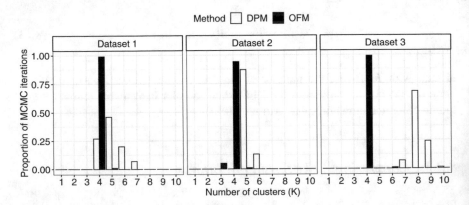

Fig. 8.3 Posterior distributions of the estimated number of clusters in Datasets 1, 2 and 3. Distributions were based on the MCMC output for the DPM (white) and OFM (black)

no support for other values of K. Uncertainty in the number of clusters among DPM models was greatest for Dataset 1, which inferred between 4 and 7 clusters with similar support across MCMC iterations. Discrepancies in the most likely number of clusters were largest for Dataset 3, with 63% of MCMC iterations proposing 8 clusters under the DPM model.

The visualisation of pairwise posterior probabilities suggested that the classification of spikes in Datasets 1 and 2 was robust to the choice of mixture model, despite evidence of differences in the true value of K (Fig. 8.4). Corresponding MAP estimates for $\mathbf{z}$ showed that the optimal clustering based on DPM models included an additional cluster, however in each case this cluster only contained a single observation (Table 8.1). Differences in pairwise posterior probabilities between models fitted to Dataset 3 were more pronounced, and were associated with a sparser clustering of spikes under the DPM model. However, additional clusters predicted by this model also had relatively low weights, representing between 0.3% and 4.3% of identified spikes.

The projection of optimal classifications onto the original data in Fig. 8.1 provided further insight into additional clusters generated under each DPM model (Fig. 8.5). For Datasets 1 and 2, the assignment of waveforms to Clusters 1, 2 and 3 was generally consistent under both approaches. Underlying spike shapes across these clusters were clearly defined, and were distinguished from one another based on minimum and maximum amplitudes. Defining features for Cluster 4 under each OFM model were less clear, and appeared to represent outlying observations; spike sorting solutions under corresponding DPM models instead attributed these observations to multiple clusters.

The assignment of outliers to singleton clusters was also observed for Dataset 3, however further inconsistencies between models indicated greater sensitivity in DPM parameter estimates. For example, spikes assigned to Cluster 3 of the OFM model varied substantially with respect to maximum amplitude. Results from the corresponding DPM model represented the same spikes by 2 smaller clusters with different maximum amplitudes.

8.5 Discussion

Using the example of spike sorting, this chapter has compared two popular approaches to mixture modelling, to assess the effect of model specification on statistical inference. Both methods represented the observed data as a mixture of multivariate Gaussian distributions and assumed that true number of clusters was unknown *a priori*.

Differences in model specification affected the estimation of K, with fitted DPM models associated with greater numbers of clusters. This outcome can be attributed to the properties of the DP when used as a prior distribution over mixture components. Unlike the OFM which assumes an upper bound on K, the DP prior assumes that observations can either be assigned to an existing cluster or be

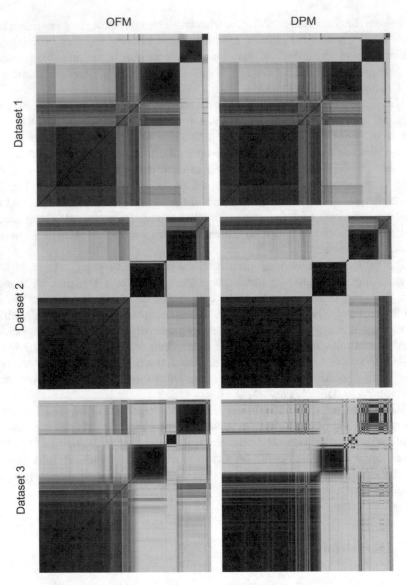

Fig. 8.4 Pairwise posterior similarity matrices for Datasets 1–3, Pairwise posterior similarity matrices for Datasets 1–3, based on MCMC output from the OFM (left column) and DPM (right column). Pairwise posterior probabilities range from 0 (light grey) to 1 (black)

Table 8.1 Frequencies of cluster membership, as determined by the optimal partition under each OFM and DPM model

Model	Dataset 1 (n = 192)			Dataset 2 (n = 211)			Dataset 3 (n = 349)		
	Cluster	Count	%	Cluster	Count	%	Cluster	Count	%
OFM	1	110	57	1	125	59	1	176	50
	2	48	25	2	44	21	2	80	23
	3	25	13	3	40	19	3	72	21
	4	9	5	4	2	1	4	20	6
DPM	1	108	56	1	127	60	1	200	57
	2	50	26	2	42	20	2	51	15
	3	25	13	3	39	18	3	38	11
	4	5	3	4	2	1	4	31	9
	5	4	2	5	1	1	5	15	4
	6	–	–	6	–	–	6	10	3
	7	–	–	7	–	–	7	2	<1
	8	–	–	8	–	–	8	1	<1

Inferred clusters under both models are labelled in decreasing order by frequency

associated with the generation of a new cluster. For results presented in Sect. 8.4, this behaviour led to the generation of additional clusters, however in most cases, these represented a single observation. In contrast, OFM models promoted a parsimonious approach to clustering, whereby outlying observations were allocated to the same cluster. This outcome can be attributed to the prior distribution specified for the unknown mixture weights, as it strongly discourages the posterior from assigning weight to clusters with limited support from the observed data. When applied to spike sorting, small clusters inferred under either approach should therefore be interpreted with care, as these are likely to represent noise as opposed to distinct source neurons.

Optimal spike sorting solutions proposed by OFM and DPM models were similar among clusters with larger weights, and performed well in capturing different waveform shapes. However, greater classification uncertainty under the DPM model reflected potential sensitivity in parameter estimation of multivariate Gaussian distributions. Whilst not considered in this chapter, the use of alternative distributions such as the multivariate-t distribution may help to address this sensitivity. Future studies in this area should therefore consider the effects of model misspecification on the performance of different mixture-based approaches.

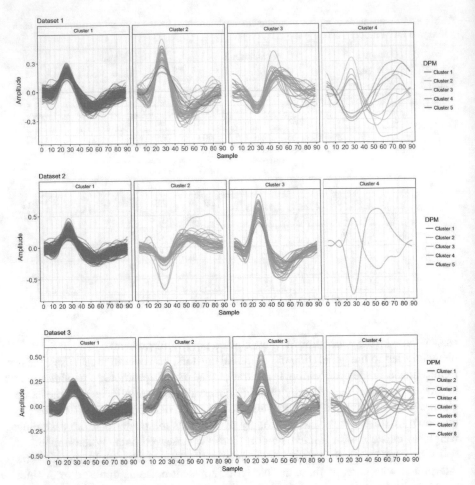

Fig. 8.5 Optimal classifications for Datasets 1, 2 and 3 based on MAP estimates produced by OFM and DPM model. For each dataset, spikes are clustered according to the OFM model. Within each OFM cluster, individual spikes are coloured based on their corresponding classifcation under the DPM model

References

1. M.S. Lewicki, A review of methods for spike sorting: the detection and classification of neural action potentials. Network **9**, R53–R78 (1998)
2. M. Sahani, Latent Variable Models for Neural Data Analysis. PhD Thesis, California Institute of Technology, Pasadena (1999)
3. M. Delescluse, C. Pouzat, Efficient spike-sorting of multi-state neurons using inter-spike intervals information. J. Neurosci. Methods **150**, 16–29 (2006)
4. C. Pouzat, M. Delescluse, P. Viot, J. Diebolt, Improved spike-sorting by modeling firing statistics and burst-dependent spike amplitude attenuation: a Markov chain Monte Carlo approach. J. Neurophysiol. **91**, 2910–2928 (2004)

5. J.C. Letelier, P.P. Weber, Spike sorting based on discrete wavelet transform coefficients. J. Neurosci. Methods **101**, 93–106 (2000)
6. E. Wood, M. Fellows, J.R. Donoghue, M.J. Black, Automatic spike sorting for neural decoding, in *The 26th Annual International Conference of the IEEE Engineering in Medicine and Biology Society*, vol. 2 (2004), pp. 4009–4012
7. S. Shoham, M.R. Fellows, R. Normann, Robust, automatic spike sorting using mixtures of multivariate t-distributions. J. Neurosci. Methods **127**, 111–122 (2003)
8. D. Görür, C.E. Rasmussen, A.S. Tolias, F. Sinz, N.K. Logothetis, Modelling spikes with mixtures of factor analysers, in *Joint Pattern Recognition Symposium* (Springer, Berlin, Heidelberg, 2004), pp. 391–398
9. D.P. Nguyen, L.M. Frank, E.N. Brown, An application of reversible-jump Markov chain Monte Carlo to spike classification of multi-unit extracellular recordings. Network **14**, 61–82 (2003)
10. A. Bar-Hillel, A. Spiro, E. Stark, Spike sorting: Bayesian clustering of non-stationary data. J. Neurosci. Methods **157**, 303–316 (2006)
11. A. Calabrese, L. Paninski, Kalman filter mixture model for spike sorting of non-stationary data. J. Neurosci. Methods **196**, 159–169 (2011)
12. F. Wood, M.J. Black, A non-parametric Bayesian approach to spike sorting. J. Neurosci. Methods. **173**, 1–12 (2008)
13. J. Gasthaus, F. Wood, D. Gorur, Y.W. Teh, Dependent Dirichlet process spike sorting, in *Advances in Neural Information Processing Systems* (2009), pp. 497–504
14. Z. van Havre, N. White, J. Rousseau, K. Mengersen, Overfitting Bayesian mixture models with an unknown number of components. PLoS One. **10**, e0131739 (2015)
15. J. Rousseau, K. Mengersen, Asymptotic behaviour of the posterior distribution in overfitted mixture models. J. R. Stat. Soc. (Ser. B) **73** 689–710 (2011)
16. M. Hubert, P.J. Rousseeuw, B.K. Vanden, ROBPCA: a new approach to robust principal component analysis. Technometrics **47**, 64–79 (2005)
17. M.A. Tanner, W.H. Wong, The calculation of posterior distributions by data augmentation. J. Am. Stat. Assoc. **82**, 528–540 (1987)
18. S. Frühwirth-Schnatter, *Finite Mixture and Markov Switching Models* (Springer, New York, 2006)
19. Y.W. Teh, Dirichlet process, in *Encyclopedia of Machine Learning*, ed. by C. Sammut, G.I. Webb (Springer, Boston, 2011)
20. J. Sethuramna, A constructive definition of Dirichlet priors. Stat. Sin. **4**, 639–650 (1994)
21. D. Blackwell, J.B. MacQueen, Ferguson distributions via polya urn schemes. Ann. Stat. **1**, 353–355 (1973)
22. S. Walker, Sampling the Dirichlet mixture model with slices. Commun. Stat. Simul. Comput. **36**, 45–54 (2007)
23. M. Stephens, Dealing with label switching in mixture models. J. R. Stat. Soc. (Ser. B) **62**, 795–809 (2000)
24. M. Medvedovic, S. Sivaganesan, Bayesian infinite mixture model based clustering of gene expression profiles. Bioinformatics **18**, 1194–1206 (2002)
25. A. Fritsch, K. Ickstadt, Improved criteria for clustering based on the posterior similarity matrix. Bayesian Anal. **4**, 367–392 (2009)

Chapter 9
Spatio-Temporal Analysis of Dengue Fever in Makassar Indonesia: A Comparison of Models Based on CARBayes

Aswi Aswi, Susanna Cramb, Wenbiao Hu, Gentry White, and Kerrie L. Mengersen

Abstract **Background:** Dengue fever is one of the world's most important vector-borne diseases and it is still a major public health problem in the Asia-Pacific region including Indonesia. Makassar is one of the major cities in Indonesia where the incidence of dengue fever is still quite high. Since dengue cases vary between areas and over time, these spatial and temporal components should be taken into consideration. However, unlike many other spatio-temporal contexts, Makassar is comprised of only a small number of areas and data are available over a relatively short timeframe. The aim of this paper is to better understand the spatial and temporal patterns of dengue incidence in Makassar, Indonesia by comparing the performance of six existing spatio-temporal models, taking into account these specific data characteristics (small number of areas and limited small number of time periods) and to select the best model for Makassar dengue dataset.

Methods: Six different Bayesian spatio-temporal conditional autoregressive (ST CAR) models were compared in the context of a substantive case study, namely annual dengue fever incidence in 14 geographic areas of Makassar, Indonesia, during 2002–2015. The candidate models included linear, ANOVA, separate spatial, autoregressive (AR), adaptive and localised approaches. The models were

A. Aswi · G. White · K. L. Mengersen (✉)
ARC Centre of Excellence for Mathematical and Statistical Frontiers, Queensland University of Technology, Brisbane, QLD, Australia
e-mail: k.mengersen@qut.edu.au

S. Cramb
ARC Centre of Excellence for Mathematical and Statistical Frontiers, Queensland University of Technology, Brisbane, QLD, Australia

Cancer Council Queensland, Brisbane, QLD, Australia

W. Hu
School of Public Health and Social Work, Queensland University of Technology, Brisbane, QLD, Australia

K. L. Mengersen et al. (eds.), *Case Studies in Applied Bayesian Data Science*, Lecture Notes in Mathematics 2259, https://doi.org/10.1007/978-3-030-42553-1_9

implemented using CARBayesST and the goodness of fit was compared using the Deviance Information Criterion (DIC) and Watanabe-Akaike Information Criterion (WAIC).

Results: The six models performed differently in the context of this case study. Among the six models, the spatio-temporal conditional autoregressive localised model had a much better fit than other options in terms of DIC, while the conditional autoregressive model with separate spatial and temporal components performed worst. However, the spatio-temporal CAR AR had a much better fit than other models in terms of WAIC. The different performance of the models may have been influenced by the small number of areas.

Conclusion: Different spatio-temporal models appeared to have a large impact on results. Careful selection of a range of spatio-temporal models is important for assessing the spatial and temporal patterns of dengue fever, especially in a context characterised by relatively few spatial areas and limited time periods.

Keywords Bayesian · Conditional autoregressive priors · CARBayesST · Spatio-temporal models

9.1   Introduction

Despite concerted efforts worldwide, dengue fever remains a serious health problem in the Asia-Pacific region including Indonesia. Makassar, the gateway to eastern Indonesia, is one of the major cities in Indonesia where the incidence of dengue fever is still quite high. However, there is substantial variation in incidence between districts and over time. Although there is strong interest in developing statistical models to estimate and predict dengue incidence, such models need to take these spatial and temporal components into account. Consideration of only the spatial component of disease can identify regions with low or high risk, but not capture anything about temporal variation of risk which is equally crucial. Similarly, focusing only on temporal variation and ignoring important spatial patterns is inadequate for effective understanding or management of the disease.

Some modelling approaches for dengue fever have been conducted in Makassar. However, these models focus on analysing the genomes of dengue viruses using phylogenetic analysis [1], predicting dengue cases using multiple regression [2] and predicting Dengue Haemorrhagic Fever (DHF) epidemics using two different models, a HR2008 model and a persistence model [3]. Spatio-temporal modelling approaches, and in particular Bayesian models, have not been explored yet for Makassar.

A variety of Bayesian spatial and spatio-temporal approaches have used in modelling dengue fever in other locations. A literature search revealed 31 journal articles about Bayesian spatial and spatio-temporal approaches to modelling dengue fever published from January 2000 to November 2017. Most studies adopted a Bayesian model with a spatially structured random effect using an intrinsic CAR prior structure to investigate the relationship between the risk of dengue and selected covariates [4]. Among the selected studies, only two studies used a generalised linear mixed model (GLMM) with spatial, temporal and spatio-temporal effects [5, 6]. An interesting feature of these studies is the wide disparity in the number of areas used to partition the region of interest; this ranged from 10 to 1490, with less than a quarter (eight studies) focusing on a small number of areas (<30). The number of periods also varied between studies, ranging from 3 months to 32 years, with 18 studies focusing on small time periods of less than 7 years.

There are specific limitations and challenges in the Makassar data, which are common to many datasets. The first challenge is that the spatial data are only available at the district level, and there are only 14 such areas in Makassar. The second challenge is that this dataset has a small number of time periods. This motivates an investigation of available spatio-temporal models that perform well in a context characterised by a small number of areas and small number of time periods. This paper provides a comparison of six existing spatio-temporal models, with the overall aim of better understanding spatial and temporal patterns of dengue incidence in Makassar, Indonesia. It is anticipated that the results of this evaluation will also inform other studies that are similarly characterised by a small number of areas and time periods.

Another consideration in choosing a statistical model for use in this case study is that the model should be easily implemented with publicly available software. This will enhance the potential for the approach to be adopted by public health agencies in Indonesia. As above, such a requirement is not unique to this case study and will have resonance with analysts and agencies in other developing countries.

9.2   Methods

9.2.1   Study Site

Makassar covers an area of 175.77 km^2 divided into 14 districts. A total of 6882 new cases of dengue were registered from 2002 to 2015, with substantial variation between districts and over time; for example, the number of cases rose sharply by 100 percent from 2012 (86 cases) to 2013 (265 cases) in a population of 1.49 million (2012) and 1.51 million (2013) [7].

9.2.2   Models

Six Bayesian spatio-temporal models with different formulations of conditional autoregressive (CAR) priors, namely linear [8, 9], ANOVA [10], separate spatial [11], AR [12], adaptive [13], and localised [14] models, were compared. These models were chosen because they fulfil the modelling requirements described above, in that they include both spatial and temporal components and they are publicly available in the CARBayesST package [9] in the statistical software package R [15]. All six models are formulated as follows,

$$y_{ij} \sim \text{Poisson}\left(e_{ij}\theta_{ij}\right)$$

$$\log\left(\theta_{ij}\right) = \psi_{ij}$$

where y_{ij} is the observed number of dengue cases in the ith district and jth time period, $i = 1, \ldots, I; j = 1, \ldots, J$; e_{ij} and θ_{ij} are, respectively, the expected number of dengue cases in area i time j and the relative risk of dengue (the underlying disease rate); and ψ_{ij} is a latent component for area i and time j involving one or more sets of spatio-temporally autocorrelated random effects. Details of each model are given below and also summarised in Table 9.1.

Models were compared using two goodness-of-fit measures, namely Deviance Information Criterion (DIC) [16] and Watanabe-Akaike Information Criterion (WAIC) [17], as well as by comparing the obtained estimates and their precision for each area.

9.2.2.1   Spatio Temporal CAR Linear Model

This model is suitable for estimating which areas have increasing or decreasing linear trends in the response over time. Here,

$$\psi_{ij} = \alpha_1 + u_i + (\beta + \delta_i)\frac{j - \overline{j}}{J}$$

where $\boldsymbol{u}$ and $\boldsymbol{\delta}$ denote normally distributed random effects that respectively describe spatial variation and the interaction between spatial and temporal effects. Thus each area i is allowed to have its own linear temporal trend, with a spatially varying intercept and slope $\alpha_1 + u_i$ and $(\beta + \delta_i)$ respectively. The random effects are assigned Leroux priors as follows:

$$(u_i|u_{-i}, \mathbf{W}) \sim \text{N}\left(\frac{\rho_{\text{int}}\sum_{k=1}^{I}\omega_{ik}u_k}{\rho_{\text{int}}\sum_{k=1}^{I}\omega_{ik} + 1 - \rho_{\text{int}}}, \frac{\tau_{\text{int}}^2}{\rho_{\text{int}}\sum_{k=1}^{I}\omega_{ik} + 1 - \rho_{\text{int}}}\right),$$

$$(\delta_i|\delta_{-i}, \mathbf{W}) \sim \text{N}\left(\frac{\rho_{\text{slo}}\sum_{k=1}^{I}\omega_{ik}\delta_k}{\rho_{\text{slo}}\sum_{k=1}^{I}\omega_{ik} + 1 - \rho_{\text{slo}}}, \frac{\tau_{\text{slo}}^2}{\rho_{\text{slo}}\sum_{k=1}^{I}\omega_{ik} + 1 - \rho_{\text{slo}}}\right).$$

Here, the elements of the adjacency matrix $\mathbf{W} = (\omega_{ik})$ represent the closeness between areas i and k, such that

$$\omega_{ik} = 1 \text{ if } i, k \text{ are adjacent}, \omega_{ik} = 0 \text{ otherwise};$$

Table 9.1 Summary of the spatio-temporal structure for the six models

Models	Structure ψ_{ij}	Prior			Additional information
		Spatial	Temporal	Spatio-temporal	
ST CAR linear	Intercept +spatial effect (for all time) +temporal effect (for all areas) +space-time interaction	CAR Leroux [18]	Linear	CAR Leroux	
ST CAR ANOVA	Spatial effect (for all time) +temporal effect (for all areas) + independent space-time interaction	CAR Leroux	CAR Leroux	Independence	
ST CAR separate spatial	Separate Spatial effect (for each time) +temporal effect (for all areas)	CAR Leroux	CAR Leroux	–	
ST CAR AR	Spatial effect (for each time)			CAR Leroux + AR (1)	-Temporal autocorrelation is induced via the mean $\rho_T u_j - 1$, Spatial autocorrelation is induced by the variance $\tau^2 \mathbf{Q}(\mathbf{W}, \rho_S)^{-1}$ -Single level ρ_S using CAR Leroux
ST CAR adaptive	Spatial effect (for each time)			CAR Leroux + AR (1)	-Model structure is the same as CAR AR, but this allows for localised spatial autocorrelation ρ_S using CAR Leroux
ST CAR localised	Spatial effect (for each time) + cluster component λ			ICAR +AR (1)	Model structure is the same as CAR AR except a component λ and random effects $\mathbf{u}$ are modelled with $\rho_S = 1$ (ICAR)

AR(1) first order autoregressive, ICAR intrinsic conditional autoregressive, ST spatio-temporal

α_1 is the mean log incidence over all areas; β is the mean linear temporal trend over all areas; τ_{int}^2 and τ_{slo}^2 are precision terms associated respectively with the intercept and slope of the regression; and ρ_{int}, ρ_{slo} are parameters of spatial dependence with values in the interval [0,1]. The Bayesian model is completed by specifying priors for the hyperparameters; in this study, the default priors in the CARBayes package were evaluated and deemed to be suitable, i.e., τ_{int}^2, $\tau_{slo}^2 \sim$ Inverse-Gamma (1, 0.01); ρ_{int}, $\rho_{slo} \sim$ Uniform (0, 1); $\beta \sim$ N(0, 1000).

9.2.2.2   Spatio Temporal CAR ANOVA Model

This model is suitable for estimating overall temporal trends and spatial patterns. Here,

$$\psi_{ij} = u_i + \delta_j + \gamma_{ij}$$

where u denotes the spatial random effects over all time periods; δ denotes the temporal random effect over all spatial units and γ denotes the space-time random interaction. The priors for the first two of these terms are as follows:

$$(u_i|u_{-i}, \mathbf{W}) \sim N\left(\frac{\rho_S \sum_{k=1}^I \omega_{ik}u_k}{\rho_S \sum_{k=1}^I \omega_{ik} + 1 - \rho_S}, \frac{\tau_S^2}{\rho_S \sum_{k=1}^I \omega_{ik} + 1 - \rho_S}\right),$$

$$(\delta_j|\delta_{-j}, \mathbf{D}) \sim N\left(\frac{\rho_T \sum_{k=1}^J d_{jk}\delta_k}{\rho_T \sum_{k=1}^J d_{jk} + 1 - \rho_T}, \frac{\tau_T^2}{\rho_T \sum_{k=1}^J d_{jk} + 1 - \rho_T}\right),$$

where the adjacency matrix $\mathbf{D} = (d_{jk})$ represents the closeness between times j and k, where k is the time immediately before or after j so that $d_{jk} = 1$ if $|k\text{-}j| = 1$ and $d_{jk} = 0$ otherwise. An independent normal prior is assigned to the last term,

$$\gamma_{ij} \sim N\left(0, \tau_\gamma^2\right).$$

As above, default priors were used for the remaining parameters, so that

$$\tau_S^2, \tau_T^2, \tau_\gamma^2 \sim \text{Inverse-Gamma}(1, 0.01),$$

$$\rho_S, \rho_T \sim \text{Uniform}(0, 1).$$

9.2.2.3 Spatio Temporal CAR Separate Spatial Model

This model is suitable for estimating overall temporal trends and to what extent the spatial variation has changed over time. Here,

$$\psi_{ij} = u_{ij} + \delta_j,$$

$$\left(u_{ij} | u_{-ij}, \mathbf{W} \right) \sim N \left(\frac{\rho_S \sum_{k=1}^{I} \omega_{ik} u_{kj}}{\rho_S \sum_{k=1}^{I} \omega_{ik} + 1 - \rho_S}, \frac{\tau_j^2}{\rho_S \sum_{k=1}^{I} \omega_{ik} + 1 - \rho_S} \right),$$

$$\left(\delta_j | \delta_{-j}, \mathbf{D} \right) \sim N \left(\frac{\rho_T \sum_{k=1}^{J} d_{jk} \delta_k}{\rho_T \sum_{k=1}^{J} d_{jk} + 1 - \rho_T}, \frac{\tau_T^2}{\rho_T \sum_{k=1}^{J} d_{jk} + 1 - \rho_T} \right),$$

$u_j = (u_{1j}, u_{2j}, \ldots u_{Ij})$ are separate spatial effects at each time period j.
$\delta = (\delta_1, \delta_2, \ldots, \delta_J)$ are temporal random effects over all the spatial areas i.

$$\tau_1^2, \ldots, \tau_J^2, \tau_T^2 \sim \text{Inverse-Gamma}(1, 0.01)$$

$$\rho_S, \rho_T \sim \text{Uniform}(0, 1).$$

9.2.2.4 Spatio Temporal CAR AR Model

This model is suitable for estimating the evolution of the spatial response surface over time without forcing it to be the same for each time. It has a single level of spatial dependence controlled by ρ_S, so that

$$\psi_{ij} = u_{ij},$$

$$\left(u_j | u_{j-1}, \right) \sim N \left(\rho_T u_{j-1}, \tau^2 \mathbf{Q}(\mathbf{W}, \rho_S)^{-1} \right) \qquad j = 2, \ldots, J,$$

$$u_1 \sim N \left(\mathbf{0}, \tau^2 \mathbf{Q}(\mathbf{W}, \rho_S)^{-1} \right)$$

$$\tau^2 \sim \text{Inverse-Gamma}(1, 0.01)$$

$$\rho_S, \rho_T \sim \text{Uniform}(0, 1).$$

9.2.2.5 Spatio Temporal CAR Adaptive Model

This model is an extension of spatio temporal CAR AR to allow for spatially adaptive smoothing (localised spatial autocorrelation), noting that ST CAR AR has only a single level of spatial dependence. This model is suitable when the residual spatial autocorrelation in the response is consistent over time but has a localised structure. The model structure is the same as CAR AR but nonzero (spatial) elements of neighbourhood matrix (W) can vary locally.

9.2.2.6 Spatio Temporal CAR Localised Model

This model is suitable for identifying clusters of areas that exhibit elevated values of the response compared with their geographical and temporal neighbours. This model structure is the same as CAR AR, but there is an additional cluster component λ and random effects **u** are modelled with $\rho_S = 1$ (ICAR). This model is similar to ST adaptive, in that both avoid the restrictive assumption that two areas that are close together must have similar estimates. The differences between the ST CAR localised and ST CAR adaptive models is that ST CAR localised captures any step-changes in the response via the mean function, but ST CAR adaptive captures any step changes via the correlation structure (via **W**). Here,

$$\psi_{ij} = u_{ij} + \lambda_{Z_{ij}},$$

$$\left(u_j | u_{j-1},\right) \sim N\left(\rho_T u_{j-1},\ \tau^2 Q(W)^{-1}\right) \qquad j = 2, \ldots . J,$$

$$u_1 \sim N\left(0, \tau^2 Q(W)^{-1}\right)$$

$$\tau^2 \sim \text{Inverse-Gamma}\,(1, 0.01)$$

$$\rho_T \sim \text{Uniform}\,(0, 1)\,.$$

$\lambda_{Z_{ij}}$ is a piecewise constant clustering or intercept component,
$\lambda_k \sim \text{Uniform}\,(\lambda_{k-1}, \lambda_{k+1})$ for $k = 1, 2, \ldots, G$

$$f\left(Z_{ij} | Z_{i,j-1}\right) = \frac{\exp\left(-\delta\left[\left(Z_{ij} - Z_{i,j-1}\right)^2 + \left(Z_{ij} - G^*\right)^2\right]\right)}{\sum_{r=1}^{G} \exp\left(-\delta\left[\left(r - Z_{i,j-1}\right)^2 + (r - G^*)^2\right]\right)}$$

for $j = 2, \ldots, J$

$$f\left(Z_{i1}\right) = \frac{\exp\left(-\delta(Z_{i1} - G^*)^2\right)}{\sum_{r=1}^{G} \exp\left(-\delta(r - G^*)^2\right)}$$

$\delta \sim \text{Uniform}\,(1, 10)$ where δ is the penalty parameter.

9.2.3   Case Study

Annual dengue fever incidence data for Makassar, Indonesia (14 geographic areas) during 2002–2015 were obtained from the Health Office of Makassar, South Sulawesi Province. An ethics exemption to use these datasets was obtained from QUT (exemption number: 1700000479) as it involves the use of existing collections of data that contain only non-identifiable data about human beings.

9.3   Results

9.3.1   Dengue Data

The descriptive analysis and the plot of the number of Makassar dengue cases from 2002 to 2015 can be seen in Table 9.2 and Fig. 9.1 respectively.

All models were fit using the CARBayesST package version 2.5.1 [9] in R version 3.3.3 or [15]. Posterior estimates and inferences were based on 100,000 MCMC samples collected after a burn in of 20,000 samples.

Crude risk estimates of dengue based on a raw SIR (Standardized incidence ratio) model were calculated for each area and represent the risk of being diagnosed with

Table 9.2 Descriptive analysis of dengue cases from 2002 to 2015

Year	Min	1st Qu	Median	Mean	3rd Qu	Max	Var
2002	14.00	44.00	86.00	104.80	98.00	419.00	10622.80
2003	11.00	40.75	61.00	82.43	73.75	251.00	5233.19
2004	10.00	20.50	33.50	45.50	56.25	178.00	1929.96
2005	11.00	26.00	49.50	63.71	71.00	236.00	3747.60
2006	19.00	27.75	55.50	60.86	65.75	209.00	2278.90
2007	4.00	15.25	27.50	32.64	47.25	80.00	550.09
2008	2.00	11.50	14.00	18.93	19.75	59.00	250.53
2009	3.00	9.00	15.00	18.29	20.00	68.00	286.06
2010	1.00	6.25	8.50	13.21	16.75	45.00	147.26
2011	1.00	2.00	3.00	6.07	9.25	16.00	29.46
2012	0.00	2.00	6.00	6.14	8.75	17.00	24.44
2013	4.00	9.00	13.50	18.93	27.75	52.00	210.22
2014	0.00	3.25	6.50	9.93	13.75	41.00	112.38
2015	2.00	4.75	8.00	10.14	14.00	26.00	50.59

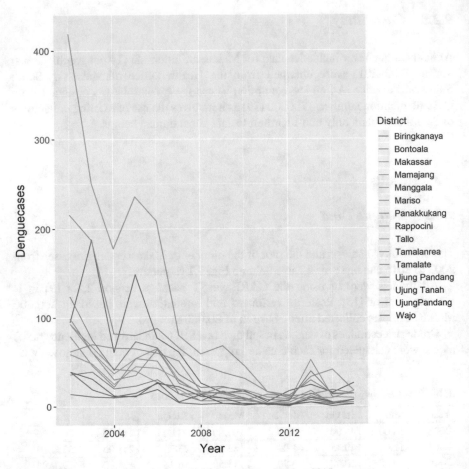

Fig. 9.1 The number of dengue cases in Makassar from 2002 to 2015

dengue fever and are depicted in Fig. 9.2. SIR is the ratio of the observed number of disease cases (y_{ij}) to the expected number of cases (e_{ij}) [19].

$$SIR_{ij} = \frac{y_{ij}}{e_{ij}}$$

It is apparent that there is substantial variation in dengue incidence between districts and over time.

The spatio-temporal CAR localised model with G = 2 had substantially better model fit as demonstrated by the smallest DIC (Table 9.3) followed by the ST CAR AR and ST CAR adaptive models. In contrast, the spatio-temporal CAR separate

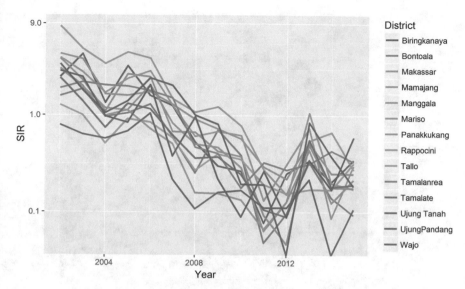

Fig. 9.2 Crude risk estimates (raw SIR model)

Table 9.3 DIC and WAIC for six models using dengue fever data for Makassar

Model	DIC	WAIC	Time (seconds)
ST CAR linear	2012.91	2230.25	44.40
ST CAR ANOVA	9374.02	45374.63	54.70
ST CAR separate spatial	9499.17	Inf	71.80
ST CAR AR	1632.36	**1884.21**	**33.00**
ST CAR Adaptive	1923.21	2111.74	162.90
ST CAR localised, G = 2	**1367.07**	1927.39	117.10
ST CAR localised, G = 3	1438.99	1892.07	128.80

The smallest DIC, WAIC of models and time to run the models are shown in bold

spatial model had the largest DIC. These two models (ST CAR localised, and ST CAR separate spatial) had very different estimates in certain regions and time periods (Fig. 9.3).

Under the preferred spatio temporal localised model with G = 2 (meaning a maximum of two clusters are allowed), most years had two clusters, but a few years had only one cluster (Fig. 9.4 and Table 9.4). Figure 9.5 shows that the overall Standardised Incidence ratios (SIR) have been decreasing over time for all areas but, as discussed above, there was a lot of variation/fluctuation from year to year.

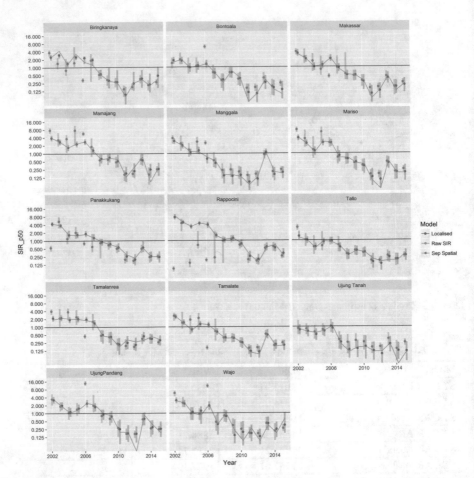

Fig. 9.3 SIR plot under the ST CAR localised model (corresponding to the smallest DIC), ST separate spatial model (corresponding to the largest DIC) for every area, with associated 95% credible intervals, and raw SIR

9.4 Discussion

Six different Bayesian spatio-temporal conditional autoregressive (ST CAR) models were compared by applying them to dengue incidence data from Makassar, Indonesia. The different structures, similarities and dissimilarities of the models have been summarized. The ST CAR localised model with $G = 2$ proposed by Lee and Lawson [14] performed the best based on the DIC goodness of fit measure, followed by the ST CAR AR and ST adaptive models. However, the ST CAR AR model proposed by Rushworth et al. [12] performed best in terms of WAIC and the computing time required. This is reasonable as the spatio-temporal random effect structure of the ST CAR localised model is the same as the ST CAR AR model except for an additional

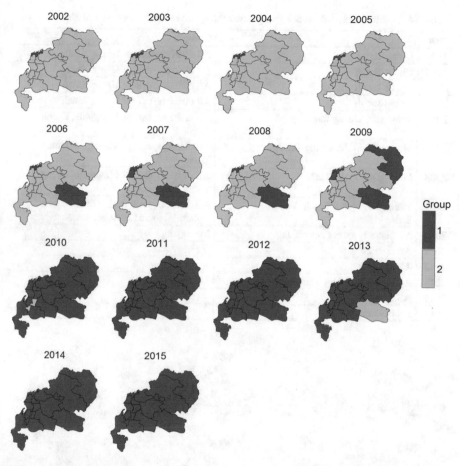

Fig. 9.4 Localised maps obtained under the spatio-temporal localised model with G = 2 local areas

Table 9.4 Districts included in each group under the spatio-temporal localised model

Year	Group 1	Group 2
2002	Ujung Tanah	All districts, except Ujung Tanah
2003	Ujung Tanah	All districts, except Ujung Tanah
2004	Ujung Tanah	All districts, except Ujung Tanah
2005	Ujung Tanah	All districts, except Ujung Tanah
2006	Manggala, Ujung Tanah	All districts, except Manggala, Ujung Tanah
2007	Manggala, Wajo, Ujung Tanah	All districts, except Manggala, Ujung Tanah
2008	Manggala, Ujung Tanah	All districts, except Manggala, Ujung Tanah
2009	Manggala, Ujung Tanah, Wajo, Biringkanaya	All districts, except Manggala, Ujung Tanah, Wajo and Biringkanaya
2010	All districts except Mamajang	Mamajang
2011	All districts	–
2012	All districts	–
2013	All districts, except Manggala	Manggala
2014	All districts	–
2015	All districts	–

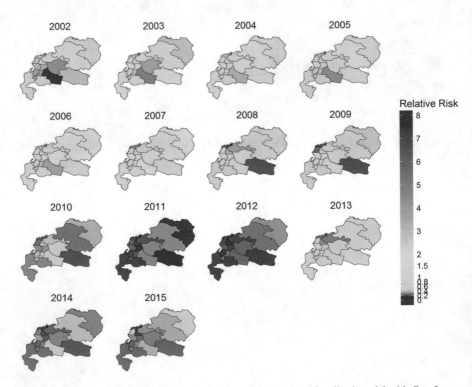

Fig. 9.5 Relative Risk maps obtained under the spatio-temporal localised model with $G = 2$

cluster component. In contrast, the ST CAR separate spatial model proposed by Napier et al. [11] performed the worst in terms of both DIC and WAIC. This may have been influenced by the small number of areas and time periods.

9.5 Conclusion

Bayesian CAR models can allow for different representations of spatial, temporal and spatio-temporal patterns. Results from the case study showed that the choice of model can have a large impact on goodness of fit, and that a spatio-temporal CAR localised model with $G = 2$ spatial groups provided the best fit in terms of DIC. Careful exploration of a range of models is important, especially when there are few areas and few time periods. The study motivates future research to provide more general insight into the behaviour of Bayesian spatio-temporal CAR models when the disease rate and degrees of spatio-temporal autocorrelation varies over different numbers of areas and time periods.

Acknowledgments This work was supported by the ARC Centre of Excellence for Mathematical and Statistical Frontiers (ACEMS) and Queensland University of Technology (QUT).

The authors acknowledge the City Health Department of Makassar, South Sulawesi Province for providing and permitting analysis of the dengue fever dataset used in the case study presented in this paper.

References

1. R.T. Sasmono, I. Wahid, H. Trimarsanto, B. Yohan, S. Wahyuni, M. Hertanto, et al., Genomic analysis and growth characteristic of dengue viruses from Makassar, Indonesia. Infect. Genet. Evol. **32**, 165–177 (2015)
2. H. Halide, P. Ridd, A predictive model for dengue hemorrhagic fever epidemics. Int. J. Environ. Health Res. **18**(4), 253–265 (2008)
3. H. Halide, Household protection against dengue hemorrhagic fever epidemics in coastal city of Makassar. J. Coast. Dev. **13**(3), 195–204 (2011)
4. A. Aswi, S.M. Cramb, P. Moraga, K. Mengersen, Bayesian spatial and spatio-temporal approaches to modelling dengue fever: a systematic review. Epidemiol. Infect. 1–14 (2018). https://doi.org/10.1017/S0950268818002807 Epub 10/29
5. S.P. Wijayanti, T. Porphyre, M. Chase-Topping, S.M. Rainey, M. McFarlane, E. Schnettler, et al., The Importance of socio-economic versus environmental risk factors for reported dengue cases in Java, Indonesia. PLoS Negl. Trop. Dis. **10**(9), e0004964 (2016). https://doi.org/10.1371/journal.pntd.0004964
6. D. Martínez-Bello, A. López-Quílez, A. Prieto, Spatiotemporal modeling of relative risk of dengue disease in Colombia. Stoch. Env. Res. Risk A. **32**(6), 1587–1601 (2018). https://doi.org/10.1007/s00477-017-1461-5
7. D. Kesehatan, *Propil Kesehatan Kota Makassar* (Dinas Kesehatan Makassar, 2012)
8. L. Bernardinelli, D. Clayton, C. Pascutto, C. Montomoli, M. Ghislandi, M. Songini, Bayesian analysis of space—time variation in disease risk. Stat. Med. **14**(21–22), 2433–2443 (1995)

9. D. Lee, A. Rushworth, G. Napier, Spatio-temporal areal unit modeling in R with conditional autoregressive priors using the CARBayesST package. J. Stat. Soft. **84**(9), 1–39 (2018)
10. L. Knorr-Held, Bayesian modelling of inseparable space-time variation in disease risk. Stat. Med. **19**(17–18), 2555–2567 (1999). https://doi.org/10.1002/1097-0258
11. G. Napier, D. Lee, C. Robertson, A. Lawson, K. Pollock, A model to estimate the impact of changes in MMR vaccine uptake on inequalities in measles susceptibility in Scotland. Stat. Methods Med. Res. **25**(4), 1185–1200 (2016). https://doi.org/10.1177/0962280216660420
12. A. Rushworth, D. Lee, R. Mitchell, A spatio-temporal model for estimating the long-term effects of air pollution on respiratory hospital admissions in Greater London. Spat. Spatio-Temporal Epidemiol. **10**(C), 29–38 (2014). https://doi.org/10.1016/j.sste.2014.05.001
13. A. Rushworth, D. Lee, C. Sarran, An adaptive spatio-temporal smoothing model for estimating trends and step changes in disease risk. J. R. Stat. Soc. Ser. C **66**(1), 141–157 (2017)
14. D. Lee, A. Lawson, Quantifying the spatial inequality and temporal trends in maternal smoking rates in Glasgow. Ann. Appl. Stat. **10**(3), 1427–1446 (2016)
15. Team RC, *R: A Language and Environment for Statistical Computing* (R Foundation for Statistical Computing, Vienna, Austria)
16. D.J. Spiegelhalter, N.G. Best, B.P. Carlin, A. Van Der Linde, Bayesian measures of model complexity and fit. J. R. Stat. Soc. Ser. B (Stat. Methodol.) **64**(4), 583–639 (2002)
17. S. Watanabe, Asymptotic equivalence of bayes cross validation and widely applicable information criterion in singular learning theory. J. Mach. Learn. Res. **11**, 3571–3594 (2010)
18. B.G. Leroux, X. Lei, N. Breslow, Statistical Models, in *Epidemiology, the Environment, and Clinical Trials*, ed. by M. E. Halloran, D. Berry, (New York, Springer, 2000)
19. A. Lawson, in *Disease Mapping with WinBUGS and MLwiN*, ed. by W. J. Browne, C. L. Vidal Rodeiro, (Wiley, Hoboken, NJ, 2003), p. 277

Chapter 10
A Comparison of Bayesian Spatial Models for Cancer Incidence at a Small Area Level: Theory and Performance

Susanna Cramb, Earl Duncan, Peter Baade, and Kerrie L. Mengersen

Abstract The increase in Bayesian models available for disease mapping at a small area level can pose challenges to the researcher: which one to use? Models may assume a smooth spatial surface (termed global smoothing), or allow for discontinuities between areas (termed local spatial smoothing). A range of global and local Bayesian spatial models suitable for disease mapping over small areas are examined, including the foundational and still most popular (global) Besag, York and Mollié (BYM) model through to more recent proposals such as the (local) Leroux scale mixture model. Models are applied to simulated data designed to represent the diagnosed cases of (1) a rare and (2) a common cancer using small-area geographical units in Australia. Key comparative criteria considered are convergence, plausibility of estimates, model goodness-of-fit and computational time. These simulations highlighted the dramatic impact of model choice on posterior estimates. The BYM, Leroux and some local smoothing models performed well in the sparse simulated dataset, while centroid-based smoothing models such as geostatistical or P-spline models were less effective, suggesting they are unlikely to succeed unless areas are of similar shape and size. Comparing results from several different models is recommended, especially when analysing very sparse data.

S. Cramb (✉) · P. Baade
Viertel Cancer Research Centre, Cancer Council Queensland, Fortitude Valley, QLD, Australia
e-mail: susanna.cramb@qut.edu.au; peterbaade@cancerqld.org.au

E. Duncan · K. L. Mengersen
ARC Centre of Excellence for Mathematical and Statistical Frontiers, Queensland University of Technology (QUT), Brisbane, QLD, Australia

© The Editor(s) (if applicable) and The Author(s), under exclusive licence to Springer Nature Switzerland AG 2020
K. L. Mengersen et al. (eds.), *Case Studies in Applied Bayesian Data Science*, Lecture Notes in Mathematics 2259, https://doi.org/10.1007/978-3-030-42553-1_10

245

10.1 Introduction

Bayesian spatial modelling continues to increase in popularity, offering a suite of models with a range of strengths in various contexts. Modelling spatial effects through a Bayesian hierarchical model has many advantages, such as being able to include a range of functions to represent outcomes over space and time, as well as the capacity to incorporate data characteristics such as rare outcomes, missing information, misclassifications, measurement error and known biases [9, 47]. Moreover, direct probabilistic statements can be made, such as the probability that an area has a higher disease risk than a comparison area [20].

A popular form for a Bayesian spatial model for disease mapping uses data aggregated by area and specifies the likelihood as:

$$Y_i \sim \text{Poisson}\left(E_i e^{\mu_i}\right) \quad \text{for } i = 1, \dots, N \text{ areas}$$

where $\{Y_1, \dots, Y_N\}$ are count data for a relatively uncommon disease, making a Poisson distribution appropriate. Other distributions are possible, including variants of Poisson such as negative binomial. The expected counts (E_i) are commonly defined using indirect standardisation to account for population size and age structure. The modelled log standardised incidence ratio (SIR) μ_i, also called log-relative risk, is often expressed as a regression equation and typically includes an overall fixed effect (intercept, denoted α), covariate effects (β) where x_i denotes a vector of covariates relating to area i, and spatial random effect(s) R_i, as follows:

$$\mu_i = \alpha + x_i^{\mathrm{T}} \beta + R_i.$$

Much of this chapter shall discuss options for modelling the spatial random effect(s), R_i. Prior distributions are then specified for each of the unknown parameters:

$$\alpha \sim p\left(\cdot | \theta_\alpha\right)$$
$$\beta \sim p\left(\cdot | \theta_\beta\right)$$
$$R_i \sim p\left(\cdot | \theta_R\right).$$

The spatial random effects are given a spatial prior, which may be assumed to follow a conditional autoregressive (CAR) or alternative prior to enable spatial correlation and smoothing [8, 10]. If the parameters θ_α, θ_β, or θ_R are unknown, then the hyperpriors represent an additional stage of the hierarchy.

Many different Bayesian spatial models have been proposed, most of which vary the representation of the spatial prior. Understanding the theoretical assumptions and appropriateness of different models is important. It is also necessary to consider how models perform in different circumstances. Therefore, this chapter discusses the theoretical underpinnings of key spatial models. Where possible and pertinent, these models were applied to typical cancer incidence mapping scenarios obtained by simulating rare and common cancer incidence data across Australia. This nation has more than 2100 small areas, with large differences in population size, demographic structure, land area size and shape.

10.2   Bayesian Spatial Models

Fourteen Bayesian spatial models used in disease mapping are considered. These can be divided into two broad types, namely 'global' spatial smoothing models that have a common spatial correlation term across the region, and 'local' spatial smoothing models that allow for differential spatial correlation depending on neighbourhood characteristics.

10.2.1   Global Spatial Smoothing

Global spatial smoothing means that the same correlation parameters are applied consistently across the entire region [26]. Although the global CAR-based models are relatively easy to implement in a range of software, disadvantages of global models include the potential to obscure genuine deviations in the underlying spatial patterns (i.e. to over-smooth), as discontinuities between adjacent areas are smoothed over.

10.2.1.1   Intrinsic CAR and BYM Models

The most commonly used prior for enabling spatial correlation within a Bayesian model is the intrinsic CAR distribution. This approach allows for smoothing of estimates over neighbouring areas, but it assumes a common variance for the smoothing term (and therefore a smooth spatial trend) over the whole region.

The intrinsic CAR (ICAR) model specifies the following set of conditional distributions for the spatial random effect parameter:

$$R_i = S_i$$

$$S_i | s_{\backslash i} \sim \mathcal{N}\left(\frac{\sum_j w_{ij} s_j}{\sum_j w_{ij}}, \frac{\sigma_s^2}{\sum_j w_{ij}}\right)$$

or in matrix notation

$$S_i | s_{\backslash i} \sim \mathcal{N}\left(\left\{\mathbf{D}^{-1}\mathbf{W}s\right\}_i, \sigma_s^2 \left\{\mathbf{D}^{-1}\right\}_{ii}\right)$$

where w_{ij} is the element of a spatial weights matrix $\mathbf{W}$ corresponding to row i and column j [6, 10], and $\mathbf{D}$ is a diagonal matrix with elements diag $\left\{\sum_j w_{ij}\right\}$. The term $\mathbf{W}$ determines the spatial proximity between the random effects, and it is most commonly defined as a binary, first-order, adjacency matrix, whereby

$$w_{ij} = \begin{cases} 1 & \text{if areas } i \text{ and } j \text{ are adjacent} \\ 0 & \text{otherwise.} \end{cases} \tag{10.1}$$

This model implies that the conditional expectation of S_i is equal to the mean of the random effects at neighbouring locations.

The S_i can be regarded as structured spatial random effects. If $R_i = S_i + U_i$, so that unstructured spatial random effects $U_i \sim N(0, \sigma_U^2)$ are also included, the resulting model is referred to as the convolution model, or the BYM model in honour of Besag et al. [8]. However, the two separate random effects components cannot be individually identified—only their sum is identifiable [15]. Note that for all CAR-based models, the strength of the partial autocorrelation depends on the number of neighbouring areas rather than on any underlying relationship [27]. The BYM remains the most popular approach to incorporating spatial smoothing, in part due to its computational synergy with fairly standard MCMC approaches [47] and ease of implementation.

10.2.1.2 Proper CAR Model

The full conditionals for the ICAR prior are proper, but the joint distribution is improper since the precision matrix is singular [7]. The impropriety of the ICAR prior can be overcome by redefining the precision matrix

$$\mathbf{T} = \frac{1}{\sigma_s^2} (\mathbf{D} - \mathbf{W})$$

to

$$\mathbf{T} = \frac{1}{\sigma_s^2} (\mathbf{D} - \phi \mathbf{W})$$

such that the conditional distributions for the spatial random effect are:

$$S_i | s_{\backslash i} \sim N\left(\frac{\phi \sum_j w_{ij} s_j}{\sum_j w_{ij}}, \frac{\sigma_s^2}{\sum_j w_{ij}}\right)$$

with the constraint $|\phi| < 1$, where ϕ represents the expected proportional 'reaction' of S_i to $\sum_j w_{ij} s_j / \sum_j w_{ij}$ [5]. This ensures that the covariance matrix $\mathbf{T}^{-1}$ is positive definite and S has a proper joint distribution [19]. The proper CAR prior may have certain disadvantages, including potentially limiting the breadth of the posterior spatial pattern. Moreover, ϕ will likely need to be very close to 1 for there to be a reasonable amount of spatial association [5].

10.2.1.3   Leroux CAR Model

Another variant of the BYM model was proposed by Leroux et al. [29],

$$S_i | s_{\backslash i} \sim \mathcal{N} \left(\frac{\rho \sum_j w_{ij} s_j}{\rho \sum_j w_{ij} + 1 - \rho}, \frac{\sigma_s^2}{\rho \sum_j w_{ij} + 1 - \rho} \right)$$

which only requires a single set of random effects [24]. This avoids the difficulties in identifiability, and also the selection of hyperpriors (given that in the BYM model, the S_i variance are conditional on neighbouring areas, while the U_i have a marginal variance term) [41].

The precision matrix can be expressed as

$$\mathbf{T} = \frac{1}{\sigma_s^2} [\rho (\mathbf{D} - \mathbf{W}) + (1 - \rho)].$$

This mixture representation consists of correlated smoothing of the neighbouring random effects (weighted by ρ) as well as uncorrelated smoothing to a global mean of zero (weighted by $(1 - \rho)$) [26]. Thus S_i has a conditional expectation based on a weighted average of both the independent random effects and the spatially structured random effects. The ICAR prior is therefore a limiting case of both the proper CAR and Leroux CAR models when ρ is set to 1. The spatial autocorrelation parameter ρ is typically given either a continuous [19, 25] or a discrete [24] uniform prior

$$\rho \sim \text{Uniform}(0, 1),$$

where the discrete case offers gains in computational efficiency [24], although other priors have been suggested such as a diffuse Gaussian prior on the logit scale [27].

10.2.1.4   Geostatistical Model

Here, the residual spatial structure is modelled as a Gaussian process using a geostatistical design [11]. Because this model incorporates distance, counts are assumed to be located in the centroid of an area.

$$R_i \sim \mathcal{N}(S_i, \sigma^2)$$
$$S_i = \exp\left(-(\lambda d_{ij})^k\right), \quad \lambda > 0$$

where λ controls the rate of decay, k is the "degree of spatial smoothing", and d_{ij} is the distance between points (e.g. centroids of areas) i and j [11]. This expression is the exponential decay function with the addition of the power k. Rather than fix decay parameter λ *a priori*, a hyperprior is specified as a fourth stage of the hierarchy:

$$\lambda \sim \text{Uniform}(0.1, 6).$$

The justification for the bounds 0.1 and 6 were based on the minimum and maximum separating distance in decimal degrees between area centroids to ensure that the spatial correlation was able to be high at the minimum distance, and likely to be low at the maximum distance. This choice is also able to give near zero correlation for distances within the study region, which is vital to avoid non-identifiability of the mean and correlation parameters [10].

Alternative functions are possible, including the disc model [40] (a linear decrease with increasing distance, where two discs of common radius are centred on centroids, and the correlation is proportional to the disc intersection area), or combining two parametric functions to obtain different shapes of decrease, such as the Matern class [10]. Note that often limited information is available to guide the choice of functional form, or correlation parameters, especially as complexity increases [10]. Because the covariance matrix is inverted at each iteration, these models can be computationally intensive and slow to run in a naïve algorithm, although this can be mitigated to some extent with the use of sparse matrix algebra.

10.2.1.5 Global Spline Models

The spline model also assumes that the cases are all located at the centroid of each area [17].

There are two main methods: smoothing splines and P-splines [32]. Smoothing splines are penalised splines which have knots on all data points. P-splines allow for a smaller number of knots, and are commonly formulated as a penalised spline regression under a 'difference penalty' based on the coefficients of adjacent B-spline bases or other spline bases [32].

The correlation between areas i and j can be modelled by a two-dimensional smooth surface [17]. First, define the longitude and latitude pairs representing the centroid of each area, denoted (c_{1i}, c_{2i}). Then

$$R_i = f(c_{1i}, c_{2i})$$

where the smooth function $f(\cdot)$ is expressed as

$$f(c_{1i}, c_{2i}) = \theta_1 B_1(c_{1i}, c_{2i}) + \cdots + \theta_k B_k(c_{1i}, c_{2i})$$

which is estimated using P-splines with B-spline bases $B_1, \ldots, B_k$. The terms $\theta_1, \ldots, \theta_k$ are unknown coefficients which are penalised to control for "wiggliness" through a penalty matrix, and k depends on the number of knots and the degree of the B-spline bases.

Define $c_1 = (c_{11}, \ldots, c_{1N})^T$ and $c_2 = (c_{21}, \ldots, c_{2N})^T$ and univariate B-spline bases $\mathbf{B}_1 = \{B_{11}(c_1), \ldots, B_{1k_1}(c_1)\}$ and $\mathbf{B}_2 = \{B_{21}(c_2), \ldots, B_{2k_2}(c_2)\}$. The bivariate B-spline basis is then constructed as the row-wise Kronecker product (denoted by $\boxtimes$) of the marginal B-spline bases:

$$\mathbf{B} = \mathbf{B}_2 \boxtimes \mathbf{B}_1$$
$$= \left(\mathbf{B}_2 \otimes \mathbf{1}_{k_1}^T\right) \odot \left(\mathbf{1}_{k_1}^T \otimes \mathbf{B}_1\right).$$

The basis $\mathbf{B}$ is of dimension $N \times k$ where $k = k_1 k_2$, the symbols $\otimes$ and $\odot$ represent the Kronecker product and "element-wise" matrix product respectively, and $\mathbf{1}_{k_1}$ and $\mathbf{1}_{k_2}$ are column vectors of ones of length k_1 and k_2 [17].

Overall this model provides a relatively smooth surface, as the covariance structure is impacted by long distance effects that influence the smoothing. This is in contrast to the covariance structure of the CAR model where an area's estimate depends on the mean of its neighbours [17].

The formulation of the P-spline model using the row-wise Kronecker product, or tensor product, is better suited to data which lie on a regular grid, or at least have similar distances between the centroids.

An alternative formulation [42] is to define the B-spline bases in terms of the distances,

$$z_{ik} = \exp\left(-\frac{d_{ik}}{\Delta}\right)\left(1 + \frac{d_{ik}}{\Delta}\right)$$

where d_{ik} is the distance between the ith area and the kth knot, and Δ is a constant used to normalise the distances so that the values of $\mathbf{B}$ are more evenly spread between the lower and upper limits. This version of the P-spline uses a radial basis function which achieves rotational invariance [42].

10.2.2   Local Spatial Smoothing

In contrast to the global smoothing models, local smoothing is focused on allowing nearby areas to potentially have different amounts of spatial smoothing. Many of these are based on modifying the CAR prior to allow for discontinuous surfaces.

10.2.2.1 CAR Dissimilarity Models

Lee and Mitchell [26] based this model on the Leroux CAR prior, with ρ set to be 0.99 to ensure strong global spatial smoothing which could then be altered locally through estimating $\{w_{ij}|i \sim j\}$. Here, the elements in $\mathbf{W}$ are modelled so the partial autocorrelations can be reduced between certain adjacent random effects. This approach can have binary or non-binary elements in $\mathbf{W}$.

The similarity between areas is determined by including non-negative dissimilarity metrics in the model, i.e. $z_{ij} = (z_{ij1}, \ldots, z_{ijq})$ where $z_{ijk} = |z_{ik} - z_{jk}|/\sigma_k$ and σ_k is the standard deviation of $|z_{ik} - z_{jk}|$ over all pairs of contiguous areas.

The set of w_{ij} are determined using regression parameters $\boldsymbol{\alpha} = (\alpha_1, \ldots, \alpha_q)$. These can be based on social or physical factors. Physical boundaries (e.g. river/railway line, or the distance between centroids) can be used if the aim is to explain the spatial pattern in the response and include covariates in the model. Alternatively, covariate information can be used to construct the dissimilarity metrics if the aim is to identify the locations of any boundaries [25].

$$R_i = S_i$$

$$S_i|s_{\setminus i} \sim \mathcal{N}\left(\frac{0.99\sum_j w_{ij}(\boldsymbol{\alpha})s_j + 0.01\mu_0}{0.99\sum_j w_{ij}(\boldsymbol{\alpha}) + 0.01}, \frac{\sigma_s^2}{0.99\sum_j w_{ij}(\boldsymbol{\alpha}) + 0.01}\right).$$

The default binary formulation is:

$$w_{ij}(\boldsymbol{\alpha}) = \begin{cases} 1 & \text{if } \exp\left(-\sum_{k=1}^q z_{ijk}\alpha_k\right) \geq 0.5 \text{ and } i \sim j \\ 0 & \text{otherwise} \end{cases}$$

$$\alpha_k \sim \text{Uniform}(0, M_k) \quad \text{for } k = 1, \ldots, q$$

where M_k is fixed so that a maximum of 50% of borders could be defined as boundaries [26]. The non-binary formulation (which does not allow identification of hard boundaries, but does allow for localised smoothing) is:

$$w_{ij}(\boldsymbol{\alpha}) = \exp\left(-\sum_{k=1}^q z_{ijk}\alpha_k\right)$$

$$\alpha_k \sim \text{Uniform}(0, 50) \quad \text{for } k = 1, \ldots, q.$$

10.2.2.2 Localised Autocorrelation

The spatially smooth random effects in this model are augmented with a piecewise constant intercept (cluster model). This allows for large jumps in the risk surface between adjacent areas if they are in different clusters. The approach by Lee and Sarran [28] partitions the I areas into a maximum of G clusters, each with their own intercept term $(\lambda_1, \ldots, \lambda_G)$. The model is thus given by:

$$R_i = S_i + \lambda_{z_i}$$

$$S_i | s_{\backslash i} \sim \mathcal{N}\left(\frac{\sum_j w_{ij} s_j}{\sum_j w_{ij}}, \frac{\sigma_s^2}{\sum_j w_{ij}}\right)$$

$$\lambda_g \sim \text{Uniform}\left(\lambda_{g-1}, \lambda_{g+1}\right) \quad \text{for } g = 1, \ldots, G$$

$$f(Z_i) = \frac{\exp\left(-\delta(Z_i - G^*)^2\right)}{\sum_{r=1}^{G} \exp\left(-\delta(r - G^*)^2\right)}$$

$$\delta \sim \text{Uniform}(1, M)$$

where $f(Z_i)$ denotes a shrinkage prior on Z_i which shrinks extreme values towards the middle intercept value. Label switching is prevented by ordering the cluster means $(\lambda_1, \ldots, \lambda_G)$ so that $\lambda_1 < \lambda_2 < \cdots < \lambda_G$. The penalty term $\delta(Z_i - G^*)^2$ where $G^* = (G + 1)/2$ means that if G is odd then each data point will be shrunk towards a single intercept λ_{G^*}, but if G is even, there may be two different intercept terms used even if there is a spatially smooth residual structure. Lee and Sarran [28] thus recommend setting G to be a small odd number, such as 3 or 5. Area i is assigned to one of the G intercepts by $Z_i \in \{1, \ldots, G\}$, and there is no spatial smoothing imposed on the indicator vector $\mathbf{Z}$. The clustering is purely non-spatial, and it is the CAR prior on the S_i term that accounts for spatial autocorrelation [28].

10.2.2.3 Locally Adaptive Model

The locally adaptive model takes a similar approach to the above dissimilarity model, except that here the boundaries are not identified by the use of additional information and the modelled w_{ij} are binary only. Lee and Mitchell [27] again based this on the Leroux CAR model:

$$S_i | s_{\backslash i} \sim \mathcal{N}\left(\frac{\rho \sum_j w_{ij} s_j}{\rho \sum_j w_{ij} + 1 - \rho}, \frac{\sigma_s^2}{\rho \sum_j w_{ij} + 1 - \rho}\right).$$

Here ρ can be estimated in the model, or fixed at a specified value. (Lee and Mitchell [27] recommend 0.99.)

The spatial weights matrix starts out as the binary, first-order, adjacency matrix given by Eq. (10.1) and is subsequently updated at each iteration which allows the weights corresponding to neighbours to be estimated as either 1 or 0 (with w_{ij} fixed at zero for non-neighbouring areas). Because only weights corresponding to neighbouring areas are estimated, this approach should be more computationally feasible than areal wombling [30] where all values in $\mathbf{W}$ are estimated.

The matrix $\mathbf{W}$ is estimated as follows. For adjacent areas i and j: if the marginal 95% credible intervals (CIs) of s_i and s_j overlap, then set $w_{ij} = 1$; else set $w_{ij} = 0$. It is therefore not a 'fully' Bayesian method of estimation for these terms, as they are not considered to be random variates. For further details, refer to Lee and Mitchell [27], who implemented this using INLA.

10.2.2.4 Weighted Sum of Spatial Priors

The BYM model with its spatially structured component S_i and its unstructured spatial component U_i was extended by Lawson and Clark [23] to be able to incorporate discontinuities:

$$R_i = p_i S_i + (1 - p_i) Z_i + U_i. \tag{10.2}$$

The Z component models abrupt discontinuities between areas. Although a range of options is possible, Lawson and Clark [23] based the prior for this parameter on the total absolute difference in risk between neighbouring areas, i.e.

$$\pi (Z_1, \ldots, Z_N) \propto \frac{1}{\sqrt{\lambda}} \exp \left(-\frac{1}{\lambda} \sum_{i \sim j} |Z_i - Z_j| \right)$$

where λ acts as a constraining term.

Note that if $p_i = 1$ in Eq. (10.2), then the model reverts to the BYM model. Conversely, if $p_i = 0$, then the model is entirely discontinuous.

10.2.2.5 Leroux Scale Mixture Model

Using a scale mixture model within a Leroux prior also enables detection of abrupt changes between areas, with the advantage over the above approaches of incorporating non-normality (heavy tailed distributions). This was proposed by Congdon [12] as follows:

$$S_i | s_{\backslash i} \sim N \left(\frac{\rho \sum_j w_{ij} s_j}{\rho \sum_j w_{ij} + 1 - \rho}, \frac{\sigma_s^2}{\kappa_i \left[\rho \sum_j w_{ij} + 1 - \rho \right]} \right).$$

If $\rho = 0$, this reduces to an unstructured iid scale mixture Student-t density, which is a heavy-tailed distribution. Small values of $\kappa_j (<1)$ will indicate areas differ from their neighbours and result in less smoothing between neighbouring areas. The scale mixture is implemented by $\kappa_i \sim \text{Gam}(0.5v, 0.5v)$, where v is a hyperparameter.

The precision matrix has the following diagonal terms [12]:

$$\{\mathbf{T}\}_{ii} = \frac{1}{\sigma_s^2} \kappa_i \left[(1 - \rho) + \rho \sum_{j \neq i} w_{ij} \right]$$

and off-diagonal terms:

$$\{\mathbf{T}\}_{ij} = \frac{1}{\sigma_s^2} \rho \kappa_i \kappa_j \mathbb{I}(i \sim j).$$

10.2.2.6   Skew-Elliptical Areal Spatial Model

Another approach that focused on incorporating skewness was introduced by Nathoo and Ghosh [36]. Here

$$R_i = \eta_i^{-0.5} \left(\delta |Z_i| + S_i \right)$$

where $\delta |Z_i|$ is the skewing component where Z_i is a set of skewing variables each independently drawn from a standard normal distribution, η provides the scale mixing and S_i is from the CAR model, i.e.

$$S_i | s_{\backslash i} \sim N \left(\kappa \frac{\sum_j w_{ij} s_j}{\sum_j w_{ij}}, \frac{\sigma_s^2}{\sum_j w_{ij}} \right)$$

where κ is a spatial smoothing parameter (note that if κ is set to 0 then the distribution corresponds to uncorrelated skew-t random effects) and other terms are defined as before.

Two versions were proposed by Nathoo and Ghosh [36]. The first aims to ensure each R_i has a skew-elliptical distribution, with the marginal distribution for each spatial effect belonging to the skew-t family of distributions.

The second is a semiparametric version that uses an approximation to a Dirichlet process to allow for data-driven departures from the parametric version. This accommodates uncertainty in the mixing structure, and gives greater flexibility in the tail behaviour of marginal distributions [36].

10.2.2.7   Hidden Potts Model

In contrast to the above approaches, this model is based on a hidden Markov field, so spatial correlation occurs in an additional latent hierarchy of the model [47]. This approach was proposed by Green and Richardson [18] and assigns each area to one of several risk categories. The spatial random effect is modelled on the log scale, as a K-component mixture model, where each component represents a different risk category, and the allocation of each area to a component follows a spatially correlated process. The number of components K is considered unknown and is estimated by the model.

$$R_i = \log(S_{z_i})$$

$$S_k \sim \text{Gamma}(a, b) \text{ for } k = 1, \ldots, K$$

$$K \sim \text{Uniform}(1, K_{\max}).$$

The Potts model is proposed as the allocation model,

$$p(z|\psi, K) = \exp(\psi U(z) - \delta_k(\psi))$$

where $\psi > 0$ is the interaction parameter to be estimated and $U(z) = \sum_{i \sim j} \mathbb{I}(z_i = z_j)$ is the number of like labelled pairs of neighbouring areas. This model allows for discontinuities between areas in different risk categories and also for the amount of spatial correlation to vary by risk category. However, it does require careful MCMC implementation due to having an unknown number of risk categories and unknown area allocation to these categories. It is also more often implemented in high-dimensional data rather than disease mapping, as its greater flexibility generally has more advantages as data complexity increases [47].

10.2.2.8 Spatial Partition Model

Closely related to the above Hidden Potts model are the spatial partition models [14, 21]. These also have K non-overlapping clusters of areas, each with a constant relative risk, and K is unknown [10]. The key differences are in defining the clusters and the hyperprior specifications [10]. Specifically, the spatial partition model assigns up to K areas as cluster centres, which are allocated with a uniform prior probability, and the number of clusters is chosen according to the distribution $p(K = k) \propto (1 - c)^k$ where $c \in [0, 1)$ is fixed *a priori*. Smaller values of c makes this prior less informative, with the limiting case $c = 0$ yielding a uniform distribution. The remaining $N - K$ areas are then assigned to their nearest cluster, according to the minimal number of boundaries that have to be crossed. Both this model and the above hidden Potts model have been criticised for forcing discontinuities into a surface, and for assuming constant relative risk within a cluster [23].

10.2.2.9 Local Spline Model

An extension to the global spline models described in Sect. 10.2.1 that results in a less smooth surface is the incorporation of unstructured random effects as in the penalised random individual dispersion effects (PRIDE) model, originally proposed by Perperoglou and Eilers [37]. Here

$$R_i = f(c_{1i}, c_{2i}) + \gamma_i$$

where γ_i is an area-specific random effect, whose vector follows a multivariate normal distribution [17]. This means that the covariance matrix captures the unstructured heterogeneity by containing an identity matrix multiplied by a variance component, in addition to the eigenvalues from the P-spline model component [17].

10.3   Case Study

10.3.1   Data

Since the dissemination of actual cancer data is restricted due to privacy and confidentiality requirements of the data custodians, simulated data that reflected the general distributions of actual data were generated to enable data sharing and reproduction of the presented results (see contact the authors for data and model code). Two datasets were generated that reflected the incidence of cancer types with a strong socioeconomic gradient: one with low total counts per geographical area over ten years (median of 2, range 0–19), considered a rare cancer, and one with higher counts over 10 years (median 25 cases, range 0–163), considered a common cancer. The focus on socioeconomic gradients meant we expected neighbouring areas having different socioeconomic levels would have different incidence rates.

The areas used were statistical areas 2 (SA2s) based on the 2011 Australian Statistical Geography Standard (ASGS) boundaries [4]. After excluding some areas with no/nominal resident populations, the number of areas was 2153. The median population of the included SA2s was 9055 (range: 3–50,251). Land area size varied from 0.8 to 520,000 km^2, with a median of 15.6 km^2.

10.3.2   Model Selection

Of the fourteen models introduced in Sect. 10.2 and described in Table 10.1, five were excluded from the application. Two of these were on theoretical grounds: the localised P-spline and the proper CAR models. The localised P-spline model was not investigated because implementing the P-spline had many challenges within the Australian context of vastly differing area sizes. The disadvantages of the proper CAR formulation such as the potentially limited breadth of estimates have limited appeal for spatial modelling [5]. We attempted to run a Hidden Potts model, spatial partition model and skew-elliptical areal spatial model, but were unable to successfully achieve this due to the computational complexity of the models, so they are also excluded from this section. The skew-elliptical model was unable to compile in WinBUGS [31], while the multidimensionality required for the spatial partition model and Hidden Potts model became too unwieldy.

10.3.3   Model Variants

Of the nine models successfully implemented, multiple variants were considered for the global P-spline model, CAR dissimilarity models, localised autocorrelation models, and the locally adaptive models, and these are detailed below. Specifications for the geostatistical model are also documented. These resulted in a total of 13 versions of models applied to the simulated data (Table 10.1).

Table 10.1 Software used for models applied to simulated data

Models investigated	Authors	Software used
Global spatial smoothing		
BYM (Intrinsic CAR)	Besag et al. [8]	R (CARBayes)
Proper CAR	Besag [6]	–
Leroux	Leroux et al. [29]	R (CARBayes)
Geostatistical	Clements et al. [11]	JAGS
P-spline (tensor)	Lang and Brezger [22]	JAGS
P-spline (radial)	Ruppert et al. [42]	JAGS
Local spatial smoothing		
CAR dissimilarity model (binary)	Lee and Mitchell [26]	R (CARBayes)
CAR dissimilarity model (non-binary)	Lee and Mitchell [26]	R (CARBayes)
Localised autocorrelation ($G = 3$)	Lee and Sarran [28]	R (CARBayes)
Localised autocorrelation ($G = 5$)	Lee and Sarran [28]	R (CARBayes)
Locally adaptive model (ρ estimated)	Lee and Mitchell [27]	R (INLA)
Locally adaptive model ($\rho = 0.99$)	Lee and Mitchell [27]	R (INLA)
Weighted sum of spatial priors	Lawson and Clark [23]	WinBUGS
Leroux scale mixture	Congdon [12]	WinBUGS
Skew-elliptical areal spatial	Nathoo and Ghosh [36]	–
Hidden Potts	Green and Richardson [18]	–
Spatial partition	Denison and Holmes [14], Knorr-Held and Raßer [21]	–
Local spline	Goicoa et al. [17], Perperoglou and Eilers [37]	–

10.3.3.1   Global P-spline Model

Two formulations of the global P-spline model were implemented: the first uses a tensor product (refer to Sect. 10.2.1.5 for a definition) to define the basis, and the second uses a radial basis based on distances. No further modifications were made to the tensor product version.

The radial P-spline model had the knots evenly spaced at intervals of 5 degrees of latitude and longitude, as shown in Fig. 10.1. Knots which were too distant from the centroids of SA2 areas were subsequently dropped. A total of 47 knots were retained for modelling. Based on these knots, Δ was set to 500.

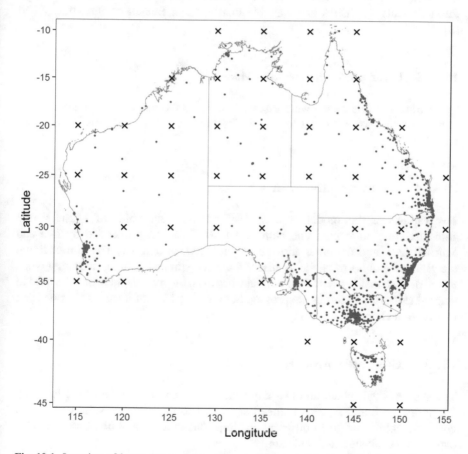

Fig. 10.1 Location of knots (crosses) in relation to SA2 centroids (dots) for the P-spline radial model

10.3.3.2 CAR Dissimilarity Model

The CAR dissimilarity model can also be applied in a variety of forms. As discussed in Sect. 10.2.2.1, the weighting matrix can be binary or non-binary, and the dissimilarity measure can be based on distance, geographical features (such as railways or mountains), or covariate information. Here we examine both binary and non-binary forms of this model based on the Socioeconomic Indexes for Areas (SEIFA) dissimilarity. This gives a continuous score for each area which is designated based on a range of socioeconomic measures, including house ownership, car ownership, employment and internet access. Several indices are available, and we used the Index of Relative Socioeconomic Disadvantage. Further details on SEIFA are available in Australian Bureau of Statistics [ABS] [3].

10.3.3.3 Localised Autocorrelation Models

Two variants of this model were assessed based on the value of G, the maximum number of clusters, being set to 3 or 5. See Sect. 10.2.2.2 for discussion of these choices.

10.3.3.4 Locally Adaptive Models

Two variants of this model were assessed based on the value of ρ, the spatial autocorrelation parameter, one being set to 0.99 (as recommended by Lee and Mitchell [27]) and the other allowed to vary between 0 and 1. The aim of fixing the value of ρ close to one is to ensure there is spatial smoothing occurring when $w_{ij} > 0$. Note that if $\rho = 0$ then w_{ij} vanishes from the model and cannot be used to determine if discontinuities are present. Setting ρ to 1 is not ideal, as the precision matrix would become singular.

10.3.3.5 Geostatistical Model

The geostatistical model had two adjustments made to provide a better fit. First, the priors for λ and k were changed according to the possible values of spatial correlation observed given different combinations of λ, k, and distances d_{ij}. This exploratory analysis suggested using

$$\lambda \sim \text{Uniform}(0.01, 1)$$

$$k \sim \text{Uniform}(0.1, 20).$$

To allow for further flexibility, λ and k were replaced by one of $\{\lambda_1, \ldots, \lambda_5\}$ and $\{k_1, \ldots, k_5\}$ respectively according to the remoteness of the area (major city, inner regional, outer regional, remote, and very remote) to allow the degree of smoothing to vary between the five levels of remoteness. Second, to make this model computationally feasible, the distance matrix $\{d\}_{ij}$ was modified by imposing a remoteness-specific radius of influence $\{r_1, \ldots, r_5\}$ on each area, such that areas beyond this threshold are not considered neighbours. These radii were $\{50, 100, 200, 400, 800\}$ km respectively. This induces a Markov random field (MRF) structure which should have only a negligible effect on parameter estimation while greatly increasing computational efficiency. Some remote and very remote areas are relatively close to major city and inner regional areas, which can lead to some areas having more than 1000 neighbouring SA2s, thereby drastically reducing any computational gains. Therefore, the imposed MRF was further modified to exclude major city areas as neighbours of remote areas, and to exclude both major city and inner regional areas as neighbours of very remote areas. This is also sensible given the differences in cancer incidence and underlying influences between these areas [13]. This was achieved by setting the distances to these excluded areas to infinity. The result of these adjustments lead to

$$\{S\}_{ij} = \begin{cases} \exp\left(-(\lambda_{z_i} d_{ij})^{k_{z_i}}\right) & \text{if } d_{ij} \leq r_{z_i} \\ 0 & \text{if } d_{ij} > r_{z_i} \end{cases}$$

$$S_i = f(S) = \frac{1}{N_i} \sum_{j=1}^{N_r} \{S\}_{ij}$$

where N_i is the number of areas within a radius of r_{z_i} units from the centroid of area i (including area i), $N_r = \max_i \{N_i\}$, and z_i represents the degree of remoteness for area i, where $z_i = 1$ corresponds to an area in a major city.

10.3.4   Statistical Software

Code for implementing the models in freely available software (Table 10.1) is available on request, as are the data sets.

The main software used to implement the statistical models were WinBUGS [31] and JAGS [38], which were run via R [39] using the packages R2WinBUGS [44] and R2jags [45] respectively, and also the R package CARBayes [25]. R-INLA [33] was also used for one model.

10.3.5   Model Comparison

Models were compared using several criteria, which are described below. The posterior SIR was calculated as $\exp(\mu_i) = \exp(\alpha + R_i)$, as no covariates were included in these models. The median, lower and upper bounds of the 80% CIs were calculated as the 50th, 10th and 90th percentiles of the posterior, respectively.

10.3.5.1   Convergence

Convergence was predominately based on calculating the Geweke convergence diagnostic [16] for each area's posterior SIR. A p-value for the test statistic below 0.01 was interpreted as suggestive evidence of non-convergence for that area. The trace and density plots for a subsample of areas were also examined for convergence.

10.3.5.2   Plausibility of Estimates

To determine how plausible the posterior SIR estimates were, the CI width was visually inspected, with unreasonably large CIs (with many of the 80% CIs spanning $\pm 5000\%$ or more of the median estimate) providing evidence the estimate was not well-defined; while very precise estimates (the majority within $\pm 4\%$) were evidence that uncertainty was not appropriately included. The magnitude of smoothing of the median posterior SIRs in comparison to the raw SIRs was also visually examined. A smoothed SIR which was very similar to the raw SIR was suggestive of under-smoothing, particularly in areas with small populations.

10.3.5.3   Model Goodness-of-Fit

Three model goodness-of-fit measures were considered: Deviance information criterion (DIC) [43], Watanabe-Akaike information criterion (WAIC) [48] and Moran's I on the residuals [35].

DIC and WAIC are both useful for comparing the predictive accuracy between models. Although DIC is a commonly used measure to compare Bayesian models, WAIC has several advantages over DIC, including that it closely approximates Bayesian cross-validation, it uses the entire posterior distribution and it is invariant to parameterisation [46]. For both these measures, smaller values indicate a better fitting model.

Moran's I was applied to the model residuals to determine if spatial autocorrelation was present after fitting the models. This measure can be quite sensitive to the spatial weights matrix used to define the spatial dependencies between areas, and while a range of spatial weights matrices (inverse-distance, third-order neighbours etc) were considered, we used a matrix based on first-order neighbours. As values of Moran's I close to 0 indicate very low or no residual spatial autocorrelation, here we consider values above 0.2 to be suggestive of some positive spatial autocorrelation. The closer Moran's I is to zero, the better the model accounts for spatial autocorrelation [2].

10.3.5.4   Computational Time

The `microbenchmark` R package [34] was used to monitor computational time to run each model. The models were run on two different computers. However, the specifications of these computers were similar and any differences should have a negligible influence on computation time.

10.4   Results and Discussion

Substantive differences in the posterior estimates were observed between the 13 model variants applied, especially for the rare cancer (Table 10.2, Figs. 10.2, 10.3, 10.4, and 10.5). Depending on the model chosen, the modelled SIR estimates for the same geographical area could range from well below to well above the Australian average (Figs. 10.3 and 10.5).

While small numbers in geographical areas require smoothing, it remains possible that the neighbouring areas may have genuinely different incidence rates. These differences would be obscured during the smoothing process. Detecting these differences is problematic, and even many of the models designed to allow for local variation gave results similar to the BYM and Leroux models (Figs. 10.2, 10.3, 10.4, and 10.5), suggesting there was insufficient statistical power to adequately detect local differences. Of the models that obtained greater variation in the median SIR estimates between areas and less smoothing, there was often excessive uncertainty around these estimates, such as the localised autocorrelation model results (Figs. 10.2 and 10.4).

The number of area-specific SIR estimates that had evidence of non-convergence (based on Geweke p-value < 0.01) did vary between models and with the extent of data sparseness. In many cases, very wide CIs were symptomatic of non-convergence. For instance, the localised autocorrelation ($G = 3$) model for the rare cancer had 86% of area-specific SIRs with significant Geweke p-values, suggesting lack of convergence, and this model had among the widest CIs (Table 10.2). In contrast, models which had implausibly narrow CIs generally had very few/no areas with small Geweke p-values. However, overly narrow CIs are equally problematic as they over-exaggerate confidence in the plausibility of the estimates, which may actually be over- or under-smoothed.

In general, especially as data sparsity increased, our application of these models suggested that global models with more smoothing tended to have 'well-behaved',

Table 10.2 Quantitative summary of results across model criteria

	Posterior median smoothed SIR	N (%) Geweke $p < 0.01$	DIC	WAIC	Moran's I on residuals	Computation time (s)
Rare cancer						
BYM	0.95 (0.38, 1.34)	93 (4.3%)	14,511.3	7432.6	0.047	420.8
Leroux	0.95 (0.39, 1.34)	100 (4.6%)	14,523.9	7429.1	0.047	309.5
Geostatistical	0.94 (0.51, 1.84)	3 (0.1%)	14,646.9	7331.3	0.204	76,772.8
P-spline (tensor)	0.95 (0.13, 1.87)	0 (0.0%)	14,515.4	7279.9	0.205	8117.2
P-spline (radial)	0.93 (0.34, 1.97)	4 (0.2%)	14,760.3	7402.6	0.241	1960.1
CAR dissimilarity (binary)	0.94 (0.09, 4.27)	55 (2.6%)	14,360.3	7354.9	0.042	8462.3
CAR dissimilarity (non-binary)	0.94 (0.16, 3.62)	50 (2.3%)	14,315.4	7361.2	0.031	5316.0
Localised autocorrelation ($G = 3$)	0.96 (0.00, 1.71)	1858 (86.3%)	13,571.9	8107.0	0.054	1035.6
Localised autocorrelation ($G = 5$)	0.99 (0.00, 1.61)	151 (7.0%)	13,129.9	7435.9	0.068	1081.4
Locally adaptive (ρ estimated)	0.95 (0.38, 1.34)				0.067	1580.7
Locally adaptive ($\rho = 0.99$)	0.95 (0.38, 1.34)				0.064	264.9
Weighted sum of spatial priors	0.98 (0.09, 2.03)	64 (3.0%)	14,421.8	7414.2	0.094	6815.7
Leroux scale mixture	Would not run					
Common cancer						
BYM	1.00 (0.74, 1.62)	25 (1.2%)	23,680.8	12,148.9	0.131	419.4
Leroux	1.00 (0.74, 1.61)	33 (1.5%)	23,686.6	12,135.1	0.122	308.1
Geostatistical	0.98 (0.69, 1.62)	14 (0.7%)	22,213.4	11,218.7	0.220	98,412.3
P-spline (tensor)	0.97 (0.71, 1.57)	0 (0.0%)	22,173.0	11,110.1	0.228	7640.8

P-spline (radial)	0.97 (0.71,2.38)	0 (0.0%)	22,167.1	11,120.7	0.254	1831.6
CAR dissimilarity (binary)	1.00 (0.74,1.61)	26 (1.2%)	23,681.4	12,132.1	0.121	4617.5
CAR dissimilarity (non-binary)	0.98 (0.65,2.53)	51 (2.4%)	22,947.1	11,698.9	0.120	9188.9
Localised autocorrelation ($G = 3$)	1.00 (0.74,1.62)	27 (1.3%)	23,690.2	12,131.6	0.124	898.4
Localised autocorrelation ($G = 5$)	1.00 (0.74,1.61)	40 (1.9%)	23,690.9	12,132.2	0.124	907.2
Locally adaptive (ρ estimated)	1.00 (0.74,1.61)				0.148	556.0
Locally adaptive ($\rho = 0.99$)	1.00 (0.74,1.61)				0.147	138.2
Weighted sum of spatial priors	1.00 (0.75,1.62)	28 (1.3%)	23,761.2	12,218.1	0.182	6553.6
Leroux scale mixture	1.00 (0.78,1.52)	19 (0.9%)	23,983.4	12,311.3	0.198	9044.0

BYM Besag, York and Mollié, *DIC* Deviance Information Criterion, *SIR* Standardised incidence ratio, *WAIC* Watanabe-Akaike Information Criterion

Notes:

- The posterior median smoothed SIR is the median SIR estimate over all MCMC iterations for each area, then the median over all areas
- The lowest values in each column by type of cancer are shaded light grey. Any Moran's I above 0.2 are shaded a darker grey, as this is considered suggestive of some spatial correlation existing in the residuals
- The Leroux scale mixture model would not run for the rare cancer
- % Geweke, DIC and WAIC are unavailable for the locally adaptive results (which used INLA via custom functions), as it was not possible to obtain the necessary posterior samples to use in our calculations

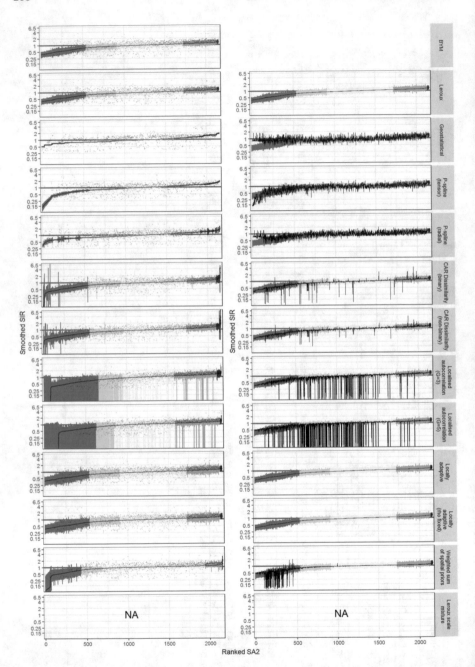

Fig. 10.2 Graphs of posterior SIR results by model, rare cancer. Note: Axes are consistent. Column 1 shows the 80% CI (shaded as per the tones on the maps in Fig. 10.3), the black line is the median SIR (in ascending order), the dots are the raw SIRs and the horizontal line at 1 represents the national average. For column 2, the 80% CIs are the BYM model, and the SA2s are ordered according to the BYM median SIR. The black line is the median estimate for the model named

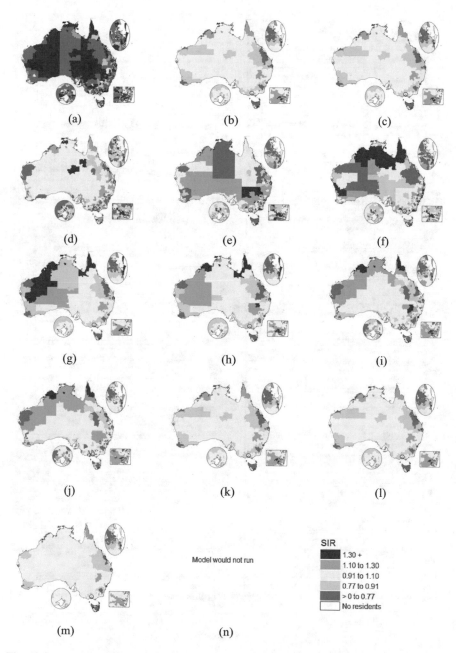

Fig. 10.3 Rare cancer median posterior SIR mapped by model. (**a**) Raw (observed/expected), (**b**) BYM, (**c**) Leroux, (**d**) Geostatistical, (**e**) P-spline (tensor), (**f**) P-spline (radial), (**g**) CAR dissimilarity (binary), (**h**) CAR dissimilarity (non-binary), (**i**) Localised autocorrelation ($G = 3$), (**j**) Localised autocorrelation ($G = 5$), (**k**) Locally adaptive (ρ estimated), (**l**) Locally adaptive ($\rho = 0.99$), (**m**) Weighted sum of spatial priors, (**n**) Leroux scale mixture

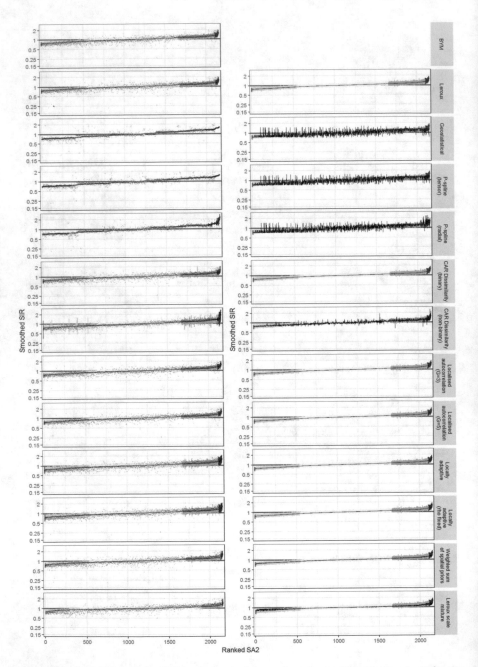

Fig. 10.4 Graphs of posterior SIR results by model, common cancer. Note: Axes are consistent. Column 1 shows the 80% CI (shaded as per the tones on the maps in Fig. 10.5), the black line is the median SIR (in ascending order), the dots are the raw SIRs and the horizontal line at 1 represents the national average. For column 2, the 80% CIs are the BYM model, and the SA2s are ordered according to the BYM median SIR. The black line is the median estimate for the model named

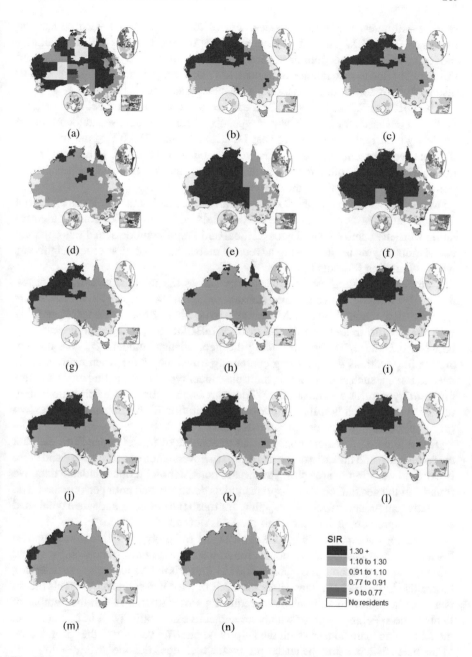

Fig. 10.5 Common cancer median posterior SIR mapped by model. (**a**) Raw (observed/expected), (**b**) BYM, (**c**) Leroux, (**d**) Geostatistical, (**e**) P-spline (tensor), (**f**) P-spline (radial), (**g**) CAR dissimilarity (binary), (**h**) CAR dissimilarity (non-binary), (**i**) Localised autocorrelation ($G = 3$), (**j**) Localised autocorrelation ($G = 5$), (**k**) Locally adaptive (ρ estimated), (**l**) Locally adaptive ($\rho = 0.99$), (**m**) Weighted sum of spatial priors, (**n**) Leroux scale mixture

reliable estimates, while local models tended to struggle in producing plausible estimates (Figs. 10.2 and 10.4). The estimates for the binary CAR dissimilarity model (based on socioeconomic differences) in our study were often unreliable, and this is likely due to its tendency to remove too many neighbours. This is expected to also apply to other formulations, such as distance-based models.

The DIC and WAIC (Table 10.2) measures of goodness of model fit were generally in consensus for a given cancer type, apart from the localised autocorrelation models which had among the lowest DIC, but highest WAIC. Some models fit the data well for one type of simulated data, but not the other. For example, the geostatistical and P-spline models fit the common cancer quite well, but resulted in poor to average model fit for the rare cancer.

Moran's I statistic (Table 10.2) generally indicated that the residual spatial autocorrelation is quite small. The only models with noticeable remaining correlation were the centroid based geostatistical and P-spline models, and this apparent correlation may result from using a weights matrix based on first-order neighbours when calculating Moran's I.

Computational time varied substantially across the models, with times for the rare cancer ranging from 5 minutes (Leroux model in CARBayes) to over 20 hours (geostatistical model in JAGS). Models able to be run in CARBayes were generally very fast, while models run in JAGS or WinBUGS took longer (approximately between 0.5 and 2.5 hours, excluding the geostatistical model). Of note though, are the implications these varying computing times may have when many models need to be run, such as considering multiple cancer types, or repeating models to test different hyperprior specifications. While increasing computing specifications may reduce these times, it is still an important consideration when choosing between two (or more) otherwise well performing models.

It is a tenet of statistical research that the choice of model depends on the data characteristics and the aims of the analysis. However, when data are sparse and there is extreme variation in area size, such as are consistent with our simulated data, we found that the geostatistical or P-spline models generally had poor performance. The geostatistical model is prohibitively slow for these type of data, and when combined with the unpredictable model fit, this model is not recommended.

A previous comparison by Adin et al. [1] of the global P-spline model against the moving average and CAR models found the P-spline performed well for sparse disease mapping, although Goicoa et al. [17] found it to be more prone to detecting more false high-risk areas than either the CAR or a local P-spline model. This model is also rather complex to implement, requiring a penalty matrix and the number of knots to be specified, both of which are subjective and can have a large impact on model fit. The main concern with the P-spline model, however, was the specification of the basis matrix using the tensor product, which does not adequately address the fact that the SA2s are irregular in shape and the distances between their centroids can be vastly different. The radial basis version of the P-spline model was designed to address this, but aside from being computationally faster, it provided similar levels of smoothing and a worse model fit.

The BYM and Leroux models may be prone to over-smoothing when neighbouring areas have abrupt differences [23, 27], but they generally converged, provided a reasonable model fit with plausible estimates and were computationally efficient to implement. The Leroux model may be preferred over the BYM model to avoid the inability of the BYM model to identify both the structured and unstructured spatial random effects separately, but we found that in some cases it struggled to achieve convergence for its mixing parameter.

The locally adaptive models provided results similar to that of the BYM model, with slightly wider credible intervals. The main disadvantage was the difficulty in obtaining samples from the posterior due to the script calling up INLA from within another function.

A non-binary dissimilarity model may also provide an adequate fit, as this smooths more than a P-spline but less than BYM or Leroux. The non-binary dissimilarity formulation using the SEIFA covariate worked quite well for both cancer types, with noticeably less constraining of modelled SIR estimates than under BYM or Leroux. Whether these SIR estimates are appropriate or are under-smoothed will depend on data characteristics and the aims of the analysis.

Note that the final specification of each model requires additional sensitivity analyses to determine the influence of the priors and hyperpriors, the topic of which was outside the scope of this chapter.

10.5   Conclusion

The number of Bayesian spatial models available continues to increase, along with the capacity of the computing software and hardware. Determining the optimal amount of smoothing in spatial analyses remains difficult, but our study demonstrates the benefits of running a range of model types and provides insights into the relative merits of the different models for the study dataset. Comparing estimates from several different model types is important to assess consistency of results when conducting a spatial analysis

In summary, in sparse data contexts, the BYM, Leroux, locally adaptive, non-binary CAR dissimilarity models or some versions of localised autocorrelation models may outperform the other models examined. We suggest considering using centroid-based smoothing models only when areas are of similar size and shape.

References

1. A. Adin, M.A. Martinez-Beneito, P. Botella-Rocamora, T. Goicoa, M.D. Ugarte, Smoothing and high risk areas detection in space-time disease mapping: a comparison of P-splines, autoregressive, and moving average models. Stoch. Environ. Res. Risk Assess. **31**(2), 403–415 (2017). https://doi.org/10.1007/s00477-016-1269-8

2. C. Anderson, L. Ryan, A Comparison of spatio-temporal disease mapping approaches including an application to ischaemic heart disease in New South Wales, Australia. Int. J. Environ. Res. Public Health **14**(2), 146 (2017). https://doi.org/10.3390/ijerph14020146

3. Australian Bureau of Statistics [ABS], Census of Population and Housing: Socio-Economic Indexes for Areas (SEIFA), Australia, 'Statistical Area Level 2, Indexes, SEIFA 2011', data cube: Excel spreadsheet, cat. no. 2033.0.55.001. (2013). www.abs.gov.au/AUSSTATS/abs.nsf/DetailsPage/2033.0.55.0012011

4. Australian Bureau of Statistics [ABS], Australian Statistical Geography Standard (ASGS): Volume 1 - Main structure and greater capital city statistical areas, July 2011, cat. no. 1270.0.55.001 (2013). www.abs.gov.au/AUSSTATS/abs.nsf/DetailsPage/1270.0.55.001July%202011

5. S. Banerjee, B.P. Carlin, A.E. Gelfand, *Hierarchical Modeling and Analysis for Spatial Data*. Monographs on Statistics and Applied Probability, 2nd edn., vol. 135 (CRC Press/Chapman & Hall, Boca Raton, 2014)

6. J. Besag, Spatial interaction and the statistical analysis of lattice systems. J. R. Stat. Soc. Ser. B (Stat. Methodol.) **36**(2), 192–236 (1974)

7. J. Besag, C. Kooperberg, On conditional and intrinsic autoregressions. Biometrika **82**(4), 733–746 (1995). https://doi.org/10.1093/biomet/82.4.733

8. J. Besag, J. York, A. Mollié, Bayesian image restoration with application in spatial statistics. Ann. Inst. Stat. Math. **43**(1), 1–20 (1991). https://doi.org/10.1007/BF00116466

9. N. Best, S. Cockings, J. Bennett, P. Wakefield, Ecological regression analysis of environmental benzene exposure and childhood leukaemia: sensitivity to data inaccuracies, geographical scale and ecological bias. J. R. Stat. Soc. Ser. A (Stat. Soc.) **164** (1), 155–174 (2001). https://doi.org/10.1111/1467-985x.00194

10. N. Best, S. Richardson, A. Thomson, A comparison of Bayesian spatial models for disease mapping. Stat. Methods Med. Res. **14**(1), 35–59 (2005). https://doi.org/10.1191/0962280205sm388oa

11. A.C.A. Clements, N.J.S. Lwambo, L. Blair, U. Nyandindi, G. Kaatano, S. Kinung'hi, J.P. Webster, A. Fenwick, S. Brooker, Bayesian spatial analysis and disease mapping: tools to enhance planning and implementation of a schistosomiasis control programme in Tanzania. Trop. Med. Int. Health **11** (4), 490–503 (2006). https://doi.org/10.1111/j.1365-3156.2006.01594.x

12. P. Congdon, Representing spatial dependence and spatial discontinuity in ecological epidemiology: a scale mixture approach. Stoch. Environ. Res. Risk Assess. **31**(2), 291–304 (2017). https://doi.org/10.1007/s00477-016-1292-9

13. S.M. Cramb, K.L. Mengersen, P.D. Baade, Identification of area-level influences on regions of high cancer incidence in Queensland, Australia: a classification tree approach. BMC Cancer **11**, 311 (2011). https://doi.org/10.1186/1471-2407-11-311

14. D.G.T. Denison, C.C. Holmes, Bayesian partitioning for estimating disease risk. Biometrics, **57**(1), 143–149 (2001). https://doi.org/10.1111/j.0006-341x.2001.00143.x

15. L.E. Eberly, B.P. Carlin, Identifiability and convergence issues for Markov chain Monte Carlo fitting of spatial models. Stat. Med. **19**(17), 2279–2294 (2000). https://doi.org/10.1002/1097-0258(20000915/30)19:17/18<2279::aid-sim569>3.0.co;2-r

16. J. Geweke, Evaluating the accuracy of sampling-based approaches to the calculation of posterior moments, in *Bayesian Statistics*, vol. 4, ed. by J.M. Bernardo, J. Berger, A.P. Dawid, A.F.M. Smith (Oxford University Press, Oxford, 1992), pp. 169–193

17. T. Goicoa, M.D. Ugarte, J. Etxeberria, A.F. Militino, Comparing CAR and P-spline models in spatial disease mapping. Environ. Ecol. Stat. **19**(4), 573–599 (2012). https://doi.org/10.1007/s10651-012-0201-8

18. P.J. Green, S. Richardson, Hidden Markov models and disease mapping. J. Am. Stat. Assoc. **97**(460), 1055–1070 (2002). https://doi.org/10.1198/016214502388618870

19. C. Kandhasamy, K. Ghosh, Relative risk for HIV in India–an estimate using conditional autoregressive models with Bayesian approach. Spatial Spatio-temporal Epidemiol. **20**, 13–21 (2017). https://doi.org/10.1016/j.sste.2017.01.001

20. S.Y. Kang, S.M. Cramb, N.M. White, S.J. Ball, K.L. Mengersen, Making the most of spatial information in health: a tutorial in Bayesian disease mapping for areal data. Geospat. Health **11**(2), 428 (2016). https://doi.org/10.4081/gh.2016.428

21. L. Knorr-Held, G. Raßer, Bayesian detection of clusters and discontinuities in disease maps. Biometrics **56**(1), 13–21 (2000). https://doi.org/10.1111/j.0006-341x.2000.00013.x

22. S. Lang, A. Brezger, Bayesian P-Splines. J. Comput. Graph. Stat. **13**(1), 183–212 (2004). https://doi.org/10.1198/1061860043010

23. A.B. Lawson, A. Clark, Spatial mixture relative risk models applied to disease mapping. Stat. Med. **21**(3), 359–370 (2002). https://doi.org/10.1002/sim.1022

24. D. Lee, A comparison of conditional autoregressive models used in Bayesian disease mapping. Spatial Spatio-temporal Epidemiol. **2**(2), 79–89 (2011). https://doi.org/10.1016/j.sste.2011.03.001

25. D. Lee, *CARBayes Version 4.7: An R package for Spatial Areal Unit Modelling with Conditional Autoregressive Priors* (University of Glasgow, Glasgow, 2017). https://CRAN.R-project.org/package=CARBayes

26. D. Lee, R. Mitchell, Boundary detection in disease mapping studies. Biostatistics **13** (3), 415–426 (2012). https://doi.org/10.1093/biostatistics/kxr036

27. D. Lee, R. Mitchell, Locally adaptive spatial smoothing using conditional auto-regressive models. J. R. Stat. Soc. Ser. C (Appl. Stat.) **62**(4), 593–608 (2013). https://doi.org/10.1111/rssc.12009

28. D. Lee, C. Sarran, Controlling for unmeasured confounding and spatial misalignment in long-term air pollution and health studies. Environmetrics **26**(7), 447–487 (2015). https://doi.org/10.1002/env.2348

29. B.G. Leroux, X. Lei, N. Breslow, Estimation of disease rates in small areas: a new mixed model for spatial dependence. Stat. Models Epidemiol. Environ. Clin. Trials **116**, 179–191 (2000). https://doi.org/10.1007/978-1-4612-1284-3_4

30. H. Lu, C. Reilly, S. Banerjee, B. Carlin, Bayesian areal wombling via adjacency modelling. Environ. Ecol. Stat. **14**(4), 433–452 (2007). https://doi.org/10.1007/s10651-007-0029-9

31. D.J. Lunn, A. Thomas, N. Best, D. Spiegelhalter, WinBUGS–a Bayesian modelling framework: concepts, structure, and extensibility. Stat. Comput. **10**(4), 325–337 (2000). https://doi.org/10.1023/a:1008929526011

32. Y.C. MacNab, Spline smoothing in Bayesian disease mapping. Environmetrics **18**(7), 727–744 (2007). https://doi.org/10.1002/env.876

33. T.G. Martins, D. Simpson, F. Lindgren, H. Rue, Bayesian computing with INLA: new features. Comput. Stat. Data Anal. **67**, 68–83 (2013). https://doi.org/10.1016/j.csda.2013.04.014

34. O. Mersmann, microbenchmark: accurate timing functions. R package version 1.4-6 (2018). http://CRAN.R-project.org/package=microbenchmark

35. P.A.P. Moran, Notes on continuous stochastic phenomena. Biometrika **37**(1/2), 17–23 (1950). https://doi.org/10.2307/2332142

36. F.S. Nathoo, P. Ghosh, Skew-elliptical spatial random effect modeling for areal data with application to mapping health utilization rates. Stat. Med. **32**(2), 290–306 (2013). https://doi.org/10.1002/sim.5504

37. A. Perperoglou, P.H.C. Eilers, Penalized regression with individual deviance effects. Comput. Stat. **25**(2), 341–361 (2010). https://doi.org/10.1007/s00180-009-0180-x

38. M. Plummer, JAGS version 4.3.0 user manual (2017). https://CRAN.R-project.org/package=CARBayes

39. R Core Team, R: A language and environment for statistical computing. R Foundation for Statistical Computing (2018). http://www.R-project.org/

40. S. Richardson, Statistical methods for geographical correlation studies, in *Geographical and Environmental Epidemiology*, ed. by P. Elliot, J. Cuzick, D. English, R. Stern (Oxford University Press, Oxford, 1996), pp. 181–204

41. A. Riebler, S.H. Sørbye, D. Simpson, H. Rue, An intuitive Bayesian spatial model for disease mapping that accounts for scaling. Stat. Methods Med. Res. **25**(4), 1145–1165 (2016). https://doi.org/10.1177/0962280216660421

42. D. Ruppert, M.P. Wand, R.J. Carroll, *Semiparametric Regression* (Cambridge University Press, Cambridge, 2003)
43. D.J. Spiegelhalter, N.G. Best, B.P. Carlin, V. der Linde, Bayesian measures of model complexity and fit (with discussion). J. R. Stat. Soc. Ser. B (Stat. Methodol.) **64** (4):583–640 (2002). https://doi.org/10.1111/1467-9868.00353
44. S. Sturtz, U. Ligges, A. Gelman, R2WinBUGS: a package for running WinBUGS from R. J. Stat. Softw. **12**(3), 1–16 (2005). https://doi.org/10.18637/jss.v012.i03
45. Y.-S. Su, M. Yajima, R2jags: using R to run 'JAGS'. R package version 0.5-7 (2015). https://CRAN.R-project.org/package=R2jags
46. A. Vehtari, A. Gelman, J. Gabry, Practical Bayesian model evaluation using leave-one-out cross-validation and WAIC. Stat. Comput. **27**(5), 1413–1432 (2017). https://doi.org/10.1007/s11222-016-9709-3
47. L.A. Waller, B.P. Carlin, Disease mapping, in *Handbook of Spatial Statistics*, ed. by A.E. Gelfand, P.J. Diggle, P. Guttorp, M. Fuentes (CRC Press, Boca Raton, 2010)
48. S. Watanabe, Asymptotic equivalence of Bayes cross validation and widely applicable information criterion in singular learning theory. J. Mach. Learn. Res. **11**, 3571–3594 (2010)

Chapter 11
An Ensemble Approach to Modelling the Combined Effect of Risk Factors on Age at Parkinson's Disease Onset

Aleysha Thomas, Paul Wu, Nicole M. White, Leisa Toms, George Mellick, and Kerrie L. Mengersen

Abstract Ensemble approaches to statistical modelling combine multiple statistical methods to form a comprehensive analysis. They are of increasing interest for problems that involve diverse data sources, complex systems and subtle outcomes of interest. An example of such an ensemble approach is described in this chapter, in the context of a substantive case study that aimed to tease out factors affecting the age at onset of the neurodegenerative medical condition, Parkinsons Disease (PD), with a particular focus on the role of a particular potential risk factor, pesticide exposure.

Keywords Bayesian network · Parkinson's disease · Organochlorine pesticide · Risk factors · Combined effect

11.1 Introduction

Parkinson's disease (PD) is the second most common neurodegenerative disorder in the world. The aetiology and pathogenesis of PD is not well understood [1]. Genetic studies have identified multiple genes and genetic variations associated with the disease [2–10] and research on twins have discussed the larger role of non-genetic risk factors on the incidence of PD [11–17]. The association between the

A. Thomas · P. Wu · K. L. Mengersen (✉)
ARC Centre of Excellence for Mathematical and Statistical Frontiers, Parkville, VIC, Australia

Queensland University of Technology, Brisbane, QLD, Australia
e-mail: p.wu@qut.edu.au; K.Mengersen@qut.edu.au

N. M. White · L. Toms
School of Public Health and Social Work, Institute of Health and Biomedical Innovation,
Queensland University of Technology, Brisbane, QLD, Australia

G. Mellick
Griffith Institute for Drug Discovery, Griffith University, Brisbane, QLD, Australia

K. L. Mengersen et al. (eds.), *Case Studies in Applied Bayesian Data Science*, Lecture Notes in Mathematics 2259, https://doi.org/10.1007/978-3-030-42553-1_11

incidence of PD and non-genetic risk factors has been well studied [17–19]. Some of the risk factors that have been analysed for an association with the incidence of PD include family history [18, 20, 21], smoking [22, 23], coffee [23, 24], alcohol intake [25–27], pesticide exposure [28–31], prior head injury [32–34], stroke [35] as well as red meat consumption [36]. Pesticide exposure, particularly organochlorine pesticides (OCPs) have been known to be associated with the incidence of PD [37, 38]. There have been studies on the association between PD and combined effect of smoking, coffee as well as non-steroidal anti-inflammatory drugs [39, 40], however the combined effect of other risk factors on PD has not been well explored.

A notable characteristic of PD is its gradual onset. The disease is often mistaken for normal ageing [41, 42]. There is usually a lag of two to three years between the time of the first symptom and clinical diagnosis [41]. A systematic understanding of the association between the age at onset and risk factors could lead to early diagnosis of PD in patients. This can contribute to timely disease and symptom management [42]. However, the combined effect of risk factors on the age at PD onset is not well studied or understood. Studies have focused on separate effects of risk factors on the age at PD onset. OCP exposure [43, 44], smoking [45–47], alcohol [45, 46], coffee [47], hydrocarbon exposure [48], plasma nitrate concentrations [49, 50], ferritin iron [51], exercise [45], red meat consumption [46], the use of multivitamins [46], prior head injuries [32, 45, 46] and family history [52, 53] are some risk factors that are associated with the age at onset of PD. However, risk factors for a feature like age at onset for a complex disease such as PD rarely occur in isolation and it is essential to investigate the combined effects of multiple risk factors on the age at PD onset. To the best of our knowledge there have been no such studies on this association. This gap in the literature could be a result of the lack of understanding of the age at onset of PD. Financial limitations and the effort required to conduct a population based study on multiple risk factors could also be reasons not to pursue such a study. Modelling can help to address this gap and inform future studies to better understand the age at PD onset.

This chapter retrospectively analyses the combined effect of non-genetic risk factors and quantitative serum OCP concentrations on the early age at PD onset. As it is expensive and time consuming to attain a single data source on patient information, risk factor exposure as well as serum OCP concentrations, we integrate multiple data sources and available literature. The source of data for PD patient information and risk factors is the Queensland Parkinson's Project (QPP) [54]. Serum OCP concentrations are taken from a general population study in South-East Queensland [55]. Estimates of the association between risk factors and the age at PD onset are sourced from the published literature.

An ensemble model approach, comprised of a meta-analysis and Bayesian Network (BN), is applied to this study. Meta-analysis is a common method where the results of multiple research studies are combined to understand inferences from a study in the context of related studies [56]. A meta-analysis is adopted to determine an overall estimate of the association between risk factors and age at onset from published literature. The method is also applied to the general population study to estimate the combined OCP exposure for each age group and gender. Estimates of the PD patient population and their risk factors, estimates from a systematic survey of the published literature as well as estimates of OCP exposure are integrated into a BN. A BN is a directed acyclic graphical model that are useful for combining uncertain knowledge under a probabilistic framework [57]. This approach is the most appropriate method for integrating the available data sources and estimates into a single model that infers the combined effects of risk factors with uncertainty. The structure of the BN model represents the relationship between risk factors and the age at PD onset, and a conditional probability table (CPT) is determined for each variable [58].

11.2   Data

11.2.1   Queensland Parkinson's Project

The risk factor information was collected on a cohort of 350 PD patients as part of the Queensland Parkinson's Project (QPP). The subjects were recruited from three specialist movement disorder clinics in Brisbane, Australia between 2002 and 2008 [54]. All subjects provided informed consent for the study and were de-identified. At the time of the survey, all patients had a diagnosis of idiopathic PD and no previous treatment. Ethics exemption for the use of this data source was granted by the Queensland University of Technology (QUT) Ethics Committee (Ethics Exemption Number: 1700000480).

11.2.2   OCP Concentration

Concentrations of five organochloride pesticides, namely hexachlorobenzene (HCB), β-hexachlorocyclohexane (β-HCH), trans-nonachlor, p,p'-dichlorodiphenyldichloroethylene (p,p-DDE) and p,p'-dichlorodiphenyltrichloroethane (p,p'-DDT), were measured from pooled samples of human blood serum from males and females collected in Brisbane, Australia in 2002/03, 2006/07, 2008/09, 2010/11 and 2012/13 across the age groups 5–15, 16–30, 31–45, 46–60 and >60 years [55]. The age group 0–4 years was available but excluded from the analysis due to missing data for the samples collected in 2002/2003. Further details on data collection and a summary of the data are provided in Thomas et al. [55]. Ethics approval for this study was granted by The University of Queensland (UQ) Medical Research Ethics Committee and Queensland University of Technology (QUT) Ethics Committee (Ethics Approval Number: 2013000317).

11.2.3 Previous Literature

Estimates of the association between risk factors and age at onset were taken from literature resulting from two unpublished systematic surveys. The systematic survey on the association between the age at PD onset and risk factors yielded 16 studies and the survey on the association between the age at PD onset and OCP exposure resulted in 1 study (presented in Supporting Information). The literature search took place between 3rd and 10th July 2017. The survey was based on protocols described in the Cochrane Handbook for Systematic Reviews of Interventions, Preferred Reporting Items for Systematic Reviews (PRISMA) and Critical Appraisal Skills Programme (CASP). Research articles that reported an odds ratio (OR) estimate and a 95% confidence interval (95% CI) were included in the meta-analysis. ORs for smoking, alcohol, head injury and family history were extracted from articles authored by Tsai et al. [45], Stern et al. [44] and Rybicki et al. [53]. It is acknowledged that these studies span more than a decade, so there is an implicit assumption that the relationships between factors and the age at onset remained similar over this time period.

As the definition of early age at onset varies between the research articles identified in the systematic survey [43–45, 53], we adopted a standard mean age at onset (50 years) as a threshold to interpret the outcome node of the BN.

11.3 Methods

11.3.1 Meta-analysis

A meta-analysis is a model to combine estimates of multiple scientific studies on the same effect of interest [56]. A hierarchical linear model was adopted and implemented using two data sources, namely to combine log OR estimates on the association between an earlier age at PD onset and risk factors from previous literature (Tables 11.5 and 11.6, see Appendix) and combine OCP concentrations for each combination of age group and gender (estimates presented in Thomas et al. [55]).

The following meta-analysis model was adopted to parameterise part of the BN.

$$y_i \sim N(\theta_i, \sigma_i^2)$$

$$\theta_i \sim N(\theta_0, \sigma_0^2)$$

$$\theta_0 \propto 1$$

$$\sigma_0^2 \sim U(0, 100) \qquad (11.1)$$

As the meta-analyses of the effects of smoking and of prior head injury on age at onset of PD involved the combination of only two studies each, the between-study variance in the above model was set to $\sigma_0^2 = 0$. This is equivalent to reversion from a random effects model to a fixed effects model [56].

For each ith OCP, y_i was the concentration estimate, and the OCP-specific mean (θ_i) and variance (σ_i^2) were reported from the original hierarchical model presented in Thomas et al. [55]. Each θ_i was further considered to have been generated from a Normal distribution with an overall mean θ_0 and overall variance σ_0^2. The prior distributions for θ_0 and σ_0 were chosen to represent a lack of knowledge about the overall estimates. Separate models were run for each combination for age group and gender. The population average of OCP concentrations (θ_0) for each age group and gender were used to parameterise part of the BN.

11.3.2   Bayesian Network

BNs can integrate diverse data sources that may include different variables collected over different time periods. Here, the BN modelled the associations between selected non-genetic risk factors, including OCP exposure, and the age at PD onset. Nodes represent risk factors while arcs/links denote direct dependencies between nodes. The dependencies between nodes are parameterised by conditional probability tables (CPTs). The conditional probability table for each node X_i describes the probability of that node given its parent nodes or $P(X_i|parents(X_i))$. Once the BN is specified, the joint probability distribution of a collection of nodes can be determined by the local CPTs.

11.3.2.1   Building the BN

The main outcome of interest in this analysis was an early age at PD onset. The BN was applied to study the combined impact of selected non-genetic risk factors on an early age at PD onset (Fig. 11.1 with definitions in Table 11.1). The risk factors included smoking, alcohol, head injury, family history, age, gender and OCP exposure. As there is no single data source that explicitly captures interactions between all risk factors, we developed a network model to combine data sources as well as the effects of risk factors using latent nodes, lifestyle effect and medical history effect. These nodes reflected the effect of the relevant risk factors that are conducive to an early age at onset. The latent nodes were conceptualised to describe the combined effect of lifestyle, medical history and OCP exposure on an early age at onset.

The inclusion of the latent nodes also managed the size of the CPT for the terminal node as the CPT size scales quickly with the number of parent nodes. Fewer states in a node also reduced the size of the CPT and avoided over-parameterisation

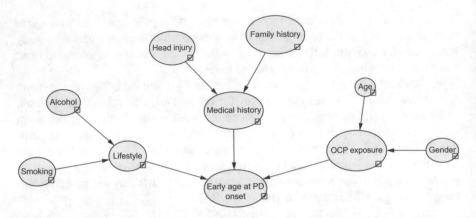

Fig. 11.1 Bayesian network. Graphical model of multiple risk factors and the final outcome—an early age at onset given that the patient will develop PD. The risk factors in the BN are smoking and alcohol which lead to an overall lifestyle effect, family history and head injury which lead to an overall medical history effect as well as age and gender which determine the OCP exposure levels

given the availability of data. In this BN, all nodes except 'Age' were limited to two states (Table 11.1).

We quantified the CPTs for the BN using the systematic survey of previous literature as well as the QPP and OCP population study. The CPTs for the root nodes, smoking, alcohol, head injury, family history, age and gender were quantified with the QPP data source. The probabilities were estimated using the proportion of patients with these risk factors. The meta-analysis method was adopted to parameterise the CPT for the OCP exposure node as described in Sect. 11.3.1. Thresholds were applied to the posterior distribution of θ_0 (described in Eq. 11.1) to obtain the conditional probability for 'High' or 'Low' OCP concentration for each age and gender combination. The threshold values were taken from the quantiles of the posterior distribution of overall OCP concentrations, irrespective of age and gender. The CPTs for the lifestyle and medical history effect as well as the early age at PD onset are parameterised from OR estimates obtained from the meta-analysis on previous studies.

Let Y represent early age at onset and X represent the presence of one or more risk factors, which include smoking, alcohol, head injury or family history. Let X^c represent the absence of risk factors and define E as,

$$E = Y \cap (X) \tag{11.2}$$

where E can be a lifestyle or medical history effect conducive to an early age at onset, such that when E represents lifestyle, $X = \{X_1 = \text{smoking}, X_2 = \text{alcohol}\}$. When E represents medical history, $X = \{X_1 = \text{head injury}, X_2 = \text{family history}\}$.

We wish to quantify the conditional probabilities $P(E|X)$ using the data available in the form of ORs and marginal probabilities derived from the meta-

Table 11.1 Summary of nodes, states, definitions and data sources for BN

Node	State	Definition	Data source	CPT input
Age	16–30/31–45/46–60/>60	Patient's age group	QPP	The proportion of patients in an age group. Age group limits were derived from the OCP population study.
Gender	Male/Female	Gender of patient	QPP	The proportion of patients by gender
Smoking	Yes/No	Has the patient smoked daily/weekly for one or more years?	QPP	The proportion of patients who have smoked for daily/weekly for one or more years
Alcohol	Yes/No	Has the patient had alcohol daily weekly for one or more years?	QPP	The proportion of patients who have had alcohol daily/weekly for one or more years
Head injury	Yes/No	History of prior head injury that led to unconsciousness or the patient admitted to hospital	QPP	The proportion of patients with a history of head injury prior to PD onset that led to unconsciousness or the hospital
Family history	Yes/No	History of PD among patient's blood relatives	QPP	The proportion of patients who have a family history of PD among blood relatives
OCP exposure	High/Low	Probability of a patient, of a specified age group and gender, having an overall OCP concentration above or below a defined threshold	OCP population study	The probability of having an overall OCP concentration level above/below a defined threshold
Lifestyle effect	Yes/No	A lifestyle effect/risk factors that are conducive to an early age at onset	Systematic literature survey	Approximated from OR estimates using Eq. 11.5
Medical history effect	Yes/No	A medical history effect/risk factors that are conducive to an early age at onset	Systematic literature survey	Approximated from OR estimates using Eq. 11.5
Early age at PD onset	Yes/No	An early age at onset given that the patient will develop PD	Systematic literature survey	Approximated using Eq. 11.7

analyses. These ORs take the following form,

$$OR(E|X) = \frac{odds(E|X)}{odds(E|X^c)} = \frac{\frac{P(E|X)}{1 - P(E|X)}}{\frac{P(E|X^c)}{1 - P(E|X^c)}} \tag{11.3}$$

Equation 11.3 can be rearranged to express $OR(E|X)$ entirely in terms of $P(E|X)$ and its marginal probabilities, $P(X)$, $P(X^c)$, $P(E)$ and $P(E^c)$. Solving for $P(E|X)$, we obtain

$$P(E|X) = \frac{OR(E|X)P(X) + OR(E|X)P(E) - P(X^c) + P(E) \pm \sqrt{\psi}}{2[OR(E|X)P(X) + P(X)]}$$

$$\text{where } \psi = [OR(E|X)P(X) + OR(E|X)P(E) + P(X^c) + P(E)]^2$$

$$- 4[OR(E|X)P(X) + P(X)][OR(E|X)P(E)] \tag{11.4}$$

In the presence of more than one risk factor, due to the absence of data covering different combinations of risk factors, we assume conditional independence. For example, to estimate the lifestyle effects CPT $P(E|X_1, X_2)$, assuming smoking (X_1) and alcohol (X_2) are conditionally independent, the CPT is quantified as,

$$P(E|X_1, X_2) = \frac{P(E|X_1)P(E|X_2)}{P(E)} \tag{11.5}$$

where $P(E|X_1)$ and $P(E|X_2)$ can be evaluated as per Eq. 11.4. From Eq. 11.2 where we have the presence of at least one risk factor, $P(E)$ is,

$$P(E) = P(Y, X_1, X_2) + P(Y, X_1, X_2^c) + P(Y, X_1^c, X_2) \tag{11.6}$$

Full details on the derivation of Eqs. 11.5 and 11.6 are provided in Supplementary Material.

Using the QPP data source, we derived an OR for early onset given pesticide exposure using logistic regression. We assume this is approximately equal to the OR for early onset given OCP, $OR(Y|OCP)$ where OCP represents exposure to OCPs. Thus, we apply Eq. 11.4 to estimate $P(Y|E_{OCP})$ where E_{OCP} represents the effect of OCP exposure conducive to an early age at onset.

To estimate the conditional probabilities of the terminal node $P(Y|E_L, E_M, E_{OCP})$, we assume that E_L, E_M and E_{OCP} are conditionally independent of each other due to the lack of existing studies on the combined effects of lifestyle, medical history and OCP exposure on an early age at onset of PD.

$$P(Y|E_L, E_M, E_{OCP}) = \frac{P(Y|E_L)P(Y|E_M)P(Y|E_{OCP})}{P(Y)^2} \tag{11.7}$$

We can obtain $P(Y|E_L)$ and equivalently $P(Y|E_M)$ based on their relevant risk factors.

$$P(Y|E_L) = P(Y|S \vee A) = \frac{P(Y, S, A) + P(Y, S, A^c) + P(Y, S^c, A)}{P(S, A) + P(S, A^c) + P(S^c, A)} \qquad (11.8)$$

where S and A represent smoking and alcohol respectively and S^c and A^c represent the absence of smoking and alcohol respectively.

Additional detail can be found in supplementary information. The resultant CPTs for the BN are presented in Tables 11.7, 11.8, 11.9, 11.10 and 11.11.

11.3.2.2 Network Interrogation and Outcomes

The BN was created, quantified and analysed in *GeNIe 2.0* [59]. The BN was conditional on patient-only cases. A sensitivity analysis as well as strength of influence analysis were applied to observe the impact of risk factors on the terminal node. An additional sensitivity analysis was performed to determine the effect of varying OCP concentration thresholds for the OCP exposure node. The posterior marginal probability distribution for the terminal node was observed when the BN was updated with no set evidence. The probability distribution of the final node was also observed when evidence was set for smoking, alcohol, head injury, family history, age and gender. This was compared to results when evidence was set for the same nodes as well as OCP exposure.

11.4 Results

11.4.1 Meta-analysis

The results of the meta-analysis are given in Table 11.2. The overall odds ratios were estimated to be 0.878 for smoking and 1.094 for head injury; however these were not substantively different from 1 given the large standard deviations.

Table 11.2 Summary of observed log odds ratio with corresponding 95% confidence intervals and combined log odds ratio with corresponding standard deviation on the association between age at onset of PD and smoking as well as head injury

Risk factor	Study reference	Observed Log OR (95%CI)	Combined Log OR (SD)
Smoking	Stern	−0.22 (−0.92, 0.47)	−0.13 (1.10)
	Tsai	−0.08 (−1.51, 1.34)	
Head injury	Stern	1.10 (0.18, 2.03)	0.09 (1.18)
	Tsai	1.50 (0.05, 2.97)	

Table 11.3 Summary of posterior θ_0 and σ_0 of five OCPs (HCB, β-HCH, transnonachlor, p,p'-DDE and p,p'-DDT) for each combination of age group and gender

Age	Gender	θ_0			σ_0		
		Mean	2.5%	97.5%	Mean	2.5%	97.5%
16–30	Male	1.9554	−0.6976	4.522	1.726	0.9304	6.768
16–30	Female	2.0903	−0.5872	4.7044	1.756	0.9517	6.91
31–45	Male	2.4085	−0.1175	4.8258	1.598	0.8536	6.308
31–45	Female	2.4258	−0.1482	4.9215	1.656	0.888	6.605
46–60	Male	2.9588	0.0729	5.7329	1.886	1.0251	7.289
46–60	Female	3.1691	0.3034	5.86	1.825	0.9839	7.104
>60	Male	3.6816	0.7535	6.5348	1.917	1.037	7.561
>60	Female	4.0266	1.0022	6.9247	1.969	1.0621	7.744

The posterior mean OCP exposure (in ng/g lipid) across five OCPs, four age groups and two genders was 2.854 (2.297, 3.402) and the posterior standard deviation was 1.614 (1.292, 2.121). A summary of posterior mean OCP exposure (in ng/g lipid) across five OCPs for each combination of age group and gender is summarised in Table 11.3.

11.4.2 Bayesian Network

11.4.2.1 Sensitivity Analyses

The results of the sensitivity analysis identified that the terminal node was susceptible to changes in medical history effect, family history, head injury, age and OCP exposure (Fig. 11.2). The analysis demonstrated that based on the quantification of the nodes, the posterior marginal probabilities of an early age at onset could range from 0.037 to 0.708. The terminal node was most sensitive to OCP exposure, followed by medical history, followed by lifestyle.

11.4.2.2 Strength of Influence

Results of the strength of influence in the BN showed that OCP exposure had the strongest direct influence on the terminal node (Table 11.4). Medical history also had a strong influence on the terminal node. OCP exposure and medical history were strongly influenced by age and family history, respectively. Alcohol also had a strong influence on lifestyle however the latter had a weak direct effect on the terminal node.

The sensitivity analysis and strength of influence reconciled the influence of OCP exposure, medical history and lifestyle on the terminal node. The direct influence of

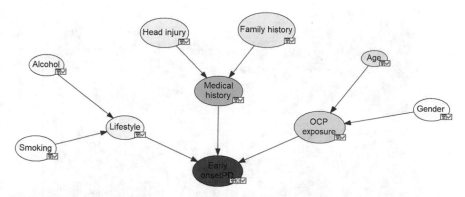

Fig. 11.2 Illustration of sensitivity results on the BN. This illustration provides results of a sensitivity analysis on the BN when the node of interest is an early age at onset. Darker shades of red for a node indicate higher sensitivity of the terminal node to the risk factor

Table 11.4 BN strength of influence

Parent node	Daughter node	Strength of influence
Family history	Medical history effect	0.244
Age	OCP exposure	0.178
Alcohol	Lifestyle effect	0.119
Head injury	Medical history effect	0.115
OCP exposure	Early onset|PD	0.096
Medical history effect	Early onset|PD	0.082
Smoking	Lifestyle effect	0.034
Lifestyle effect	Early onset|PD	0.031
Gender	OCP exposure	0.024

age on OCP exposure and family history on medical history is also observed in both analyses.

11.4.2.3 Altering BN Evidence

When the evidence for OCP exposure was altered in the BN along with smoking, alcohol, head injury, family history, age and gender, the posterior marginal probability of an early age at onset varied with all risk factors (Fig. 11.3). The probability of an early age at onset did not differ by age or gender due to d-separation of the age and gender nodes from an early age at onset. The probability of an early age at onset when OCP exposure was 'High' was larger compared to 'Low' OCP exposure. The difference in 'High' or 'Low' OCP exposure for outcome probabilities was approximately 0.06 if there was no history of head injury or family history. On the other hand, the difference was approximately 0.08 if there was a history of head injury and family history. There was also a small

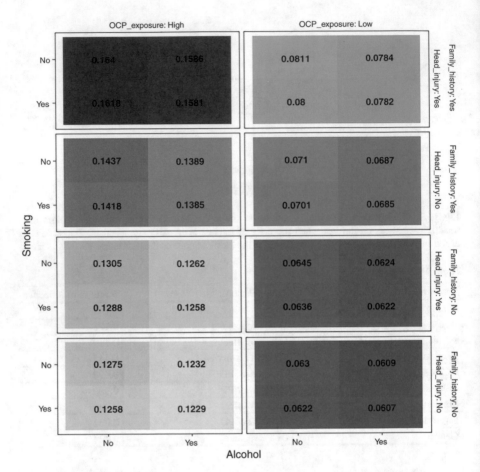

Fig. 11.3 Heat map of the posterior probability of an early age at onset with OCP exposure as input. Illustration of the posterior probability of an early age at onset given PD when the evidence for smoking, alcohol, head injury, family history, age, gender and OCP exposure was altered. Values in the plot are the posterior probabilities of an early age at onset. The range of colours represent the probability where red and green indicate a higher and lower probabilities of an early age at onset respectively

variation in the probability of an early age at onset for smoking and alcohol. The absence of one or both medical history risk factors resulted in a smaller probability of an early age at onset than the presence of both head injury and family history.

When the evidence for smoking, alcohol, head injury, family history, age and gender was altered while keeping OCP exposure constant, the posterior marginal probability of an early age at onset varied by all risk factors that were altered (Fig. 11.4). Similar to the previous results, the probability of an early age at onset varied by smoking, alcohol, head injury and family history. The conditions for the

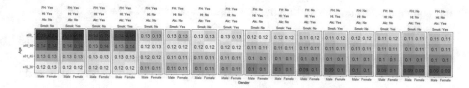

Fig. 11.4 Heat map of the posterior probability of an early age at onset with input from risk factors excluding OCP exposure. Illustration of the posterior probability of an early age at onset when the evidence for smoking (Smok), alcohol (Alc), head injury (HI), family history (FH), age and gender was altered. Values in the plot are the posterior probabilities of an early age at onset. The range of colours represent the probability where red and green indicate a higher and lower probabilities of an early age at onset respectively

smallest probabilities were smoking, alcohol as well the lack of head injury and family history. The presence of head injury and family history without smoking and alcohol had the highest probabilities for an early age at onset. Smoking and alcohol appeared to be protective risk factors for the disease, as demonstrated by previous literature [22, 23]. When the evidence for smoking and alcohol were positive, patients with medical history of only family history had a smaller probability (0.012) of an early age at onset compared to those with only head injury. Older age groups had higher probabilities than younger age groups; the difference was approximately 0.019–0.028 where the difference was larger for patients with exposure to both lifestyle risk factors. Women also had an incrementally higher probability of an early age at onset than men.

11.5   Discussion

This chapter has proposed an ensemble model approach to investigate the combined effects of risk factors on the age at PD onset. We combined a meta-analysis model and BN to identify the combined effect between smoking, alcohol consumption, prior head injury, family history, age, gender and cumulative serum OCP concentrations on an early age at onset of PD. Inferences from the ensemble model focused on how the probability of an early age at onset varied by the presence or absence of risk factors as well as the strength of influence of risk factors.

This analysis integrated OR estimates from a systematic survey on published literature about the association between the age at PD onset and risk factors as well as OCP exposure, QPP data on PD patient risk factors and a population study on serum OCP concentrations. The meta-analysis was adopted to determine the overall estimates for selected risk factors and combined OCP exposure. This was later incorporated into the BN. The results of the meta-analysis were converted into conditional probabilities for the BN. The application of the BN was motivated by its unique probabilistic features that allow multiple disparate data sets to be combined in a single model. The ability to alter evidence for the BN nodes was a key feature that contributed to understanding the combined effects of risk factors on the age at PD onset.

Results of the analysis highlighted that OCP exposure was an influential risk factor that led to variation in the probability of an early age at onset. Medical history effects and its associated risk factors, head injury and family history, also had a substantial effect on the age at onset.

The probability of an early age at PD onset varied widely when evidence for the selected non-genetic risk factors was altered. These results demonstrate the necessity and usefulness of understanding the combined effects of risk factors on PD age at onset and the inclusion of quantitative OCP measurements in an analysis.

When the evidence for all risk factors, including OCP exposure, was altered, there was no variation in the probability of an early age at onset due to d-separation between the terminal node and the nodes for age group. However when OCP exposure was kept constant, there was a difference in the probability for an early age at onset across age groups, genders and the presence of lifestyle and medical history risk factors. The change in probabilities in both cases is indicative of a combined effect of risk factors on the early age at onset of PD. We also observed that altering the threshold of 'High' or 'Low' OCP exposure did not substantively change the probability of an early age at onset.

These results are not meant to be interpreted as conclusive inferences due to the disparity of the data sources as well as the required approximations owing to the lack of data. The resulting inferences are intended to guide potential further studies on the combined impact of risk factors on PD. This ensemble model is not meant to replace a valid, well designed study and analysis.

The absence of a single data source is a major limitation in the study that prevents conclusive inferences. The disparity between the data sources also led to the forced integration of two separate populations (based in Brisbane, Australia) into a single analysis.

The QPP data source did not have information on control subjects. Therefore the BN model was conditional on the patients eventually being diagnosed with PD. As the focus of the paper was on the age at PD onset, it was reasonable to exclude information on controls from the model. Previous studies that conducted analyses focused on age at PD onset had also excluded controls from the study [46, 47, 60, 61].

A range of estimates was used to summarise the research articles identified in the systematic survey of literature (Tables 11.5 and 11.6). ORs were adopted for the ensemble model as they were the most frequently reported estimates on individual risk factors and the age at PD onset. The scarcity of data on the combined effects of risk factors on age at PD onset necessitated the use of assumptions and approximations from the available data sources. We assumed that estimates of ORs from published studies of the association between early onset and a risk factor could approximate the probability of an effect conducive to early onset given exposure to the risk factor. The CPT for the terminal node, early age at PD onset, was also approximated from available OR estimates as there was no information on the combined effects of smoking and alcohol on lifestyle

Table 11.5 Summary of estimates for resulting studies of systematic survey on association between age at PD onset and risk factors

Author	Year	Cases	Controls	Type of estimate	Exposure	Stratification	Estimate	95%CI	p-value
Tsai et al.	2002	90	90	Odds ratio	Ex-Smoker	Patients ≤ 40 yrs	0.92	0.22–3.8	0.76
					Current Smoker	Patients ≤ 40 yrs	2.28	0.19–26.88	0.76
					Ex-Alcohol	Patients ≤ 40 yrs	0.24	0.02–3.26	0.2
					Current Alcohol	Patients ≤ 40 yrs	2.66	0.40–17.5	0.2
					Head injury	Patients ≤ 40 yrs	4.5	1.05–19.4	0.043
					Exercise	Patients ≤ 40 yrs	0.07	0.01–0.54	0.01
Savica et al.	2009	96	196	Odds ratio	Anaemia	Early onset	2.08	1.05–4.15	0.04
						Late onset	1.95	1.14–3.34	0.02
Stern et al.	1991	149	149	Odds ratio	Well water	Patients < 40 yrs	0.9	0.4–2.1	>0.05
						Patients ≥ 60 yrs	1.2	0.6–2.7	
					Rural living	Patients < 40 yrs	1.2	0.6–2.5	>0.05
						Patients ≥ 60 yrs	2.4	1.0–5.4	
					Insecticide	Patients < 40 yrs	0.6	0.2–1.7	>0.05
						Patients ≥ 60 yrs	0.8	0.3–2.1	
					Herbicide	Patients < 40 yrs	0.9	0.5–1.7	>0.05
						Patients ≥ 60 yrs	1.3	0.7–2.4	
					Head injury	Patients < 40 yrs	3	1.2–7.6	>0.05
						Patients ≥ 60 yrs	2.3	0.9–6.1	
					Smoking	Patients < 40 yrs	0.8	0.4–1.6	>0.05
						Patients ≥ 60 yrs	0.5	0.2–1.0	
					Education	Patients < 40 yrs	1	0.5–2.1	>0.05
						Patients ≥ 60 yrs	2.7	1.2–6.2	

(continued)

Table 11.5 (continued)

Author	Year	Cases	Controls	Type of estimate	Exposure	Stratification	Estimate	95%CI	p-value
Maher et al.	2002	203	–	Generalised estimating equation	Multivitamin use	–	3.2 yrs older than mean AAO	–	0.007
					Smoking		Later mean AAO	–	0.0001
					Head injury		3.3 yr earlier mean AAO	–	0.03
					Alcohol	–	No difference	–	0.44
					Coffee	–	No difference	–	0.79
					Red meat	–	No difference	–	0.42
					Pesticides	–	No difference	–	0.31
					Insecticide	–	No difference	–	0.77
					Herbicide	–	No difference	–	0.93
					Rodenticide	–	No difference	–	0.74
					Fungicide	–	No difference	–	0.78
Molina et al.	1994	68	68	Correlation	Plasma nitrate	–	No estimate	–	–
Sanyal et al.	2010	80	80	Correlation	Plasma nitrate	Patients > 50 yrs	correlation coefficient = 0.26	–	0.0191
Pezzoli et al.	2000	188	188	Mean AAO and SE	Hydrocarbon exposure	Early onset	55.2	9.8	0.014
						Late onset	58.6	10	
Molina et al.	1992	37	37	Correlation	Serum lipid peroxide	–	-0.4	–	<0.01
Bartzokis et al.	2004	12	14	Other	Ferritin iron	–	Refer to Tables 1A and 1B in paper	–	–

Study	Year	n	n2	Measure	Factor	Subgroup	Estimate	CI/range	p
Wilk and Lash	2007	2559	–	Beta coefficients	Coffee	All patients	4	1.3–6.7	–
						Patients ≥ 40 yrs	2.1	0–4.1	–
					Ex-smoker	All patients	0.7	–1.3–2.6	–
						Patients ≥ 40 yrs	0.6	–1.1–2.2	–
					Current smoker	All patients	–6.4	–9.9 to –2.9	–
						Patients ≥ 40 yrs	–4.2	–7.4 to –1.1	–
Goldman et al.	2006	93	–	Other	Head injury	–	No estimate	–	>0.05
Jimenez-Jimenez et al.	1992	128	256	Other	Well water drinking, pesticides	–	No estimate	–	–
Ferraz et al.	1996	118	–	Proportions	Rural living	Patients ≤ 40 yrs	5.6%	–	>0.05
					Urban living	Patients ≤ 40 yrs	12.8%	–	–
Marder et al.	2003	477	409	Relative risk	Risk among first degree relatives	Patients ≤ 50 yrs	2.9	1.6–5	0.0002
					Risk among first degree relatives	Patients > 50 yrs	2.7	1.6–4.4	0.0002
Rybicki et al.	1999	144	464	Odds ratio and 95% CI	Family history	Patients < 70 yrs	8.8	3.4–22.8	–
					Family history	Patients > 70 yrs	2.8	3.4–19.2	–
Kuopio et al.	2001	119	238	Mean AAO and SD	Family history	–	61.8	9.1	–
					No family history	–	63.2	9.4	–
De Reuck et al.	2005	512	–	Mean AAO and SD	Never smoked	–	63.9	11.4	0.002
					Current smoker	–	67	9.7	–

Table 11.6 Summary of OR estimates for resulting study of systematic survey on association between age at PD onset and OCP exposure

Author	Year	Cases	Controls	Population	Results	Exposure	Stratification	Estimate	95%CI	p-value
Elbaz et al.	2009	224	557	Mutualite Socialte Agricole	Organochlorine insecticides had a significant association to men with older onset (>65 years) PD	Insecticides	Men	2.2	1.1–4.5	0.03
							Women	1.4	0.5–3.8	0.49
							Men AAO ≤ 65 yrs	1.2	0.5–2.8	0.67
							Men AAO > 65 yrs	4.9	1.4–17.3	0.01
						Carbamate	Men	1.6	0.7–3.7	>0.05
							Men AAO > 65 yrs	1.2	0.3–4.7	>0.05
						Organochlorine	Men	2.4	1.2–5	<0.05
							Men AAO > 65 yrs	4.2	1.5–11.9	<0.05
						Organophosphate	Men	1.8	0.9–3.7	>0.05
							Men AAO > 65 yrs	2.9	0.9–9.1	>0.05
						Pyrethoid	Men	1.4	0.7–2.8	>0.05
							Men AAO > 65 yrs	1.8	0.6 0 5.1	>0.05
						Arsenic	Men	1.5	0.6–3.7	>0.05
							Men AAO > 65 yrs	1.6	0.5–5.1	>0.05

effects conducive to an early age at onset or head injury and family history on medical history effects conducive to an early age at onset. These assumptions made use of the data sources available and incorporated feasible information into the terminal node and latent nodes, lifestyle effect and medical history effects. The resultant CPTs for the BN are presented in Tables 11.7, 11.8, 11.9, 11.10 and 11.11.

Table 11.7 Conditional probability tables for smoking, alcohol, head injury, family history, age and gender

Risk factor	State	Proportion
Smoking	Yes	0.49
	No	0.51
Alcohol	Yes	0.42
	No	0.58
Head injury	Yes	0.13
	No	0.87
Family history	Yes	0.24
	No	0.76
Age group	16–30	0.00
	31–45	0.01
	46–60	0.19
	>60	0.80
Gender	Male	0.65
	Female	0.35

Table 11.8 Conditional probability table for lifestyle effect conducive to an early onset age in PD patients

Smoking	Alcohol	Lifestyle effect	Probability
Yes	Yes	Yes	0.02
Yes	Yes	No	0.98
Yes	No	Yes	0.12
Yes	No	No	0.88
No	Yes	Yes	0.04
No	Yes	No	0.96
No	No	Yes	0.17
No	No	No	0.83

Table 11.9 Conditional probability table for a medical history effect conducive to an early onset age in PD patients

Head injury	Family history	Medical history	Probability
Yes	Yes	Yes	0.50
Yes	Yes	No	0.50
Yes	No	Yes	0.07
Yes	No	No	0.93
No	Yes	Yes	0.17
No	Yes	No	0.83
No	No	Yes	0.03
No	No	No	0.97

Table 11.10 Conditional probability table for OCP exposure at each threshold

Threshold number	Age	Gender	OCP exposure	Probability
1	a16_30	Male	High	0.50
1	a16_30	Male	Low	0.50
1	a16_30	Female	High	0.53
1	a16_30	Female	Low	0.47
1	a31_45	Male	High	0.61
1	a31_45	Male	Low	0.39
1	a31_45	Female	High	0.61
1	a31_45	Female	Low	0.39
1	a46_60	Male	High	0.70
1	a46_60	Male	Low	0.30
1	a46_60	Female	High	0.75
1	a46_60	Female	Low	0.25
1	a60_	Male	High	0.81
1	a60_	Male	Low	0.19
1	a60_	Female	High	0.85
1	a60_	Female	Low	0.15
2	a16_30	Male	High	0.36
2	a16_30	Male	Low	0.64
2	a16_30	Female	High	0.39
2	a16_30	Female	Low	0.61
2	a31_45	Male	High	0.45
2	a31_45	Male	Low	0.55
2	a31_45	Female	High	0.46
2	a31_45	Female	Low	0.54
2	a46_60	Male	High	0.58
2	a46_60	Male	Low	0.42
2	a46_60	Female	High	0.62
2	a46_60	Female	Low	0.38
2	a60_	Male	High	0.71
2	a60_	Male	Low	0.29
2	a60_	Female	High	0.77
2	a60_	Female	Low	0.23
3	a16_30	Male	High	0.30
3	a16_30	Male	Low	0.70
3	a16_30	Female	High	0.33
3	a16_30	Female	Low	0.67
3	a31_45	Male	High	0.39
3	a31_45	Male	Low	0.61
3	a31_45	Female	High	0.40
3	a31_45	Female	Low	0.60

(continued)

Table 11.10 (continued)

Threshold number	Age	Gender	OCP exposure	Probability
3	a46_60	Male	High	0.52
3	a46_60	Male	Low	0.48
3	a46_60	Female	High	0.57
3	a46_60	Female	Low	0.43
3	a60_	Male	High	0.67
3	a60_	Male	Low	0.33
3	a60_	Female	High	0.72
3	a60_	Female	Low	0.28
4	a16_30	Male	High	0.25
4	a16_30	Male	Low	0.75
4	a16_30	Female	High	0.28
4	a16_30	Female	Low	0.72
4	a31_45	Male	High	0.33
4	a31_45	Male	Low	0.67
4	a31_45	Female	High	0.34
4	a31_45	Female	Low	0.66
4	a46_60	Male	High	0.47
4	a46_60	Male	Low	0.53
4	a46_60	Female	High	0.51
4	a46_60	Female	Low	0.49
4	a60_	Male	High	0.62
4	a60_	Male	Low	0.38
4	a60_	Female	High	0.68
4	a60_	Female	Low	0.32
5	a16_30	Male	High	0.15
5	a16_30	Male	Low	0.85
5	a16_30	Female	High	0.17
5	a16_30	Female	Low	0.83
5	a31_45	Male	High	0.20
5	a31_45	Male	Low	0.80
5	a31_45	Female	High	0.21
5	a31_45	Female	Low	0.79
5	a46_60	Male	High	0.34
5	a46_60	Male	Low	0.66
5	a46_60	Female	High	0.38
5	a46_60	Female	Low	0.62
5	a60_	Male	High	0.49
5	a60_	Male	Low	0.51
5	a60_	Female	High	0.56
5	a60_	Female	Low	0.44

Table 11.11 Conditional probability table for final outcome node, an early onset age given PD (EO|PD)

Lifestyle effect	Medical history effect	OCP exposure	EO|PD	Probability
Yes	Yes	High	Yes	0.40
Yes	Yes	High	No	0.60
Yes	Yes	Low	Yes	0.23
Yes	Yes	Low	No	0.77
Yes	No	High	Yes	0.22
Yes	No	High	No	0.78
Yes	No	Low	Yes	0.13
Yes	No	Low	No	0.87
No	Yes	High	Yes	0.32
No	Yes	High	No	0.68
No	Yes	Low	Yes	0.19
No	Yes	Low	No	0.81
No	No	High	Yes	0.18
No	No	High	No	0.82
No	No	Low	Yes	0.10
No	No	Low	No	0.90

The BN model in this paper could be further extended to incorporate information on more risk factors. This would provide a more comprehensive understanding of combined risk factor effects on PD age at onset. The inclusion of genetic data in the BN would provide a quantitative understanding of the influence of family history on the age at onset. It would also be valuable to have longitudinal information on patients to match to the cumulative serum OCP concentrations over time to observe any change in probability of an early age at onset over time and altered risk factors.

Appendix: Supporting Information

Let Y represent early age at onset and X represent the presence of one or more risk factors, which include smoking, alcohol, head injury or family history. Let X^c represent the absence of risk factors and define E as,

$$E = Y \cap (X) \tag{11.9}$$

where E can be a lifestyle or medical history effect conducive to an early age at onset, such that when E represents lifestyle, $X = \{X_1 = \text{smoking}, X_2 = \text{alcohol}\}$. When E represents medical history, $X = \{X_1 = \text{head injury}, X_2 = \text{family history}\}$.

We wish to quantify the conditional probabilities $P(E|X)$ and the data available is in the form of ORs and marginal probabilities as derived from the literature and from meta-analyses. These ORs take the following form,

$$OR(E|X) = \frac{odds(E|X)}{odds(E|X^c)} = \frac{\dfrac{P(E|X)}{1 - P(E|X)}}{\dfrac{P(E|X^c)}{1 - P(E|X^c)}} \tag{11.10}$$

Equation 11.10 can be rearranged to express $OR(E|X)$ entirely in terms of $P(E|X)$ and its marginal probabilities,

$$OR(E|X) = \frac{\dfrac{P(E|X)}{1 - P(E|X)}}{\dfrac{P(E) - P(E|X)P(X)}{P(X^c)}}$$

$$1 - \left(\frac{P(E) - P(E|X)P(X)}{P(X^c)} \right)$$

$$= \frac{P(E|X)\left(1 - \dfrac{P(E) - P(E|X)P(X)}{P(X^c)}\right)}{(1 - P(E|X))\left(\dfrac{P(E) - P(E|X)P(X)}{P(X^c)}\right)}$$

$$= \frac{P(E|X) - P(E|X)\left(\dfrac{P(E) - P(E|X)P(X)}{P(X^c)}\right)}{(1 - P(E|X))\left(\dfrac{P(E) - P(E|X)P(X)}{P(X^c)}\right)}$$

$$= \frac{\dfrac{P(E|X)P(X^c)}{P(X^c)} - \dfrac{P(E|X)(P(E) - P(E|X)P(X))}{P(X^c)}}{(1 - P(E|X))\left(\dfrac{P(E) - P(E|X)P(X)}{P(X^c)}\right)}$$

$$OR(E|X) = \frac{P(E|X)P(X^c) - P(E|X)(P(E) - P(E|X)P(X))}{(1 - P(E|X))(P(E) - P(E|X)P(X))} \tag{11.11}$$

We solve Eq. 11.11 for $P(E|X)$, which involves solving for the roots of the quadratic on $P(E|X)$, to obtain the following expression for the CPT,

$$P(E|X) = \frac{OR(E|X)P(X) + OR(E|X)P(E) + P(X^c) + P(E) \pm \sqrt{\psi}}{2[OR(E|X)P(X) + P(X)]}$$

where $\psi = [OR(E|X)P(X) + OR(E|X)P(E) + P(X^c) + P(E)]^2$

$$- 4[OR(E|X)P(X) + P(X)][OR(E|X)P(E)] \tag{11.12}$$

In the presence of more than one risk factor, due to the absence of data covering different combinations of risk factors, we assume conditional independence. For example, to estimate the lifestyle effects CPT $P(E|X_1, X_2)$, assuming smoking (X_1) and alcohol (X_2) are conditionally independent, the CPT is quantified as,

$$
\begin{aligned}
P(E|X_1, X_2) &= \frac{P(E, X_1, X_2)}{P(X_1, X_2)} \\
&= \frac{P(E, X_1, X_2)}{P(X_1)P(X_2)} \\
&= \frac{P(X_1|E, X_2)P(E, X_2)}{P(X_1)P(X_2)} \\
&= \frac{P(X_1|E)P(E, X_2)}{P(X_1)P(X_2)} \\
&= \frac{P(E|X_1)P(X_1)P(E, X_2)}{P(E)P(X_1)P(X_2)} \\
&= \frac{P(E|X_1)P(E, X_2)}{P(E)P(X_2)} \\
&= \frac{P(E|X_1)P(E|X_2)P(X_2)}{P(E)P(X_2)} \\
P(E|X_1, X_2) &= \frac{P(E|X_1)P(E|X_2)}{P(E)}
\end{aligned}
\tag{11.13}
$$

where $P(E|X_1)$ and $P(E|X_2)$ can be evaluated as per Eq. 11.12. From Eq. 11.9 where we have the presence of atleast one risk factor, $P(E)$ is,

$$
P(E) = P(Y, X_1, X_2) + P(Y, X_1, X_2^c) + P(Y, X_1^c, X_2)
\tag{11.14}
$$

In the QPP data source, we derived an OR for early onset given pesticide exposure using logistic regression. We assume this is approximately equal to the OR for early onset given OCP, $OR(Y|OCP)$ where OCP represents exposure to OCPs. Thus, we apply Eq. 11.12 to estimate $P(Y|E_{OCP})$ where E_{OCP} represents the effect of OCP exposure conducive to an early age at onset.

To estimate the conditional probabilities of the terminal node $P(Y|E_L, E_M, E_{OCP})$, we assume that E_L, E_M and E_{OCP} are conditionally independent of each other due to the lack of existing studies on the combined effects of lifestyle, medical history and OCP exposure on an early age at PD onset. Here, E_L represents lifestyle effect and E_M represents medical history effect.

$$
\begin{aligned}
P(Y|E_L, E_M, E_{OCP}) &= \frac{P(Y, E_L, E_M, E_{OCP})}{P(E_L, E_M, E_{OCP})} \\
&= \frac{P(E_L|Y, E_M, E_{OCP})P(Y, E_M, E_{OCP})}{P(E_L)P(E_M)P(E_{OCP})}
\end{aligned}
$$

$$= \frac{P(E_L|Y)P(Y, E_M, E_{OCP})}{P(E_L)P(E_M)P(E_{OCP})}$$

$$= \frac{P(E_L|Y)P(E_M|Y)P(Y, E_{OCP})}{P(E_L)P(E_M)P(E_{OCP})}$$

$$= \frac{P(E_L|Y)P(E_M|Y)P(Y|E_{OCP})}{P(E_L)P(E_M)}$$

$$= \frac{P(E_L, Y)P(E_M, Y)P(Y|E_{OCP})}{P(E_L)P(E_M)P(Y)^2}$$

$$P(Y|E_L, E_M, E_{OCP}) = \frac{P(Y|E_L)P(Y|E_M)P(Y|E_{OCP})}{P(Y)^2} \tag{11.15}$$

We can obtain $P(Y|E_L)$ and equivalently $P(Y|E_M)$ based on their relevant risk factors as informed by the QPP data source.

$$P(Y|E_L) = P(Y|S \vee A) = \frac{P(Y, S, A) + P(Y, S, A^c) + P(Y, S^c, A)}{P(S, A) + P(S, A^c) + P(S^c, A)} \tag{11.16}$$

where S and A represent smoking and alcohol respectively and S^c and A^c represent the absence of smoking and alcohol respectively.

References

1. W. Dauer, S. Przedborski, Parkinson's disease: mechanisms and models. Neuron **39**, 889–909 (2003)
2. M.J. Farrer, Genetics of Parkinson disease: paradigm shifts and future prospects. Nat. Rev. Genet. **7**, 306–318 (2006)
3. A.A. Hicks, H. Pétursson, T. Jonsson, H. Stefánsson, H.S. Johannsdottir, J. Sainz, M.L. Frigge, A. Kong, J.R. Gulcher, K. Stefansson, et al., A susceptibility gene for late-onset idiopathic Parkinson's disease. Ann. Neurol. **52**, 549–555 (2002)
4. E.R. Martin, W.K. Scott, M.A. Nance, et al., Association of single-nucleotide polymorphisms of the Tau gene with late-onset Parkinson's disease. J. Am. Med. Assoc. **286**, 2245–2250 (2001)
5. T.H. Hamza, et al., Common genetic variation in the HLA region is associated with late-onset sporadic Parkinson's disease. Nat. Genet. **42**, 781–785 (2010)
6. C.B. Lücking, A. Dürr, V. Bonifati, J. Vaughan, G. De Michele, T. Gasser, B.S. Harhangi, G. Meco, P. Denéfle, N.W. Wood, Association between early-onset Parkinson's disease and mutations in the Parkin gene. New Engl. J. Med. **342**, 1560–1567 (2000)
7. M. Periquet, M. Latouche, E. Lohmann, N. Rawal, G. De Michele, S. Ricard, H. Teive, V. Fraix, M. Vidailhet, D. Nicholl, Parkin mutations are frequent in patients with isolated early-onset Parkinsonism. Brain **126**, 1271–1278 (2003)
8. S. Hague, E. Rogaeva, D. Hernandez, C. Gulick, A. Singleton, M. Hanson, J. Johnson, R. Weiser, M. Gallardo, B. Ravina, Early-onset Parkinson's disease caused by a compound heterozygous DJ-1 mutation. Ann. Neurol. **54**, 271–274 (2003)
9. V. Bonifati, P. Rizzu, M.J. van Baren, O. Schaap, G.J. Breedveld, E. Krieger, M.C. Dekker, F. Squitieri, P. Ibanez, M. Joosse, Mutations in the DJ-1 gene associated with autosomal recessive early-onset Parkinsonism. Science, **299**, 256–259 (2003)

10. E.M. Valente, P.M. Abou-Sleiman, V. Caputo, M.M. Muqit, K. Harvey, S. Gispert, Z. Ali, D. Del Turco, A.R. Bentivoglio, D.G. Healy, Hereditary early-onset Parkinson's disease caused by mutations in PINK1. Science **304**, 1158–1160 (2004)

11. R.C. Duvoisin, R. Eldridge, A. Williams, J. Nutt, D. Calne, Twin study Of Parkinson disease. Neurology **31**, 77–77 (1981)

12. C.D. Ward, R.C. Duvoisin, S.E. Ince, J.D. Nutt, R. Eldridge, D.B. Calne, Parkinson's disease in 65 pairs of twins and in a set of quadruplets. Neurology **33**, 815–815 (1983)

13. C. Marsden, Parkinson's disease in twins. J. Neurol. Neurosurg. Psychiatry **50**, 105–106 (1987)

14. R. Marttila, J. Kaprio, M. Koskenvuo, U. Rinne, Parkinson's disease in a nationwide twin cohort. Neurology, **38**, 1217–1217 (1988)

15. T. Zimmerman, M. Bhatt, D. Calne, R. Duvoisin, Parkinson's disease in monozygotic twins: a follow-up. Neurology **41**, 255 (1991)

16. P. Vieregge, K. Schiffke, H. Friedrich, B. Müller, H. Ludin, Parkinson's disease in twins. Neurology **42**, 1453–1453 (1992)

17. C. Tanner, R. Ottman, S. Goldman, et al., Parkinson disease in twins: an etiologic study. J. Am. Med. Assoc. **281**, 341–346 (1999)

18. J.M. Gorell, E.L. Peterson, B.A. Rybicki, C.C. Johnson, Multiple risk factors for Parkinson's disease. J. Neurol. Sci. **217**, 169–174 (2004)

19. H. Checkoway, K. Powers, T. Smith-Weller, G.M. Franklin, W.T. Longstreth, P.D. Swanson, Parkinson's disease risks associated with cigarette smoking, alcohol consumption, and caffeine intake. Am. J. Epidemiol. **155**, 732–738 (2002)

20. H. Payami, K. Larsen, S. Bernard, J. Nutt, Increased risk of Parkinson's disease in parents and siblings of patients. Ann. Neurol. **36**, 659–661 (1994)

21. C.A. Taylor, M.H. Saint-Hilaire, L.A. Cupples, C.A. Thomas, A.E. Burchard, R.G. Feldman, R.H. Myers, Environmental, medical, and family history risk factors for Parkinson's disease: a New England-based case control study. Am. J. Med. Genet. **88**, 742–749 (1999)

22. J.A. Driver, G. Logroscino, J.M. Gaziano, T. Kurth, Incidence and remaining lifetime risk of Parkinson disease in advanced age. Neurology **72**, 432–438 (2009)

23. M.A. Hernán, B. Takkouche, F. Caamaño Isorna, J.J. Gestal-Otero, A meta-analysis of coffee drinking, cigarette smoking, and the risk of Parkinson's disease. Ann. Neurol. **52**, 276–284 (2002)

24. A. Ascherio, S.M. Zhang, M.A. Hernán, I. Kawachi, G.A. Colditz, F.E. Speizer, W.C. Willett, Prospective study of caffeine consumption and risk of Parkinson's disease in men and women. Ann. Neurol. **50**, 56–63 (2001)

25. A.E. Lang, C.D. Marsden, J.A. Obeso, J.D. Parkes, Alcohol and Parkinson disease. Ann. Neurol. **12**, 254–256 (1982)

26. M.A. Hernán, H. Chen, M.A. Schwarzschild, A. Ascherio, Alcohol consumption and the incidence of Parkinson's disease. Ann. Neurol. **54**, 170–175 (2003)

27. N. Palacios, X. Gao, E. O'Reilly, M. Schwarzschild, M.L. McCullough, T. Mayo, S.M. Gapstur, A.A. Ascherio, Alcohol and risk of Parkinson's disease in a large, prospective cohort of men and women. Mov. Disord. **27**, 980–987 (2012)

28. A. Ascherio, H. Chen, M.G. Weisskopf, E. O'Reilly, M.L. McCullough, E.E. Calle, M.A. Schwarzschild, M.J. Thun, Pesticide exposure and risk for Parkinson's disease. Ann. Neurol. **60**, 197–203 (2006)

29. J.R. Richardson, S.L. Shalat, B. Buckley, et al., Elevated serum pesticide levels and risk of Parkinson's disease. Arch. Neurol. **66**, 870–875 (2009)

30. A.F. Hernández, B. González-Alzaga, I. López-Flores, M. Lacasaña, Systematic reviews on neurodevelopmental and neurodegenerative disorders linked to pesticide exposure: methodological features and impact on risk assessment. Environ. Int. **92**, 657–679 (2016)

31. M. Weisskopf, P. Knekt, E. O'Reilly, J. Lyytinen, A. Reunanen, F. Laden, L. Altshul, A. Ascherio, Persistent organochlorine pesticides in serum and risk of Parkinson disease. Neurology **74**, 1055–1061 (2010)

32. S.M. Goldman, C.M. Tanner, D. Oakes, G.S. Bhudhikanok, A. Gupta, J.W. Langston, Head injury and Parkinson's disease risk in twins. Ann. Neurol. **60**, 65–72 (2006)

33. A. Hofman, H. Collette, A. Bartelds, Incidence and risk factors of Parkinson's disease in The Netherlands. Neuroepidemiology **8**, 296–299 (1989)

34. K. Rugbjerg, B. Ritz, L. Korbo, N. Martinussen, J.H. Olsen, Risk of Parkinson's disease after hospital contact for head injury: population based case-control study. Br. Med. J. **337**, a2494 (2008)

35. R.L. Levine, J.C. Jones, N. Bee, Stroke and Parkinson's disease. Stroke **23**, 839–842 (1992)

36. X. Gao, H. Chen, T.T. Fung, G. Logroscino, M.A. Schwarzschild, F.B. Hu, A. Ascherio, Prospective study of dietary pattern and risk of Parkinson disease. Am. J. Clin. Nutr. **86**, 1486–1494 (2007)

37. A. Priyadarshi, S.A. Khuder, E.A. Schaub, S. Shrivastava, A meta-analysis of Parkinson's disease and exposure to pesticides. Neurotoxicology **21**, 435–440 (2000)

38. E.E. Ntzani, M. Chondrogiorgi, G. Ntritsos, E. Evangelou, I. Tzoulaki, Literature review on epidemiological studies linking exposure to pesticides and health effects. EFSA Supporting Publication (2013)

39. D.B. Hancock, E.R. Martin, J.M. Stajich, R. Jewett, M.A. Stacy, B.L. Scott, J.M. Vance, W.K. Scott, Smoking, caffeine, and nonsteroidal anti-inflammatory drugs in families with Parkinson disease. Arch. Neurol. **64**, 576–580 (2007)

40. K.M. Powers, D.M. Kay, S.A. Factor, C.P. Zabetian, D.S. Higgins, A. Samii, J.G. Nutt, A. Griffith, B. Leis, J.W. Roberts, E.D. Martinez, J.S. Montimurro, H. Checkoway, H. Payami, Combined effects of smoking, coffee, and NSAIDs on Parkinson's disease risk. Move. Disord. **23**, 88–95 (2008)

41. A.J. Lees, J. Hardy, T. Revesz, Parkinson's disease. Lancet **373**, 2055–2066 (2009)

42. A.J. Noyce, A.J. Lees, A.-E. Schrag, The prediagnostic phase of Parkinson's disease. J. Neurol. Neurosurg. Psychiatry **87**, 871–878 (2016). jnnp–2015

43. A. Elbaz, J. Clavel, P.J. Rathouz, F. Moisan, J.-P. Galanaud, B. Delemotte, A. Alperovitch, C. Tzourio, Professional exposure to pesticides and Parkinson disease. Ann. Neurol. **66**, 494–504 (2009)

44. M. Stern, E. Dulaney, S.B. Gruber, L. Golbe, M. Bergen, H. Hurtig, S. Gollomp, P. Stolley, The epidemiology of Parkinson's disease: a case-control study of young-onset and old-onset patients. Arch. Neurol. **48**, 903–907 (1991)

45. C. Tsai, S. Lo, L. See, H. Chen, R. Chen, Y. Weng, F. Chang, C. Lu, Environmental risk factors of young onset Parkinson's disease: a case-control study. Clin. Neurol. Neurosurg. **104** (2002), 328–333

46. N.E. Maher, et al., Epidemiologic study of 203 sibling pairs with Parkinson's disease: the GenePD study. Neurology **58**, 79–84 (2002)

47. J.B. Wilk, T.L. Lash, Risk factor studies of age-at-onset in a sample ascertained for Parkinson disease affected sibling pairs: a cautionary tale. Emerg. Themes Epidemiol. **4**, 1 (2007)

48. G. Pezzoli, M. Canesi, A. Antonini, A. Righini, L. Perbellini, M. Barichella, C. Mariani, F. Tenconi, S. Tesei, A. Zecchinelli, K. Leenders, Hydrocarbon exposure and Parkinson's disease. Neurology **55**, 667–673 (2000)

49. J. Molina, F.J. Jiménez-Jiménez, J. Navarro, E. Ruiz, J. Arenas, F. Cabrera-Valdivia, A. Vázquez, P. Fernández-Calle, L. Ayuso-Peralta, M. Rabasa, et al., Plasma levels of nitrates in patients with Parkinson's disease. J. Neurol. Sci. **127**, 87–89 (1994)

50. J. Sanyal, B.N. Sarkar, T.K. Banerjee, S.C. Mukherjee, B.C. Ray, V.R. Rao, Plasma level of nitrates in patients with Parkinson's disease in West Bengal. Neurol. Asia **15**, 55–59 (2010)

51. G. Bartzokis, T.A. Tishler, I.S. Shin, P.H. Lu, J.L. Cummings, Brain ferritin iron as a risk factor for age at onset in neurodegenerative diseases. Ann. N. Y. Acad. Sci. **1012**, 224–36 (2004)

52. K. Marder, G. Levy, E.D. Louis, H. Mejia-Santana, L. Cote, H. Andrews, J. Harris, C. Waters, B. Ford, S. Frucht, et al., Familial aggregation of early- and late-onset Parkinson's disease. Ann. Neurol. **54**, 507–513 (2003)

53. B.A. Rybicki, C.C. Johnson, E.L. Peterson, G.X. Kortsha, J.M. Gorell, A family history of Parkinson's disease and its effect on other PD risk factors. Neuroepidemiology **18**, 270–278 (1999)

54. G.T. Sutherland, G.M. Halliday, P.A. Silburn, F.L. Mastaglia, D.B. Rowe, R.S. Boyle, J.D. O'Sullivan, T. Ly, S.D. Wilton, G.D. Mellick, Do polymorphisms in the familial Parkinsonism genes contribute to risk for sporadic Parkinson's disease? Move. Disord. **24**, 833–838 (2009)

55. A. Thomas, L.-M.L. Toms, F.A. Harden, P. Hobson, N.M. White, K.L. Mengersen, J.F. Mueller, Concentrations of organochlorine pesticides in pooled human serum by age and gender. Environ. Res. **154**, 10–18 (2017)

56. M. Borenstein, L.V. Hedges, J. Higgins, H.R. Rothstein, *Introduction To Meta-analysis* (Wiley Online Library, Hoboken, 2009)

57. J. Pearl, Bayesian networks: a model of self-activated memory for evidential reasoning, in *Proceedings of the 7th Conference of the Cognitive Science Society*, University of California, Irvine, CA, USA, pp. 15–17 (1985)

58. M.J. Druzdzel, F.J. Díez, Combining knowledge from different sources in causal probabilistic models. J. Mach. Learn. Res. **4**, 295–316 (2003)

59. M.J. Druzdzel, GeNIe: a development environment for graphical decision-analytic models, in *Proceedings of the AMIA Symposium* (1999), p. 1206

60. H.B. Ferraz, L.A. Andrade, V. Tumas, L.C. Calia, V. Borges, Rural or urban living and Parkinson's disease. Arquivos de neuro-psiquiatria **54**, 37–41 (1996)

61. J. De Reuck, M. De Weweire, G. Van Maele, P. Santens, Comparison of age of onset and development of motor complications between smokers and non-smokers in Parkinson's disease. J. Neurol. Sci. **231**, 35–39 (2005)

Chapter 12
Workplace Health and Workplace Wellness: Synergistic or Disconnected?

G. Davis, E. Moloney, M. da Palma, Kerrie L. Mengersen, and F. Harden

Abstract Workplace health and wellness is paramount in many businesses and industries, for economic and social reasons. Workplace wellness programs have emerged to meet this need. This paper pursues a deeper understanding of the relationship between workplace health and workplace wellness initiatives in Australia. Based on a survey of published literature, Bayesian networks are developed to describe and quantify factors that contribute to each of these components of workplace efficiency. Workplace health was found to be a complex system of acute and chronic occupational medical conditions, as well as lifestyle factors. Successful wellness programs were found to be those that have a high level of participation and positive financial impacts, and are integrated into business strategy and company culture. It was observed that many workplace wellness programs tend to target non-occupational health risks and that there is an opportunity to address other critical components of worker health risk factors. The outputs of the Bayesian networks can provide an interrogative monitor of workplace health and the potential impact of corresponding wellness initiatives, facilitating the development of more targeted and cost-effective programs.

12.1 Introduction

Occupational health is increasing as a priority in workplaces around the world [44]. In Australia, for example, healthy workers are almost three times more productive in the workforce than their unhealthy counterparts [18]. Chronic disease is recognised as one of the key causes of absenteeism and presenteeism, early retirement and lost productivity in the workforce, and acute disease and adverse mental health are also

G. Davis · E. Moloney · M. da Palma · K. L. Mengersen (✉) · F. Harden
Queensland University of Technology, Brisbane, QLD, Australia
e-mail: K.Mengersen@qut.edu.au

K. L. Mengersen et al. (eds.), *Case Studies in Applied Bayesian Data Science*,
Lecture Notes in Mathematics 2259, https://doi.org/10.1007/978-3-030-42553-1_12

known concerns [7, 8, 15]. Despite government strategies over the past 15 years, many of these health outcomes have not declined [57, 59].

Unhealthy workers and those with chronic diseases are a significant strain on corporate and national spending [7, 24, 48]. This has resulted in an international surge in workplace wellness initiatives that target various health risks [3, 22, 24, 39, 72, 78]. This is not a new phenomenon [27]. For example, a study published in 2003 reported that 66% of more than 1200 organisations in 47 countries offered a formal wellness strategy [74].

Notwithstanding the popularity of workplace wellness programs, their effectiveness has been questioned [22, 56]. This is in part due to the multi-factorial nature of occupational diseases and the many non-occupational contributing factors [8], but also because there is great variability in activities offered through the programs, substantive differences in employee outcomes and little consistency in methods to identify, monitor and evaluate the outcomes and benefits of the programs [24, 54]. Indeed, there is increasing concern that wellness programs can have negative consequences for companies, for example by biasing health data collection [1].

In this paper, creating and maintaining a 'healthy worker' workforce is seen as a complex system, comprising not only health factors but their interaction with personal, social, economic, external and other factors. A model of this system is developed, based on a survey of the occupational health risks including musculoskeletal disorders, cardiovascular diseases, obesity, noise-induced hearing loss, cancers, and respiratory diseases, as well as lifestyle factors, particularly the effect of excessive alcohol consumption on work. A similar systems perspective is taken for workplace wellness, based on a survey of a range of wellness programs with focus on their various definitions, program characteristics, target health areas, methods of evaluation, and documented successes and failures. The systems models for healthy workers and workplace wellness are developed as Bayesian Networks. The models are probabilistically quantified and are then used to provide scenario evaluations and interrogations to develop a deeper understanding of these two areas. The article concludes with a discussion of the strengths and limitations of research on healthy workers and workplace wellness programs, the apparent points of synergy and disconnection and the potential utility of the proposed systems model as a method for integrating and analysing this body of research.

12.2   Methods

The first stage of this study comprised a substantive literature survey which aimed to identify key occupational health risks and factors that contribute to the occupational diseases discussed above. For specificity and scope, a primary focus was on the Australian context. Journal articles were the primary information source, as well as papers published by respected organisations. Several official government documents, such as the National Occupational Health and Safety (OHS) Strategy 2001–2012, were also included. The Safe Work Australia website provided government reports on occupational diseases prevalent in Australia. The Australian Health Survey (AHS) conducted by the Australian Bureau of Statistics (ABS) was used to gauge the current overall health of Australia's adult population. The AHS is a combination of the current National Health Survey (NHS), the National Aboriginal and Torres Strait Islander Health Survey, the National Health Measures Survey (NHMS) and the National Nutrition and Physical Activity Survey (NNPAS). It is the largest, most comprehensive health survey ever conducted in Australia. Given the relatively low unemployment rate (5.4% of Australians over the age of 18 were as of December 2012) [6], it was assumed that the data provided by the AHS on adult Australians is acceptably representative of the general health of Australia's working force.

A similar survey of literature on workplace wellness programs was undertaken, along with a collection of case studies. Both published and unpublished (grey) material was examined. Articles were located through the reference lists of other articles and through keyword searches in Google scholar. The most common searches included "workplace wellness programs" along with other words such as "types", "characteristics", "definitions", "successes", "failures", "diet", "exercise" and other specific phrases. The unpublished or grey material was obtained primarily from similar searches of Google web and from government published studies. Case studies were classified by country and then broken into three main components in order to analyse their effectiveness. These were purpose of the program, methods undertaken, and achievements of the program.

The second stage of the study involved the construction of systems models based on the healthy worker and workplace wellness surveys. The models were developed as Bayesian networks. A Bayesian network (BN) is a representation of a complex or complicated system, which can be graphically depicted as a set of factors (nodes) and their relationships (directed arrows). Hence a network is comprised of a set of nodes, each of which is influenced by 'parent' node/s and which in turn may influence 'daughter' nodes. These connections flow through the system to a final target node. Each node in the network is then probabilistically quantified, taking into account its parent node/s. This quantification can be achieved using diverse data sources. The resultant model provides an overall probability of the target outcome, given the various contributing factors. It can also be interrogated to identify most influential factors that impact on the target outcome, and quantitatively evaluate 'what-if' scenarios involving changes to these factors. BNs have been used in a wide variety of contexts, including many problems in health and industry [9, 36, 55, 76].

Two BNs were created in this study. The first represents the various identified health factors that contribute to the overall target outcome of a 'healthy worker'. The second represents the various identified factors that contribute to a successful workplace wellness initiative. The systems models were quantified in a nominal manner, with a future intention in mind of enabling a workplace to input and assess their own program factors. Each node was categorised as binary, for example 'yes' or 'no', or 'positive' or 'negative'. The model was quantified using a generic approach, as described in Appendix 3. This provides a platform which can be modified for specific workplaces, workforces or wellness programs.

The BNs were critically surveyed by an expert panel comprising an occupational physician, a health scientist, a workplace wellness expert and a statistician. The networks were presented to employer and employee groups in moderated meetings, and feedback was incorporated.

12.3 Results

12.3.1 Healthy Worker Survey

Results of the healthy worker survey are presented in Appendix 1. The survey identified a range of associations between health outcomes and workplace conditions and exposures. The health outcomes included musculoskeletal disorders, cardiovascular disease, noise-induced hearing loss, respiratory illness, cancers and mental health disorders.

Occupational risk factors associated with musculoskeletal disorders included one-time traumatic events, repetitive use, excessive loading, excessive workloads, insufficient rest breaks, fatigue, poor posture and stress. Factors associated with cardiovascular disease included exposure to air pollutants, occupational stress, and lack of social support. Excessive noise was also associated with elevated blood pressure, reduced performance, sleeping difficulties, annoyance and stress, temporary shift in the hearing threshold and tinnitus. Workplace agents that were reported to exacerbate or cause respiratory disease include pesticides, herbicides, dust, lead, fumes, chemicals, gases, faulty air conditioners, particulates and gases emitted from fire and other activities, emissions from furnishings, and so on. Occupational dust was reported to be a significant contributor to the development of bronchitis, asthma, chronic obstructive pulmonary disease and other respiratory illnesses. There were also numerous respiratory carcinogens including asbestos, arsenic, radon, silica, chromium, cadmium, nickel and beryllium. Occupational exposures associated with cancer included asbestos, silica, nickel, chromium, arsenicals, vinyl chloride, and halo ethers as well as ionising and solar radiation. Finally, mental health disorders among workers were found to lead to work impairment and reduced job commitment and satisfaction. Traumatic events associated with work related incidents have been linked to mental disorders, and high mental stress levels have been associated with an increased risk of musculoskeletal diseases, cardiovascular diseases and obesity.

All of these risk factors and health outcomes were documented to contribute to absenteeism and/or presenteeim. Common lifestyle confounders included obesity and alcohol consumption. Obesity was associated with a range of occupational health complications, including an increased risk of hypertension, cardiovascular diseases, asthma, musculoskeletal disorders, and some cancers, increased potential for occupational stress, impaired immune response to chemical exposures and increased risk of disease from occupational neurotoxins. Alcohol consumption reportedly affects hepatic and pancreatic systems and has also been related to increased blood pressure, an increased risk of developing certain cancers, and cerebral dysfunction. It has also been linked with other more subtle effects such as late arrival at work and reduced promotion success. Moderate drinking has also been shown to have adverse health effects.

12.3.2   Workplace Wellness Survey

The results of the workplace wellness survey are presented in Appendix 2. A number of wellness definitions were identified, ranging from that used by the World Health organization as a "state of complete physical, mental and social-wellbeing and not merely the absence of disease or infirmity" [75] to others based on holistic notions of 'self'.

A total of 25 wellness programs that provided on-line information about their goals, methods and outcomes were identified. Of the nine Australian worksite wellness programs examined, seven detailed their purpose, all nine described their methods and seven disclosed their achievements. Of the eight programs examined from the United States, these figures were five, six and seven, respectively. Among the seven programs evaluated from the United Kingdom, none listed their purpose, but all of them reported their methods and outcomes. The studies were cross-checked with the survey findings and case studies published by Mattke et al. [44] to ensure consistency and representativeness.

Among the Australian studies, the most common aims were to improve physical health and wellbeing of employees and reduce the risk of lifestyle disease; the most common methods were employee health assessments, employee needs assessments and information sessions on healthy lifestyle and management support, and the most common reported achievement was a reduction in sedentary category. Among the US studies, the main aims included reduction of tobacco intake and reduction of stress; the only method common to all studies was a tobacco cessation program, and all seven programs that reported outcomes listed saved money as a primary success. The second most common outcome was reduced absenteeism, reported in four studies. Among the United Kingdom programs, four offered smoking cessation classes along with stress interventions, and the most common reported outcome was a decrease in absenteeism.

These companies were further delineated depending on whether they comprised primarily blue collar or white collar employers. No substantive difference was found between the programs implemented in these two groups. Program delivery modes were also surveyed. The four most common delivery methods in 2004 were printed materials, the internet, in-person methods and telephone systems.

Wellness initiatives were categorised according to three functional levels [27]: awareness programs, lifestyle modification programs, and programs to promote the sustainability of these modifications. Of the Australian case studies examined, most were level one and none were level three. The United States programs run were evenly distributed between level one and two programs, with the only level three programs being the provision of onsite exercise facilities. In the United Kingdom there was a greater proportion of level two programs, with only one level one program (blood pressure testing) and one level three program (healthy eating options at work) offered by more than one company.

Characteristics of workplace wellness programs were also surveyed. Successful wellness programs reported integration of the program into business strategy and company culture; the inclusion of a full time or part time program co-ordinator; support from upper level management; social support; the use of incentives to increase participation in the program; co-worker support, and the convenience and accessibility of the program. These were contrasted with the 2004 National Worksite Health Promotion Survey, which found that the most commonly used definitions of success involved employee feedback and participation, reductions in health-related costs and reductions in absenteeism.

12.4   Healthy Worker and Wellness Systems Models

Tables 12.1 and 12.2 provide lists of the factors included in the health worker model and workplace wellness model, respectively, based on the respective literature surveys in Appendices 1 and 2. The corresponding systems models are shown in Figs. 12.1 and 12.2.

In Fig. 12.1, the target node, Worker Health, is seen to be directly influenced by three nodes representing 'Mental', 'Chronic' and 'Acute' health outcomes, which are themselves influenced by a range of health outcomes, personal factors and occupational risks. For example, it is seen that occupational stress affects mental health, both directly and also indirectly via depression. Occupational stress is in turn affected by workload, noise level and excessive working hours. All specified occupational diseases affect chronic health, whereas only musculoskeletal diseases, liver disease, cancers, cardiovascular disease and respiratory disease affect acute health.

Table 12.1 Influential factors in Workplace Health, based on literature survey

Overall health outcomes	Occupational health outcomes	Factors
Mental	Hearing loss	Workload
Chronic	Respiratory diseases	Noise level
Acute	Depression	Excessive hours
	Cardiovascular disease	Air quality
	Hypertension	Smoking
	Liver disease	Alcohol consumption
	Musculoskeletal disease	Sun exposure
	Occupational stress	Exercise required in job
	Obesity	Recreational exercise
		Diet

Table 12.2 Influential factors in Workplace Wellness Programs, based on survey of selected published programs

	Increase in wellness—physical	Increase in wellness—behavioural	Return on investment
Extent of participation			
Full-time coordinator	Reduction in obesity	Improvement in mental health	Reduced absenteeism
Incentives offered for program participation	Increased physical activity	Reduction in substance abuse	Reduced presenteeism
Encourage of program from senior management	Reduced risk of lifestyle	Improvement in diet	Reduced workforce turnover
How well program is advertised	Improved cholesterol		
Accessibility of program to workers: Times program is run	Decreased blood pressure		
Accessibility of program to workers: On-site facilities			

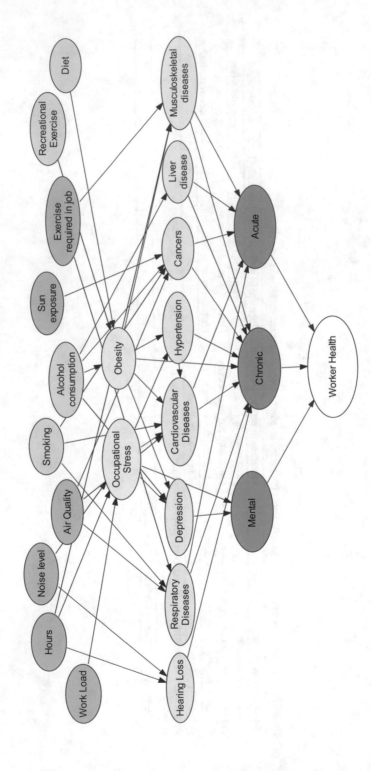

Fig. 12.1 Healthy worker systems model

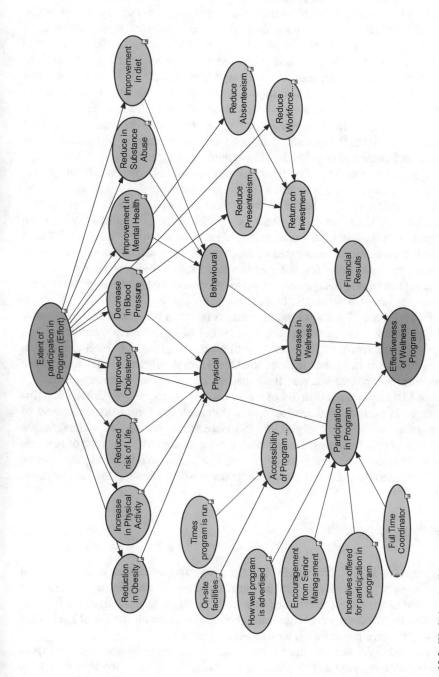

Fig. 12.2 Workplace wellness systems model

In Fig. 12.2, the target node, 'Effectiveness of Program' is directly affected by 'Participation in Program', 'Increase in Wellness' and 'Return on Investment'. These nodes are in turn affected by the factors drawn from the survey. For example, increased physical health resulting from a wellness program is influenced by a reduction in obesity, increase in physical activity, decrease in risk of lifestyle diseases, improved cholesterol and decreased blood pressure. Physical health in turn impacts directly on increase in wellness, which affects the effectiveness of the wellness program. Participation in the wellness program is seen to be central to the program success, and is affected by the five factors listed in Table 12.2.

Details of the quantification of the two systems models are provided in Appendix 3. As described there, the quantification is generic and intended for exposition, and hence is not interpreted definitively. It is helpful in its own right but if desired, the general platform provides a foundation for models tailored to specific workplaces, workforces or wellness programs in a straightforward manner.

An illustration of the quantified workplace wellness model is given in Fig. 12.3. Tables 12.3, 12.4 and 12.5 show the results of the quantification, based on the literature survey and the methods described in Appendix 3. Table 12.3 shows the relative probabilities, expressed as percentages, for each of the worker health outcomes. The table indicates that a worker has a substantially smaller probability of good mental health, compared with acute or chronic health, although these latter two outcomes are still substantially less than an optimal level of 100%.

Similarly, Table 12.4 shows that a wellness program has about an even chance (48%) of having a high participation rate and of being effective overall, but less chance of a high return on investment. While there is a relatively high chance that the program has an impact on reduced workforce turnover, there is much less chance of reducing absenteeism and very little chance (8%) of reducing presenteeism.

Table 12.5 shows the sensitivity of the outcomes of workplace wellness programs to changes in worker participation levels. Although the return on investment of the program changes only slightly (a 5% increase) as participation changes, the probability of an effective program overall more than doubles. This type of scenario or 'what-if' assessment can be used to assess most influential factors in the system, which in turn can be used as drivers to influence positive outcomes of the programs.

12.5 Discussion

This study aimed to contribute to the literature regarding workplace wellness and the effectiveness of programs that aim to improve wellness by undertaking a systematic assessment of impacting factors and their relative bearing. This is the first study to our knowledge to combine a large-scale program survey with statistical analysis in order to produce a predictive as well as analytic model.

The Healthy Worker literature survey presents strong evidence that employees are occupationally exposed to numerous risks in their workplace that may have an impact on their overall health. These include but are not limited to noise, car-

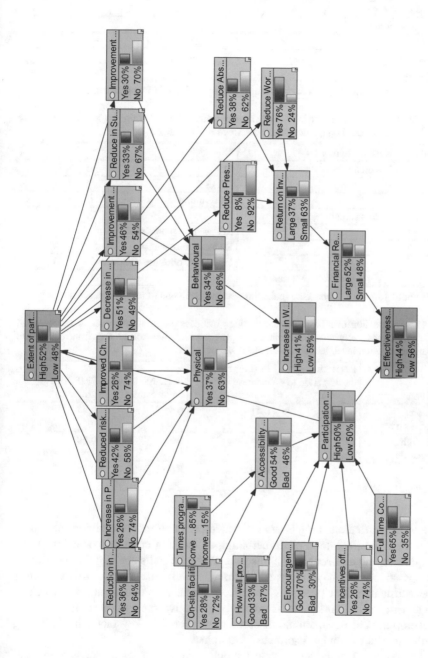

Fig. 12.3 Workplace wellness systems model quantified as a Bayesian network

Table 12.3 Results of quantified BN for worker health: overall score for selected occupational health outcomes, based on quantification described in Appendix 3

Health outcome	Score
Overall worker health	89
Mental health	50
Chronic health	86
Acute health	72

Score is out of 100, with larger score indicating greater likelihood of worker health

Table 12.4 Results of quantified BN for workplace wellness: overall score for selected workplace wellness outcomes, based on quantification described in Appendix 3

Overall effectiveness of program			48	
	Increase in wellness		41	
			Physical	37
			Behavioural	34
	Return on investment		37	
			Reduced presenteeism	8
			Reduced absenteeism	38
			Reduced workforce turnover	76
	Participation in program		50	

Score is out of 100, with larger score indicating greater likelihood of program success

Table 12.5 Evaluation of relative impact on three wellness program outcomes, namely overall increase in wellness, return on investment and program effectiveness, arising from changes in program participation level, based on workplace wellness BN

Condition	Pr (High participation in program)	Pr (Large increase in wellness)	Pr (Large return on investment)	Pr (Effective program)
Based on literature survey	0.50	0.41	0.52	0.44
100% High participation	1.00	0.44	0.55	0.60
100% Low participation	0.00	0.39	0.50	0.28

cinogens, air particulate, long hours and sedentary or strenuous activities and these may increase the likelihood of exacerbating an existing condition or heighten the risk of developing an occupational disease. The net results are that the individuals or groups concerned may suffer as a consequence of the disease, and that there can be substantive lost productivity and economic loss due to absenteeism and presenteeism in the workplace. Conversely, proactive prevention or reduction in the development of occupational diseases in workers is a key factor in sustaining a healthy, productive and cost-effective workplace.

The interplay of non-occupational factors, such as diet, exercise, smoking and alcohol consumption, can considerably influence susceptibility of developing a chronic, acute or mental disease. Nevertheless, occupational exposures are significant and employers have a duty of care to ensure that potentially modifiable occupational factors are regulated to mitigate risk. It is therefore vital to determine which factors are most likely to cause or exacerbate disease and evaluate their overall contribution to risk. Investigation into the importance of each health risk factor will provide valuable information to workplaces and enable them to target interventions or wellness programs to those individuals and or groups most in need.

Workplace health promotion programs have become increasingly popular as a way to target the growing incidence of lifestyle disease. All the programs analysed in this study have demonstrated positive results; however due to the lack of comprehensive analysis on the part of employers, it is very difficult to conclusively define a workplace health promotion program as successful. However, the literature and models reveal some common insights. For example, workplace wellness programs arguably do not have a high effectiveness rate if attendance is not satisfactory. This agrees with previous studies [56] and shows that programs that focus on encouraging attendance through workplace support, dedicated program coordinators and integration of the program into company culture are far more likely to see successful outcomes. Through the examination of wellness definitions, types of wellness programs and their characteristics along with successes and failures of workplace health promotion programs, it is clear that these programs have huge potential. A more systematic and open approach to evaluating the effectiveness of these programs would help to improve their outcomes, with positive effects for workers, workplaces and business.

The Bayesian networks presented in this paper are intended to contribute to this endeavour. The systems models of factors contributing to a healthy worker, and similarly a successful workplace wellness program, have been developed based on available literature. It is anticipated that both the structure of the models and the probabilistic inputs will change as the literature grows, and other researchers and practitioners interrogate and contribute to these models. Indeed, this dynamic nature of BNs is a positive feature, facilitating currency of knowledge and richer insights at both global and individual workplace scales.

The BN models presented in this study can be modified to enable employers to describe systems specific to their own workplace, workforce and wellness programs. The structure of the model, including the nodes and connections, as well as the definition of the target outcomes, can be modified as required. The definition of the binary quantities for the nodes can also be adapted to the situation at hand. In the BNs in this paper, the binary quantities included yes/no and present/absent depending on the context, and the internal nodes were defined in terms of a 'relative load' based on the number of parents nodes that were positive (yes, present, etc). This facilitates the evaluation of 'best' and 'worst' scenarios. Alternatively, the network probabilities could be quantified using real data, such as individual health data obtained as part of ongoing hygiene and medical assessments, or survey results on the effectiveness or otherwise of wellness programs. Such information can also be used for predicting where health and wellness programs could be best targeted and enable better use of resources to target those individuals and groups most at risk.

Appendix 1

Healthy Worker Literature Survey

The following table summarises selected literature that identifies occupational associations with health outcomes among workers.

Health outcome	Occupational associations
Obesity • Around 2/3 of Australians are overweight or obese, and the proportion is growing [5]. • The economic costs of being overweight in the workplace have been shown to be higher than those of smoking, drinking, and poverty [26, 67, 68].	• Increases risk of hypertension, cardiovascular diseases, asthma, musculoskeletal disorders, some cancers [21, 62]. • Modifies response to occupational stress, immune response to chemical exposures, and risk of disease from occupational neurotoxins [62]. • Modifies intensity of response to various occupational hazards including heat exhaustion, pesticide exposure, accidents with equipment operators, and respiratory and physiological strain during hard physical work [31]. • Reduces effectiveness of personal respirator tests, protective equipment and clothing, particularly in hot and humid conditions [62]. • Increases absenteeism [49, 62, 73]. • Increases incidence of sick leave [13, 61] • Increases presenteeism [26]. • Occupational stress and fatigue increases behaviours associated with weight gain [77]. • Augments endocrine factors related to weight gain caused by psychological strain [77].
Musculoskeletal Disorders • Encompasses a variety of inflammatory and degenerative conditions involving muscles, tendons, ligaments, joints, peripheral nerves and supporting blood vessels • Can be caused by one acute traumatic event, or by chronic stress over a period of time due to repetitive use [58].	• Muscular stress due to lifting, carrying, lowering, and handling of objects as well as from other strenuous physical movements [58]. • Repetitive movements, excessive loading, muscle overuse and vibration are specific activities often associated with a heightened risk of musculoskeletal diseases [62]. • Exacerbated by jobs that demand excessive workloads and high responsibilities, time pressures, insufficient rest breaks, and inadequate resources and workplace support and resulting in increased injury risk from fatigue, poor posture, and stress [42].

(continued)

Health outcome	Occupational associations
Cardiovascular Disease • Encompasses a range of disorders including heart disease and circulatory conditions. • A major cause of death globally [65].	• Air pollution, including short and long term exposure to gases, chemicals and particulate matter, can potentially increase the risk of heart disease. This is exacerbated by the well-established relationship between smoking, both active and passive, and heart disease and stroke [12, 30, 71]. • Occupational stress can increase the risk of cardiovascular disease [29, 35, 47]. • Impact of occupational cardiovascular disease on absenteeism and presenteeism, but this is not trivial [41]. Altering risk factors such as diet, exercise and smoking can aid in preventing the onset of various cardiovascular diseases [4, 28, 43].
Noise-induced hearing loss • Noise exposure has a range of undesirable effects including elevated blood pressure, reduced performance, sleeping difficulties, annoyance and stress, temporary shift in the hearing threshold, tinnitus and noise-induced hearing loss (NIHL) [55].	• Exposure to excessive noise at the workplace can cause NIHL, also known as industrial deafness [58]. • The extent of hearing loss can be affected by the level of noise and the length of exposure [52, 55] and by the type of exposure [46]. • NIHL impedes spoken communication and can cause social isolation and stress [55, 60]. • Hearing loss is the second most self reported occupational disease [52].
Respiratory Diseases • Various respiratory diseases have been associated with workplaces, including bronchitis, asthma, upper and lower respiratory illness, chronic obstructive pulmonary disease, lung cancer, pneumoconiosis [10, 20].	• Agents in the workplace that can exacerbate or cause several respiratory diseases include pesticides, herbicides, dust, lead, fumes, chemicals, gases, faulty air conditioners, particulates and gases emitted from fire and other activities, emissions from furnishings, and so on [10, 20]. • Occupational dust is a significant contributor to the development of bronchitis, asthma and other respiratory illnesses [20, 53]. • Exposure to dust in the workplace is also associated with chronic obstructive pulmonary disease with potential for continued development of the disease many years after exposure [10]. • There are also numerous respiratory carcinogens including asbestos, arsenic, radon, silica, chromium, cadmium, nickel and beryllium [10]. • Asbestos-related disorders and industrial bronchitis and asthma are respiratory diseases that have been amongst the most common occupational diseases, and silicosis and pneumoconiosis have been reported in US coal workers [20].

(continued)

Health outcome	Occupational associations
Cancers • A variety of cancers are reportedly associated with occupational carcinogens, including lung, bone, liver, thyroid, bladder, skin and leukaemia [33].	• Well recognised carcinogenic agents include asbestos, silica, nickel, chromium, arsenicals, vinyl chloride, and halo ethers as well as ionising and solar radiation [33].
Mental Health Disorders • Poor mental health among workers is pervasive [14]. • It can lead to work impairment [37, 40], reduced job commitment and satisfaction [38]. • Anxiety disorders can be more costly than alcohol-related disorders due to their higher frequency rate [64].	• Occupational factors that can lead to poor mental health include long hours [32, 70], shift work [17, 25], low job control and high work demand [70]. • Traumatic events associated with work related incidents, harassment, bullying and exposure to violence have also been linked to mental disorders [58]. • Occupational stress can lead to absenteeism, loss of productivity, unemployment, social impairment and a high use rate of health care [16].
Alcohol-related Disorders • Alcohol affects hepatic and pancreatic systems [23, 34] and is also related to increased blood pressure, an increased risk of developing certain cancers, and cerebral dysfunction [23].	• Excessive alcohol consumption is related to poor health, resulting in absenteeism and presenteeism. It has also been linked with late arrival at work and reduced promotion success [34].

Appendix 2

Workplace Wellness Literature Survey

Definition of Wellness

Wellness is a balance of positive mental, physical and social health. It is variously defined, for example by the World Health organization as a "state of complete physical, mental and social-wellbeing and not merely the absence of disease or infirmity" [75], or "the process and state of a quest for maximum human functioning that involves the body, mind, and spirit" [51]. Models for wellness also exist, such as the Indivisible Self Model [50] which comprises five components, namely The Essential Self, The Creative Self, The Coping Self, The Social Self and The Physical Self.

Wellness Programs: Case Studies

The following table provides a summary of the wellness programs surveyed for the purposes of developing the Workplace Wellness BN. Numbers in brackets refer to number of programs.

Country	Aims	Methods	Outcomes
Australia: No. programs surveyed (9). No. programs that detailed: - purpose (7) - methods (9) - outcomes (7)	• Improved physical health and wellbeing of employees (5) • Reduced risk of lifestyle disease (3) • Save money, improve productivity, improve staff relationships, reduce stress, reduce absenteeism, provide information to employees on benefit of a healthy lifestyle, improve quality of life, ensure employees are fit to work (≥ 2)	• Comprehensive health assessment of each employee (7) • Needs assessment and information sessions on healthy lifestyle (5) • Exercise programs, financial support to participate in community events (3) • Marketing of events outside the program, workplace audit, meditation classes, health challenges (2).	• Reduced sedentary category (4) • Improved blood pressure, energy, eating habits and staff morale, decreased stress levels (3) • Increase in employees in healthy weight range, increase in employees in ideal category for total cardiac risk, increased health knowledge, improved physical health, enhanced motivation, improved mental health, better staff relationships, increased job satisfaction (2)
USA: No. programs surveyed (8). No. programs that detailed: - purpose (5) - methods (6) - outcomes (7)	• Reduction of tobacco intake (4) • Reduction of stress (3) • Improve exercise, improve diet, manage employee weight, improve employee fitness, reduce business costs (2)	• Tobacco cessation program (6) • Nutrition classes (4) • On-site exercise facilities, weight control programs, health services, free health screenings (3) • Educational material on health, stress management programs, fitness and activity programs, vaccinations, wellness magazines or newsletters, incentive programs, fitness classes (2)	• Saved money (7) • Reduced absenteeism (4) • Reduced weight, cholesterol, smoking and blood pressure (2) • Increased productivity, improvements in nutrition and emotional health, increased exercise, decreased alcohol use (≥ 1)
UK: No. programs surveyed (7). No. programs that detailed: - purpose (0) - methods (7) - outcomes (7)	N/A	• Smoking cessation classes and stress interventions (4) • Massage sessions, healthy eating options at work, fitness classes, weight management courses, discounts at local gyms, blood pressure testing, counselling services (3) • Pedometer use (2)	• Decrease in staff absence (5) • Decrease in staff turnover, increase in corporate image, decrease in risky behaviour including smoking cessation, increased physical activity (3) • Increased employee engagement (2)

Types of Wellness Initiatives

Wellness initiatives have been categorised in many ways over many years. For example, Gebhardt and Crump [27] separated workplace wellness programs into three functional levels. Level one involves awareness programs such as newsletters, health fairs, screening sessions, posters, flyers and educational classes. These are not solely aimed at improving participants' health or instantiating long term behavioural change, but are aimed at raising awareness of the consequences of unhealthy behaviours. Level two programs aim for behaviour and lifestyle modification and include self-administered fitness programs, memberships at local fitness facilities, classes related to proper performance of physically demanding work tasks, etc. Level three programs aim to create an environment that supports the sustainability of new healthy behaviours, for example via the provision of equipment, space or locker facilities at the worksite, availability of healthy foods and the removal of unhealthy temptations.

This historic categorisation generally conforms to more recent classifications of wellness initiatives. For example, Mattke et al. [45] performed a cluster analysis of a large dataset and identified five common configurations of workplace wellness programs that offered different levels of service for health risk screening, lifestyle management to reduce health risks and encourage healthy lifestyles, and chronic disease management.

Of the Australian case studies examined, a large proportion were level one programs, with the most popular being health and needs assessments. Information sessions on healthy lifestyle were also very popular, followed by marketing of the program audits of the workplace, health risk screening and marketing of events outside the program. Level two programs in Australia most commonly included: exercise programs; financial support to participate in community events; challenges and meditation sessions. None of the Australian case studies examined included any level three wellness programs.

In the United States wellness programs run were evenly distributed between level one and two programs, with the most popular level one programs: nutrition classes, free health screenings, educational materials on health and wellness newsletters or magazines. Level two programs included tobacco cessation programs; weight control programs, fitness classes and fitness and activity programs. The only level three program was the provision of onsite exercise facilities.

In the United Kingdom studies there was a greater proportion of level two programs, with only one level one and level three program offered by more than one company. The level one program was blood pressure testing. The most common level two programs included smoking cessation classes, fitness classes, discounted bicycle purchase, weight management courses, discounts at local gym, counselling services and the provision of pedometers. The level three option included was healthy eating options at work.

Characteristics of Successful Wellness Initiatives

In order to evaluate the success of a work-based wellness programme, it is necessary to establish what is used to define a program as either a success or a failure. Success can be defined in terms of financial, health and social benefits, and within different timeframes. For example, results from the 2004 National Worksite Health Promotion (NWHP) Survey revealed that the most commonly used definition of success was employee feedback, followed by employee participation, workers' compensation costs, health care claims costs and reduced absenteeism, whereas the most common barriers to success were lack of employee interest, lack of staff resources and funding, lack of participation of high-risk employees and lack of support from upper level management.

These findings have been echoed in later surveys, for example, in the comprehensive studies of U.S. employers [45] and in U.K businesses [24]. The latter study reported reduced sickness absence (in 82% of programs surveyed), reduced staff turnover (33%), reduced accidents and injuries (29%), increased employee satisfaction (25%), reduced resource allocation (16%), increased company profile (15%), increased productivity (15%), and increased health and welfare (15%). Each of these was linked to positive economic contributions.

Similar benefits were observed among the successful wellness programs surveyed in the present paper. These included integration of the program into business strategy and company culture, the inclusion of a full time or part time program co-ordinator, support from upper level management, social support, the use of incentives to increase participation in the program and the convenience and accessibility of the program.

The following table provides a selection of references to literature supporting these findings.

Factors that increase the likelihood of success of a workplace wellness initiative:
• The program is woven into the business strategy and culture of the company; a program is more vulnerable if it is considered to be a luxury rather than a necessity [11]
• Strong management support for the program; the more an individual perceives support from their supervisor, the higher their participation [63].
• Employee involvement in the program design and implementation [69]
• A full time or part time coordinator with the ability to motivate participation [27]
• Use of incentives to increase employee participation [69]. Incentives work positively because employees prefer to feel that they are acting of their own volition rather than being forced to act by management policies [11].
• Accessibility and convenience of a program. In order to ensure maximum accessibility making a program either free or low cost to participants must be a priority [11].
• Onsite integration of wellness programs, making participation more straightforward and convenient for employees [11].
Factors that have contentious influence:
• Co-worker support is reported in some studies to be an important influential factor on participation in health-related and fitness activities for all employee subgroups [19] but not in other studies [2, 63].
• Mattke et al. [45] found little evidence of the benefit of employee participation in the management aspects of wellness programs.

A final method of determining success is to examine the purpose of the program as set out by the company and then examining the achievements that were reported by the company. To be considered successful the company must as a minimum achieve the majority of the points set out in its "purpose of program statement". For the Australian case studies, five of the nine companies detailed their purpose of program as well as the achievements. Based upon the criteria of a successful program, two of the five Australian programs that listed both their purpose of program as well as the achievements of their program can be labelled as successful. Four of the eight case studies examined from the United States detailed both their purpose of program and achievements. Three of these four programs can be classed as successful programs based on the criteria of a successful program. Three companies reportedly achieved everything they set out to achieve, saved money and improved productivity, although only one of these provided data on specific outcomes to support their claims. Success of the United Kingdom programs could not be evaluated due to the lack of recorded information on their purpose of program.

Appendix 3

Quantification of the Bayesian Networks

The Healthy Worker and Workplace Wellness BNs were quantified as follows. The quantification is intended for exposition purposes only. The resultant probabilities should not be interpreted medically, socially or economically, nor with respect to particular workplaces or wellness programs. However, the structure of the Healthy Worker network and its quantification could be targeted to a particular workplace cohort if data about the associated personal and lifestyle factors, health outcomes and workplace risks were made available. Similarly, the Workplace Wellness network could be structured and quantified using program-specific information, resulting in interpretable probabilities of success.

For the Healthy Worker model, each node was categorised as 'yes' or 'no'. All exposures and factors at the top of the network were set to equal probabilities for each category and each health outcome was assigned equal weight. Thus the probability of a health outcome was determined by the number of detrimental factors affecting the node (depicted as directed arrows to the node) divided by the total number of factors affecting the node. Exceptions to this rule were made for respiratory diseases and acute & chronic diseases. Since the survey indicated that exercise heightens the inhalation of unwanted air pollution, exercise was included as a factor when combined with poor quality air, but was ignored for good quality air. However, since exercise was reportedly not as influential a risk factor as smoking, a weight of 0.2 of developing a respiratory disease was assigned to exercise alone, 0.4 to both smoking and poor air quality, 0.6 to performing strenuous activity and poor air quality, and 1.0 if the worker also smoked. For the node representing acute & chronic diseases, it was determined that if a worker had three or more diseases then they were deemed to have a higher chronic disease rate. Having two diseases was given a 2/3 weighting; having 1 disease was given a 1/3 weighting and no disease was given a 0 weighting. While this approach simplified the quantification of the network, the many nodes linking to the chronic and acute nodes meant that it was relatively easy to obtain a poor score on the chronic, acute, and overall worker health nodes.

The systems model for the workplace wellness programs was quantified in a similar manner. Probabilities were then assigned to each node, based on these information sources and on the other nodes in the model. For each of the nodes such as participation in program, physical, behavioural, and return on investment, the probability of a positive outcome was directly proportional to the number of positive outcomes in the parent nodes, with the maximum value set to 0.95 if all parent nodes were positive and the minimum value set to 0.05 if all parent nodes were 'negative'. For those with two parent nodes, such as increase in wellness and accessibility of program, the probability of a positive outcome was set to 0.80 if both parent nodes were positive and to 0.2 if both parent nodes were negative.

References

1. I. Ajunwa, Workplace wellness programs could be putting your health data at risk. Harv. Bus. Rev. (19 January 2017)
2. B.B. Alexy, Factors associated with participation or nonparticipation in a workplace wellness center. Res. Nurs. Health **14**, 33–40 (1991)
3. O. Atilola, O. Akinyemi, B. Atilola, Taking the first step towards entrenching mental health in the workplace: insights from a pilot study among HR personnel in Nigeria. J. Natl. Assoc. Resid. Doct. Niger. **23**(1), 70–76 (2014)
4. Australian Bureau of Statistics (ABS), *Australian Health Survey: First Results, 2011-12* (Heart Stroke and Vascular Disease, Canberra, 2012a)
5. Australian Bureau of Statistics (ABS), *Australian Health Survey: First Results, 2011-12* (Key Findings, Canberra, 2012b)
6. Australian Bureau of Statistics (ABS), *Labour Fource, Australia,* Dec 2012. (Canberra, 2013)
7. Australian Institute of Health and Welfare (AIHW), Chronic Disease and Participation in Work. Report DOI: 9781740248785, 2009
8. Australian Institute of Health and Welfare (AIHW), *Australia's Health 2014.* (Canberra, 2014)
9. D.A. Beaudequin, F.A. Harden, A. Roiko, H. Stratton, C. Lemckert, K.L. Mengersen, Beyond QMRA: Modelling microbial health risk as a complex system using Bayesian networks. Environ. Int. **80**, 8–18 (2015)
10. W.S. Beckett, Occupational respiratory diseases. N. Engl. J. Med. **342**(6), 406–413 (2000)
11. L. Berry, A.M. Mirabito, W.B. Baun, What's the hard return on employee wellness programs? Harv. Bus. Rev. **88**(12), 104–112 (2010)
12. R. Brook, B. Franklin, W. Cascio, Y. Hong, G. Howard, M. Lipsett, R. Luepker, M. Mittleman, J. Samet, S. Smith Jr., I. Tager, Air pollution and cardiovascular disease: a statement for healthcare professionals from the expert panel on population and prevention science of the American Heart Association. Circulation **109**, 2655–2671 (2004)
13. A. Burdorf, Economic evaluation in occupational health–its goals, challenges, and opportunities. Scand. J. Work Environ. Health **33**(3), 161–164 (2007)

14. Business in the Community, *Mental Health at Work Report*. (National Employee Mental Wellbeing Survey Findings 2017, 2017), https://wellbeing.bitc.org.uk/system/files/research/bitcmental_health_at_work_report-2017.pdf. Accessed 8 Dec 2017
15. B.L. Callen, L.C. Lindley, V.P. Niederhauser, Health risk factors associated with presenteeism in the workplace. J. Occup. Environ. Med. **55**(11), 1312–1317 (2013)
16. W.H. Charles, E.T. Holly, C.M. Stephen, B. Nicole, The occupational burden of mental disorders in the U.S. military: psychiatric hospitalizations, involuntary separations, and disability. Am. J. Psychiatry **162**(3), 585–591 (2005)
17. L.C. Coffey, J.J.K. Skipper, F.D. Jung, Nurses and shift work: effects on job performance and job-related stress. J. Adv. Nurs. **13**(2), 245–254 (1988)
18. Comcare, *Benefits to Business: The Evidence for Investing in Worker Health and Wellbeing* (Australian Government, Canberra, 2011)
19. C.E. Crump, J.L. Earp, C.M. Kozma, I. Hert-Picciotto, Effect of organizational-level variables on differential employee participation in 10 Federal worksite health promotion programs. Health Educ. Q. **23**(2), 204–223 (1996)
20. M.R. Cullen, M.G. Cherniack, L. Rosenstock, Occupational medicine (1). N. Engl. J. Med. **322**(9), 594–601 (1990)
21. Department of Health and Ageing (DOHA), *Promoting Healthy Weight* (Australian Government Report, Canberra, 2009)
22. J.-P. Després, N. Alméras, L. Gauvin, Worksite health and wellness programs: Canadian achievements and prospects. Prog. Cardiovasc. Dis. **56**(5), 484–492 (2014)
23. M.J. Eckardt, T.C. Harford, C.T. Kaelber, E.S. Parker, L.S. Rosenthal, R.S. Ryback, G.C. Salmoiraghi, E. Vanderveen, K.R. Warren, Health hazards associated with alcohol consumption. J. Am. Med. Assoc. **246**(6), 648–666 (1981)
24. ERS Research and Consultancy (Health at Work: Economic Evidence Report. U.K., 2016), https://www.bhf.org.uk/health-at-work. Accessed 6 Dec 2017
25. M. Estryn-Behar, M. Kaminski, E. Peigne, N. Bonnet, E. Vaichere, C. Gozlan, S. Azoulay, M. Giorgi, Stress at work and mental health status among female hospital workers. Br. J. Ind. Med. **47**(1), 20–28 (1990)
26. D.M. Gates, P. Succop, B.J. Brehm, G.L. Gillespie, B.D. Sommers, Obesity and presenteeism: the impact of body mass index on workplace productivity. J. Occup. Environ. Med. **50**(1), 39–45 (2008)
27. D.L. Gebhardt, C.E. Crump, Employee fitness and wellness programs in the workplace. Am. Psychol. **45**(2), 262–272 (1990)
28. I.F. Groeneveld, K.I. Proper, A.J. van der Beek, V.H. Hildebrandt, W. van Mechelen, Lifestyle-focused interventions at the workplace to reduce the risk of cardiovascular disease–a systematic review. Scand. J. Work Environ. Health **36**(3), 202–215 (2010)
29. S.F.M. Haynes, Women, work and coronary heart disease: prospective findings from the Framingham Heart Study. Am. J. Public Health **70**, 133–141 (1980)
30. J. He, S. Vupputuri, K. Allen, M.R. Prerost, J. Hughes, P. Whelton, Passive smoking and the risk of coronary heart disease: a meta-analysis of epidemiologic studies. N. Engl. J. Med. **340**, 920–926 (1999)
31. A. Henschel, Obesity as an occupational hazard. Can. J. Public Health **58**, 491–493 (1967)
32. M.F. Hilton, H.A. Whiteford, J.S. Sheridan, C.M. Cleary, D.C. Chant, P.S. Wang, R.C. Kessler, The prevalence of psychological distress in employees and associated occupational risk factors. J. Occup. Environ. Med. **50**(7), 746–757 (2008)
33. K. Husgafvel-Pursiainen, P.W. Brandt-Rauf, A. Kannio, P. Oksa, T. Suitiala, H. Koskinen, R. Partanen, K. Hemminki, S. Smith, R. Rosenstock-Leibu, Mutations, tissue accumulations, and serum levels of p53 in patients with occupational cancers from asbestos and silica exposure. Environ. Mol. Mutagen. **30**(2), 224–230 (1997)
34. R. Jenkins, S. Harvey, T. Butler, R.L. Thomas, A six year longitudinal study of the occupational consequences of drinking over "safe limits" of alcohol. Br. J. Ind. Med. **49**(5), 369–374 (1992)
35. J. Johnson, E.M. Hall, Job strain, work place social support, and cardiovascular disease: a cross-sectional study of a random sample of the Swedish working population. Am. J. Public Health **78**(10), 1336–1342 (1988)

36. S. Johnson, K. Mengersen, Integrated Bayesian network framework for modeling complex ecological issues. Integr. Environ. Assess. Manag. Online, 1–11 (2011)
37. R.C. Kessler, R.G. Frank, The impact of psychiatric disorders on work loss days. Psychol. Med. **27**(4), 861–873 (1997)
38. H.K. Laschinger, D.S. Havens, The effect of workplace empowerment on staff nurses' occupational mental health and work effectiveness. J. Nurs. Adm. **27**(6), 42–50 (1997)
39. C. Lee, M. Chen, C. Chu, The health promoting hospital movement in Taiwan: recent development and gaps in workplace. Int. J. Public Health **58**(2), 313–317 (2013)
40. D. Lin, K. Sanderson, A. Gavin, Lost productivity among full-time workers with mental disorders. J. Mental Health Policy Econ. **3**(3), 139–146 (2000)
41. J.L.Y. Liu, N. Maniadakis, A. Gray, M. Rayner, The economic burden of coronary heart disease in the UK. Heart Br. Cardiac Soc. **88**(6), 597–603 (2002)
42. W. Macdonald, O. Evans, *Research on the Prevention of Work-related Musculoskeletal Disorders Stage 1 - Literature Review* (Safe Work Australia, 2006)
43. T.H. Marc, G.H. Deborah, W.Z. Theodore, Role of low energy expenditure and sitting in obesity, metabolic syndrome, type 2 diabetes, and cardiovascular disease. Diabetes **56**(11), 2655–2667 (2007)
44. S. Mattke, H. Liu, J.P. Caloyeras, C.Y. Huang, K.R. van Busum, D. Khodyakov, V. Shier, *Workplace Wellness Programs Study Final Report* (RAND Health, 2013). https://www.rand.org/content/dam/rand/pubs/research_reports/RR200/RR254/RAND_RR254.sum.pdf. Accessed 8 Feb 2017
45. S. Mattke, K.A. Kapinos, J. Caloyeras, E.A. Taylor, B.S. Batorsky, H.H. Liu, K.R. van Busum, S. Newberry, Workplace wellness programs. Services offered, participation and incentives. Rand Health Q. **5**(2), 7 (2015)
46. D.I. McBride, Noise-induced hearing loss and hearing conservation in mining. Occup. Med. **54**(5), 290–296 (2004)
47. J.H. Medalie, H.A. Kahn, H.N. Neufeld, E. Rise, U. Goldbourt, Five year myocardial infarction incidence: 2. Association of single variables to age and birthplace. J. Chronic Dis. **26**, 329–349 (1973)
48. C.N. Michaels, A.M. Greene, Worksite wellness: increasing adoption of workplace health promotion programs. Health Promot. Pract. **14**, 473 (2013)
49. M. Moreau, F. Valente, R. Mak, E. Pelfrene, P. de Smet, G. De Backer, M. Kornitzer, Obesity, body fat distribution and incidence of sick leave in the Belgian workforce: the Belstress study. Int. J. Obes. Relat. Metab. Disord. **28**, 574–582 (2004)
50. J.E. Myers, T.J. Sweeney, *Five Factor Wellness Inventory* (Mind Garden Inc., 2005)
51. J.E. Myers, K. Williard, Integrating spirituality into counseling and counselor training: a development, wellness approach. Couns. Values **47**(2), 142–155 (2003)
52. National Institute for Occupational Safety and Health (NIOSH), *Work Related Hearing Loss.* (Centers for Disease Control and Prevention, 2001)
53. National Institute for Occupational Safety and Health (NIOSH), *The Work-related Lung Disease Surveillance Report, 2002.* (Cincinnati, 2003)
54. National Institute for Occupational Safety and Health (NIOSH), *Promising and Best Practices in Total Worker Health.* Joint Report on workshop convened by NIOSH and the Institute of Medicine (2014). https://www.cdc.gov/niosh/twh/practices.html. Accessed 4 Dec 2017
55. D.I. Nelson, R.I. Nelson, M. Concha-Barrientos, M. Fingerhut, The global burden of occupational noise induced hearing loss. Am. J. Ind. Med. **48**(6), 446–458 (2005)
56. A. Rongen, S.J.W. Robroek, F.J. van Lenthe, A. Burdorf, Workplace health promotion: a meta-analysis of effectiveness. Am. J. Prev. Med. **44**(4), 406–415 (2013)
57. Safe Work Australia, *National OHS Strategy 2002–2012.* (Canberra, 2002)
58. Safe Work Australia, *Occupational Disease Indicators.* (Canberra, 2010)
59. Safe Work Australia, *Occupational Disease Indicators.* (Canberra, 2012)
60. R.T. Sataloff, J. Sataloff, *Occupational Hearing Loss* (CRC Press Taylor & Francis Group, Boca Raton, 2006)
61. J.K. Schmier, M.L. Jones, M.T. Halpern, Cost of obesity in the workplace. Scand. J. Work Environ. Health **32**(1), 5–11 (2006)

62. P. Schulte, Work, obesity, and occupational safety and health. Am. J. Public Health **97**(3), 428–436 (2007)
63. R.P. Sloan, J.C. Gruman, Participation in workplace health promotion programs: the contribution of health and organizational factors. Health Educ. Q. **15**(3), 269–288 (1988)
64. F. Smit, P. Cuijpers, J. Oostenbrink, N. Batelaan, R. de Graaf, A. Beekman, Costs of nine common mental disorders: implications for curative and preventive psychiatry. J. Ment. Health Policy Econ. **6**(4), 193–200 (2006)
65. J. Sowers, Obesity and cardiovascular disease. Clin. Chem. **44**(8), 1821–1825 (1998)
66. G.B. Stewart, K. Mengersen, N. Meader, Potential uses of Bayesian networks as tools for synthesis of systematic reviews of complex interventions. Res. Synth. Methods **5**(1), 1–12 (2014)
67. R. Sturm, The effects of obesity, smoking, and drinking on medical problems and costs. Health Aff. **21**, 245–253 (2002)
68. R. Sturm, K.B. Wells, Does obesity contribute as much to morbidity as poverty and smoking? J. Public Health **115**, 229–235 (2001)
69. M.S. Taitel, V. Haufle, D. Heck, R. Loeppke, D. Fetterolf, Incentives and other factors associated with employee participation in health risk assessments. J. Occup. Environ. Med. **50**(8), 863–872 (2008)
70. C. Tennant, Work-related stress and depressive disorders. J. Psychosom. Res. **51**(5), 697–704 (2001)
71. M. Thun, J. Henley, L. Apicella, Epidemiologic studies of fatal and nonfatal cardiovascular disease and ETS from spousal smoking. Environ. Health Perspect. **107**, 841–846 (1999)
72. S. Torp, H. Vinje, Is workplace health promotion research in the Nordic countries really on the right track? Scand. J. Public Health **42**, 74–81 (2014)
73. L.A. Tucker, G.M. Friedman, Obesity and absenteeism: an epidemiologic study of 10,825 employed adults. Am. J. Health Promot. **12**, 202–207 (1998)
74. W. Watson, J. Gauthier, The viability of organizational wellness programs: an examination of promotion and results. J. Appl. Soc. Psychol. *33*(6), 1297–1312 (2003)
75. World Health Organisation, *Constitution Principles* (WHO, Geneva, 1946)
76. P.P. Wu, J. Pitchforth, K. Mengersen, A hybrid queue-based Bayesian network framework for passenger facilitation modelling. Transp. Res. Part C: Emerg. Technol. **46**, 247–260 (2014)
77. Y. Yamada, M. Ishizaki, I. Tsuritani, Prevention of weight gain and obesity in occupational populations: a new target of health promotion services at worksites. J. Occup. Health **44**, 373–384 (2002)
78. S. Yamato, Y. Nakamura, Happy workplace program: workplace health promotion program driven by Thai Health Promotion Foundation. J. Occup. Health **56**(3), 87–89 (2014)

Chapter 13
Bayesian Modelling to Assist Inference on Health Outcomes in Occupational Health Surveillance

Nicholas J. Tierney, Samuel Clifford, Christopher C. Drovandi, and Kerrie L. Mengersen

Abstract *Objectives:* Occupational Health Surveillance (OHS) facilitates early detection of disease and dangerous exposures in the workplace. Current OHS analysis ignore important workplace structures and repeated measurements. There is a need to provide systematic analyses of medical data that incorporate the data structure. Although multilevel statistical models may account for features of OHS data, current applications in occupational health medicine are often not appropriate for OHS. Additionally, typical OHS data has not been analysed in a Bayesian framework, which allows for calculation of probabilities of potential events and outcomes. This paper's objective is to illustrate the use of Bayesian modeling of OHS. Three analytic aims are addressed: (1) Identify patterns and changes in health outcomes; (2) Explore the effects of a particular risk factor, smoking and industrial exposures over time for individuals and worker groups; (3) identify risk of chronic conditions in individuals. *Method*: A Bayesian hierarchical model was developed to provide individual and group level estimates and inferences for health outcomes, FEV1%, BMI, and Diastolic and Systolic blood pressure. *Results*: We identified individuals with the greatest degree of change over time for each outcome, and demonstrated how to flag individuals with substantive negative health outcome change. We also assigned probabilities of individuals moving into "at risk"

N. J. Tierney (✉)
ARC Centre of Excellence for Mathematical and Statistical Frontiers, Parkville, VIC, Australia

School of Mathematical Sciences, Queensland University of Technology, Brisbane, QLD, Australia

Department of Econometrics and Business Statistics, Monash University, Melbourne, Clayton, VIC, Australia
e-mail: nicholas.tierney@monash.edu

S. Clifford · C. C. Drovandi · K. L. Mengersen
ARC Centre of Excellence for Mathematical and Statistical Frontiers, Parkville, VIC, Australia

School of Mathematical Sciences, Queensland University of Technology, Brisbane, QLD, Australia

K. L. Mengersen et al. (eds.), *Case Studies in Applied Bayesian Data Science*, Lecture Notes in Mathematics 2259, https://doi.org/10.1007/978-3-030-42553-1_13

health categories 1 year from their last visit. *Conclusion*: Bayesian models can account for features typically encountered in OHS data, such as individual repeated measurements and group structures. We describe one way to fit these data and obtain informative estimates and predictions of employee health.

13.1 Introduction

Occupational Health Surveillance (OHS) is the systematic collection, analysis, and dissemination of employee exposure and health data to facilitate early detection of disease and dangerous exposures in the workplace [1]. Australian employers have a responsibility to identify, assess, and control risks arising from workplace hazards [2, 3]. There is a rigorous methodology for OHS data collection, but a surprising lack of agreement about analysis of these data [4]. Indeed, industry OHS data collection is often targeted for managing risk and implementing engineering controls. Consequently, many of the analyses conducted in industry focus on the likelihood of exposure rather than the impact of these risk factors on health. Moreover, current practices may ignore important data structures such as repeated measurements and workplace structures. This results in inferences not being applicable for individuals over time, or for groups with similar exposures within the workplace.

There is a need to provide systematic analyses of medical data that incorporate workplace structure, relevant to risk factors. An example of such a workplace structure is segmentation of the workplace into similar exposure groups (e.g., as in [5]). Moreover, such analyses need to incorporate typical features of OHS data, in particular where individuals have multiple health measurements, or single repeated measurements over time, or missing data. These analyses should provide both individual and group health predictions, and should improve the understanding of exposure effects on the workplace population as a whole, as well as similar exposure groups and individuals. Such analyses could flag individuals and groups for further health monitoring. The absence of such analyses in industry means that chronic disease and dangerous work environments may go unidentified and that health funding is not optimally or effectively targeted.

These features of OHS data described above can be accounted for with multilevel statistical models. These are in wide use in epidemiology [6, 7], and have been used in occupational health medicine to evaluate decline in lung function for ceramic fibre workers [8], assess impacts of asbestos [9], measure decline from cystic fibrosis [10] and model leptospirosis in abattoir employees [11].

However, applications of multilevel models in occupational health medicine do not quite mimic the analyses conducted in standard industry environments, as they might ignore individuals with only one measurement, population minority groups, or workplace structures [8, 9]. This is likely due to the fact that the goal of these papers is often to demonstrate the use of a new method [9, 10, 12], or discover new health risk factors or exposures [13–17].

In contrast, the goal of OHS analysis is to provide individual predictions for health, understand the effect of exposures on groups, and monitor exposures and health over time, so that individuals and groups at risk of some disease can be flagged for further health monitoring. Thus, ignoring cases with only one medical visit or analysing only subsets of the population for reasons such as sufficient sample size, can increase bias and/or variance of estimates.

Bayesian models provide a pragmatic framework for this research problem, as they provide simple and effective ways of analysing small effects, and provide a rich set of results that can be interpreted with probabilistic statements. Bayesian methodology also allows for direct comparison between groups and individuals, and provides probabilities on potential events and outcomes.

Bayesian techniques have been recently applied in occupational health, with [5] demonstrating the use of hierarchical models to combine monitoring data and professional judgement from occupational hygienists to facilitate decision making. Bayesian hierarchical models have also been applied to quantify chemical exposure variation in human populations [18], and to combine two data sources from animal studies and human industrial studies to create informative priors to estimate human lung function changes [19].

Industry data used in the literature typically consist of longitudinal data collected from employees in a particular industry, or set of industries. These data are used to evaluate the effect of the working conditions, such as long hours with no sleep [20, 21], metal smelting, and other exposures [8–12, 17]. Other OHS studies focus on small populations, using experiments to evaluate effects of increasing some exposure on health [22, 23], or larger cross-sectional studies using registry data, cohort studies, or surveys to evaluate the effect of an environmental exposure on diseases such as rhinitis or cystic fibrosis [10, 16].

Notwithstanding these studies, there are no examples of Bayesian hierarchical modelling and analysis of typical OHS data with applications in an industry context. This paper analyses OHS data from selected industrial sites around Australia to identify risk factors for health outcomes. The multilevel model adopted is a Bayesian hierarchical model providing individual and group level estimates and inferences.

The aims of the analysis are threefold, and are focused on the following health outcomes: lung function (as a percentage of predicted Forced Expiratory Volume in 1 s (FEV_1, $FEV_1\%$), Body Mass Index (BMI), and systolic and diastolic blood pressure. These health outcomes were selected as they are clinically substantive in the case study population and are well known in OHS. The first analytic aim is to identify patterns and changes in health outcomes. The second aim is to explore the effects of a particular risk factor: smoking and industrial exposures over time for individuals and worker groups. The third aim is to identify risk of chronic conditions (such as obesity, hypertension, and obstructive/restrictive lung disease) in individuals.

13.2 Method

13.2.1 Case Study Data

The case study considered in this paper is typical of many large companies that are involved in a range of activities, such as construction, mining, manufacturing and agriculture. For reasons of confidentiality, the particular industry and associated sites are not named here. The data are comprised of over 3000 employee medical records from nearly 2000 individuals located at a number of sites. Each observation is a medical visit, and while most employees have one or two visits, some have over ten visits over a 10 year period. Employees are typically grouped by their workplace exposure; for example, Administration employees are less likely to be exposed to environmental factors such as dust or noise, compared to maintenance exposure groups. In this way, Administration provides a useful control group to compare to the other exposure groups. Employees may change positions within the company over their career and thus may also change their exposure group. The pattern of measurements over time for individuals are illustrated in Fig. 13.1, which displays individual measurements of lung function (FEV1%) over visits for selected exposure groups.

The frequency of medical visits changes for each exposure group, as certain exposure groups require more frequent medical examinations to ensure that they

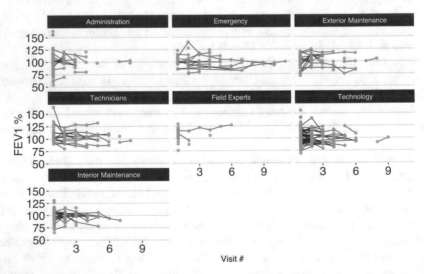

Fig. 13.1 Individual employee lung function (FEV1%) over their medical visit number (1, up to 10), for each exposure group. A sample of 50% of employees is used to reduce overplotting. Individuals are linked by a line between observations. A point on the 2nd or later visit which is not joined to previous points by a line indicates individuals who have changed exposure group. Broken lines and individual floating points without lines indicate where individuals have changed exposure group. Note that the number of days between visits varies by individual and exposure group

are fit for work. The times between visits for each worker were not equally spaced, with median number of days since first visit being 1028 (IQR = 193–3363), or 2.8 years (IQR = 0.5–9.2 years). Gender (male or female) and smoking status (ever smoker or never smoker) were also recorded. Dust data were not recorded for some dates and were interpolated using a loess model [24] fitted for each exposure group, so that the values corresponded to medical examination dates. Interpolated values should be treated with care, and explored with visual and numerical summaries.

13.2.2 Ethics

The Queensland University of Technology Human Research Ethics Committee assessed that this research met the conditions for exemption from HREC review and approval in accordance with section 5.1.22 of the Australian National Statement on Ethical Conduct in Human Research.

13.2.3 Patient and Public Involvement

The development of the research questions and outcomes were informed by discussion with health practitioners who helped collect the data. The patients were not involved in the results, design, or recruitment. The paper will be shared with the medical practitioners for their use in future designs. We thank the health practitioners and patients involved in the data collection.

13.2.4 Modelling

We construct four multilevel Bayesian hierarchical models. Each model predicts one of the four outcomes: lung function (FEV1%), Body Mass Index (BMI), systolic blood pressure, and diastolic blood pressure.

Let Y_{ij} be the ith individual's jth health observation, at a time day_{ij} after their first visit. We assume that Y_{ij} follows a normal distribution with mean μ_{ij} and variance σ_y^2. Let β_{0i} and β_{di} be respectively the individual intercept and individual health trend coefficient associated with the jth day for the ith person, these individual parameters are centered around an overall intercept β_{0c} and an overall slope β_{dc}, the effect of the number of days since arriving at the workplace. Thus β_{di} is the linear trend over time for the health characteristics of interest for the ith individual, over and above the overall population effect. Let β_g be the effect of being female (compared to being male); let β_s be the effect of being a smoker (compared to a never smoker), and let $\sum_{k=1}^{n_{exposure}-1} \beta_k I(exposure_{ij} = k)$ be the effect of a workplace exposure, where $I(.)$ indicates whether an individual i at a visit j is in exposure group k, with the baseline exposure group set to Administration. Thus the model for a particular health outcome is represented as:

$$Y_{ij} \sim \mathcal{N}(\mu_{ij}, \sigma_y^2)$$

with

$$\mu_{ij} = \beta_{0i} + \beta_{di}\,\mathrm{day}_{ij} + \beta_g\,\mathrm{gender}_{ij} + \beta_s\,\mathrm{smoke}_{ij} + \beta_p\,\mathrm{dust}_{ij} + \sum_{k=1}^{n_{\mathrm{exposure}}-1} \beta_k\,I\,(\mathrm{exposure}_{ij} = k)$$

$$\beta_{0i} \sim N(\beta_{0c}, \sigma_0^2)$$

$$\beta_{di} \sim N(\beta_{dc}, \sigma_d^2)$$

for $i = 1 \ldots n_I$, $j = 1 \ldots n_{0i}$, $k = 1 \ldots, n_E$, where n_I is the total number of individuals, n_{0i} is the number of observations for each individual, and n_E is the number of exposure groups.

In the absence of other information, all of the regression coefficients were allocated independent normal priors with a mean of 0 and a variance of $D_1 = 10^3$.

$$\beta_{0c}, \beta_{dc}, \beta_{di}, \beta_g, \beta_s, \beta_p, \beta_k \sim N(0, D_1)$$

Priors on $\sigma_0, \sigma_y, \sigma_d$ were set to a uniform distribution with bounds of zero and D2, where $D_2 = 100$ for BMI and $FEV_1\%$, and $D_2 = 50$ for Systolic and Diastolic blood pressure. D_2 is intended to better reflect the variation in BMI and $FEV_1\%$ compared to blood pressure. Note also that we do not recommend automatically choosing set values for the uniform, but to instead choose sensible bounds based on the problem at hand.

$$\sigma_y, \sigma_0, \sigma_d, \sim \mathrm{Uniform}(0, D_2)$$

Note also that the priors used for the β terms are proper priors, which produce a proper posterior. In some cases improper priors such as an infinite uniform prior might be used, but these are sometimes not valid choices (See [25] and [26] for more details). It is worthwhile to consider the choice of prior for the variance terms. Although we have used inverse gamma and uniform priors, other weakly informative priors could be considered, such as a half-t-prior (represented as a half-Cauchy) [27]. It is important to not automatically choose uniform or half-t-priors, but to explore options during model building.

Data processing and manipulation were implemented using the R statistical programming language [28] and various R packages [29–34]. To ensure reproducibility, the paper was written using rmarkdown and knitr [35, 36]. Potential outliers in the data were checked administratively and confirmed for biological plausibility in the context of the workforce under consideration. Given this, we elected to include them in the analyses. Moreover, the modelling goal is to identify those who are risk, so removing outliers seems counter to that goal. The model was run for 20,000 iterations (10,000 burnin) using JAGS [37, 38]. Thinning was applied to the analysis, removing every 20th value to assist in reducing autocorrelation and for computational storage. We note that thinning is not absolutely necessary in an analysis, and should be assessed case by case [39]. We note that other software such as STAN, WinBUGS or OpenBUGS, Nimble, and greta could also have been used [40–43]. The diagnostics for MCMC convergence were predominantly graphical and statistical [44, 45]. Graphical evaluation included expert examination of posterior density plots, traceplots and autocorrelation plots of parameters. Statistical evaluation included calculation of the Geweke diagnostic and effective sample size.

Missing values were imputed from their respective posterior conditional distributions as part of the Bayesian analysis. Posterior estimates of each parameter, including mean, 95 and 80% credible intervals, and probability of being negative were calculated after burnin. An effect was nominated as substantive if the corresponding credible interval did not contain zero. The probability of individual health outcomes reaching the threshold value of being a chronic condition was also calculated. Individuals were identified as being "at risk" if the corresponding estimates of the parameter for change over time, β_{di}, contained 0 in the 95% credible intervals, and β_{di} was far away from zero. For the purposes of exposition, individuals with 3 or more visits were selected as examples to explore further.

Patterns and trends in health outcomes were examined by exploring individuals' change over time and identifying substantive effects, addressing analytic aim 1. The effects of smoking and industrial exposures over time for individuals and exposure groups were examined by evaluating substantive effects of smoking and dust for each outcome, and finding those exposure groups substantively different from the Administration population, addressing aim 2. To identify future risk of chronic conditions, 1 year forecasts for each individual and corresponding 95% credible intervals were calculated from the respective posterior predictive distribution, and the probability of having a chronic condition in 1 year was obtained, addressing aim 3. Model fit was evaluated by examining the proportion of observed values lying within the 95% and 80% posterior predictive intervals [46, 47].

13.3   Results

13.3.1   Demographics

In the case study dataset, the population was predominantly male (86%), with the mean overall age being 35.8 years. Males were older on average, but not significantly so, compared with females. For all exposure groups there were more

Table 13.1 Percent of the population in selected exposures

Exposure	% of population
Technology	18–20
Administration	8–10
Interior maintenance	8–10
Technicians	6–7
Emergency	5–6
Exterior maintenance	4–5
Field experts	2–3

males than females, except in the Administration exposure group. The proportion of individuals in selected exposure groups is shown below in Table 13.1.

13.3.2 Model Fit

Figure 13.2 shows the posterior predictions for each outcome plotted on the y axis against the observed values on the x axis. The points represent the observed values and the corresponding posterior means with vertical lines representing the respective 95% posterior predictive intervals. A line of perfect prediction runs from the bottom left to the top right corner. The points and lines are shown in red to indicate when the observed value lies outside of the 95% posterior predictive interval.

Model fit was assessed by visual inspection of Fig. 13.2, and by assessing the percentage of observed values that lie within nominated posterior predictive intervals (Table 13.2). The models for BMI and FEV1% had very high proportions of observed values in the 95% intervals and 80% intervals, indicating reasonable model fit.

Figure 13.3 shows four selected "at risk" individuals and their posterior mean and credible intervals for the health characteristics systolic blood pressure, FEV1%, diastolic blood pressure, and BMI. The proportion of individuals "at risk" for each health outcome, and the mean and standard deviation for each health outcome for those at risk and not at risk, are shown in Table 13.3. Individuals were identified as "at risk" in this case according to whether their parameter estimates for change over time β_d were the furthermost away from zero (and did not contain zero in the credible interval) for the health characteristics systolic blood pressure, FEV1%, diastolic blood pressure, and BMI. As described in the method section, these individuals had 3 or more visits.

Table 13.4 shows the posterior mean, 95% credible interval, and probability of being negative for each of the risk factors considered, namely smoking, dust, and days, since commencement.

The number of days since first visit had a substantive effect on all outcomes, and was associated with a decrease in FEV$_1$%, and a decrease in BMI, diastolic and systolic blood pressure. Dust did not have a substantive impact on any outcomes, but was associated with an 11% chance of decreased BMI, a 22% chance of

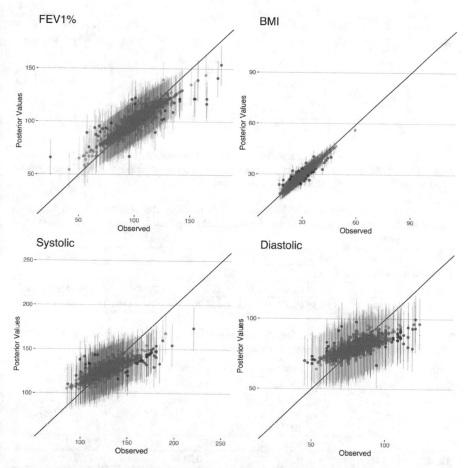

Fig. 13.2 Observed values for each health outcome plotted against the posterior mean values (point) with their respective 95% posterior predictive interval (lines). A line of perfect prediction is shown. The points and lines are shown in red to indicate when the 95% posterior predictive interval lies outside the line of perfect prediction

Table 13.2 Percent of observed values inside the 95 and 80% posterior prediction intervals and RSS for each model outcome

Outcome	Inside 95% PI	Inside 80% PI
BMI	99.09	97.52
Diastolic BP	98.38	90.18
FEV1 (%)	98.96	95.91
Systolic BP	98.32	92.15

decreased FEV$_1$%, a 31% chance of decreased systolic blood pressure, and a 73% chance of decreased diastolic blood pressure. Being a smoker was associated with substantively decreased FEV$_1$%, a 100% chance of decreased FEV$_1$ %, a 95% chance of decreased BMI, a 13% chance of increased diastolic blood pressure, and a 24% chance of increased systolic blood pressure.

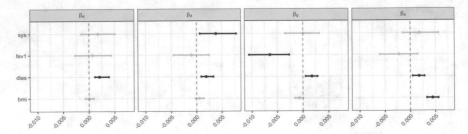

Fig. 13.3 Individuals whose parameter estimates for change over time, β_d, were the furthest away from zero (and did not contain zero in the credible interval) for the health characteristics systolic blood pressure, FEV1%, diastolic blood pressure, and BMI

Table 13.3 The proportion of individuals with 3 or more visits who were 'at risk' (95% credible interval for change over time did not include zero) and the health outcomes for the 'at risk' and 'not at risk' groups

Summary	BMI	FEV1%	Systolic	Diastolic
Proportion at risk	0.1	0.02	0.01	0.98
Mean (SD) not at risk	27.68 (3.84)	100.19 (12.36)	126.46 (13.32)	73.89 (10.98)
Mean (SD) at risk	30.55 (5.22)	102.00 (23.76)	140.77 (19.01)	80.94 (10.04)

Table 13.4 Estimated posterior mean, 95% credible intervals and probability of the effect of Day, Dust, and Smoking parameters being less than zero

Terms	BMI	FEV1%	Systolic	Diastolic
Days	6.09×10^{-4}	-8.44×10^{-4}	1.35×10^{-3}	1.78×10^{-3}
	$(4.72 \times 10^{-4}, 7.52 \times 10^{-4})$	$(-1.41 \times 10^{-3}, -2.51 \times 10^{-4})$	$(7.30 \times 10^{-4}, 1.97 \times 10^{-3})$	$(1.34 \times 10^{-3}, 2.17 \times 10^{-3})$
	0	9.98×10^{-1}	0	0
Dust	0.11	0.33	0.24	-0.21
	$(-0.06, 0.3)$	$(-0.52, 1.15)$	$(-0.72, 1.16)$	$(-0.92, 0.51)$
	0.11	0.22	0.31	0.73
Smoking	-0.33	-1.87	0.47	0.52
	$(-0.72, 0.07)$	$(-3.24, -0.55)$	$(-0.8, 1.79)$	$(-0.38, 1.44)$
	0.94	1	0.24	0.13

All exposure group effects were compared to the baseline Administration, and the effects for each outcome over all exposure groups are shown in Fig. 13.4. BMI was substantively lower in Maintenance, Emergency, and Field Experts. Diastolic blood pressure was substantively lower in Technology and Field Expert exposure groups. Technologists had substantively lower systolic blood pressure, and Technicians had substantively higher FEV$_1$%.

Figure 13.5 shows the same four selected individuals previously identified as being "at risk", from Fig. 13.3, and the observed and predicted values for health outcomes. Observed outcomes are shown as blue points and model mean posterior

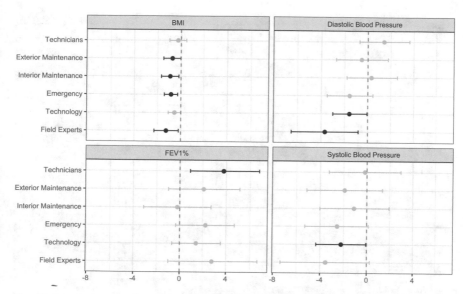

Fig. 13.4 Posterior mean and 95% credible interval for exposure group parameters, for each model. The baseline exposure group is Administration

values are shown as a blue line. The dark blue ribbon around the blue line represents an 80% credible interval, and the light blue ribbon the 95% credible interval. A 1 year forecast is shown as a red line extending from the blue line, and similarly the 80% and 95% credible intervals are displayed. A dotted line is shown for each outcome, which represents a clinically relevant threshold of chronic disease for each outcome. Individual probabilities of chronic condition in the 1 year forecast are labelled directly on Fig. 13.5.

13.4   Discussion

This paper set out to develop a Bayesian approach to analysing OHS data and to illustrate three analytic aims focussed on the health outcomes lung function (FEV$_1$%), Body Mass Index (BMI), and systolic and diastolic blood pressure. Aim 1 investigated patterns and trends in the health outcomes. Aim 2 explored the effects of smoking and industrial exposures over time for individuals and worker groups. Aim 3 identified future individual risk of chronic conditions.

Aim 1 was addressed by examining individual change over time and identifying individuals with the greatest degree of change over time for each outcome. We demonstrated how one could then assess the overall change over time for these individuals, which could be used to identify trends in other health outcomes. We also identified the proportion of individuals who fell into an "at risk" category.

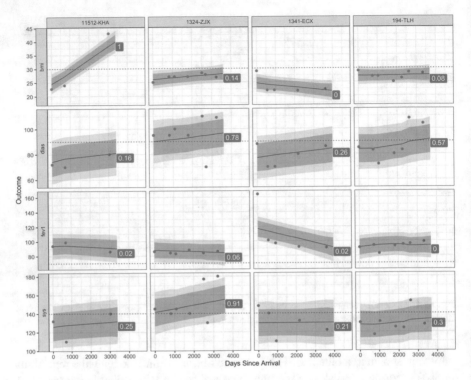

Fig. 13.5 Individuals from Fig. 13.3 and their observed (points), predicted (blue region and line), and forecasted (red region and lines) values for the health outcomes shown with 80% (darker region) and 95% credible intervals (lighter region). Labels show the probability of the individual having a clinically defined chronic disease at the forecasted timepoint

Identifying those individuals with substantive negative change in BMI, lung function, systolic, and diastolic blood pressure means that medical professionals could flag these individuals as at risk (compared to the overall worker population), and provide more frequent medical attention to better monitor their health.

Aim 2 was addressed by examining the probability that the effects of smoking and dust were different from zero for each outcome, finding those exposure groups that were substantively different from the reference group (Administration), and identifying individuals at risk based on substantive change over time in health outcomes. Smoking was associated with negative health outcomes for lung function and systolic and diastolic blood pressure. This information can be used to further support health policies, such as implementation of tobacco bans in the workplace.

The nominated industrial exposure, dust was not substantively associated with health outcomes in the workplaces in this case study. The relevant parameter estimates had quite wide credible intervals, possibly due to interpolation of the data, which was used to align dust measurement points with health measurements. This demonstrates that frequent measurements of industrial exposures of concern can provide more certainty in the measurement of effects. Interestingly, results identified that employees in Administration should be more closely monitored and perhaps should be the focus of health interventions and healthy worker programs in workplaces. Providing descriptive statistics of the outcomes for at risk and not at risk populations (Table 13.3) provides medical professionals with a measure of how meaningfully different these populations are, and facilitates more targeted health and wellness programs.

Aim 3 was addressed by calculating posterior predictions and corresponding intervals and 1 year forecasts for all individuals. This allows medical professionals to assign a probability that an individual might move into an "at risk" category 1 year from their last visit. This means that individuals may be flagged as "at risk" and further action can be taken, perhaps in the form of more frequent medical visits to more closely monitor their health measures. his demonstrates how forecasting could identify "at risk" individuals, by placing a threshold on the probability of health outcomes being medically classified as chronic or acute conditions. Some individuals might cross the threshold over the observed times, whilst others might be predicted to enter the threshold with a given probability over the next year.

The Bayesian hierarchical model accounts for important features of data such as multiple measurements for individuals, and the exposure group structure in the workplace. The definition of a Bayesian credible interval as a range of probable values for a parameter makes it easier to communicate model inferences. The model also provides probabilities of interest directly, conditional on the data. This is a useful complement to credible intervals. Forecasting of future observations in a Bayesian framework also allows for probabilistic statements based directly on the posterior predictive distribution. The Bayesian framework naturally includes additional uncertainty due to imputation of missing values. These features compare favourably to their frequentist modelling counterparts.

Extensions to the Bayesian models developed here are also straightforward. For example, an obvious next step in analysis might be to add interactions into the model, such as smoking and dust, or BMI and blood pressure. One relatively straightforward way to explore the impact of interactions is to evaluate the Bayes factors for each variable, approximated using the Savage-Dickey density ratio, which only requires samples from the posterior [48, 49]. This can add time to the model building process, in terms of deciding upon the most useful model, but is worth the effort if the practitioner is genuinely interested in one or two interaction terms. As with any working population, there may be some healthy cohort effect [50, 51], where, being employable, employees are healthier than the general population. The methods provided in this study identify employees and groups that are different from the population. Combining this information with reference chronic conditions provides a more comprehensive approach which might otherwise have missed healthy employees.

It is also possible that there may perhaps be less measurement error for long term employees; here a model that predicts the number of visits for each individual may be useful, where the number of visits n_{0i} for each individual is the outcome. Additionally, there may be some correlation between the slope and the intercept, which could be accounted for by modelling them as coming from some bivariate normal distribution [46]. It is acknowledged that the model fit is not ideal, particularly with respect to underestimation and overestimation of very high and low values, respectively. While this regression to the mean is to be expected given the random effects terms, the fit could be improved for the other outcomes, perhaps by including interactions as previously discussed.

As far as we are aware, this is the first time Bayesian methods have been applied to this kind of OHS data. It is our hope that this paper can serve as one way to fit and interpret these data, and serve as encouragement for researchers in the field of OHS, to include Bayesian approaches in their analytic toolkit. The ultimate ambition is to provide more informative evidence-based OHS assessments for a healthier workforce and more profitable workplaces.

13.5   Summary

13.5.1   Strengths and Limitations of the Study

- Strength: This is the first application of Bayesian methods to typical data found in occupational health surveillance.
- Strength: The methods used account for important features of data such as multiple measurements for individuals, and the group structure of exposure groups in the workplace.
- Strength: The model allows for groups and individuals to be flagged as "at risk", enabling proactive action on individual health.
- Strength: The definition of a Bayesian credible interval as a range of probable values for a parameter makes it easier to communicate model inferences.
- Strength: The model provides probabilities of interest directly, conditional on the data, which is a useful complement to credible intervals that makes effects and uncertainty simpler and easier to communicate to health practitioners.
- Limitation: No account was taken of the healthy worker effect, so whilst the focus of the paper is on employees rather than the general population, the analysis may be biased if healthier employees remain longer in the industry.
- Limitation: The model used vague priors, and so future work could explore the use of more informative priors, based on, for example, previous data collected in similar fields.

Acknowledgments   The authors would like to thank Dr. Xing Lee and Dr. Nicole White for their advice and collaboration with this project.

Data Sharing   For reasons of employee confidentiality and industry sensitivity, the dataset used in the case study is not available for sharing. However, the seminal features of the dataset are well described in the paper. Code used for the statistical data analysis for this work can be found at: https://github.com/njtierney/njtbatohs-chapter.

Contributors   NJT conducted literature survey, statistical analysis, created visualisations, and wrote the first draft. SC provided critical feedback, assisted in developing the statistical model and in presenting results. CCD provided assistance in developing the statistical model and in critical feedback of the paper. KLM provided initial description of the model, assisted in developing the code, and provided critical feedback of the paper and presentation of results. All authors approved the version of the paper for publishing, and agreed to respond to questions that may arise regarding integrity of the work.

Funding   This research was jointly funded by an Australian Postgraduate Award (APA), the Australian Technology Network Industry Doctoral Training Centre (IDTC), the Australian Research Council, and the ARC Centre of Excellence for Mathematical and Statistical Frontiers. CCD was supported by an Australian Research Council's Discovery Early Career Researcher Award funding scheme (DE160100741).

Competing Interests   None declared.

Ethics Approval   The QUT University Human Research Ethics Committee.

References

1. P.A. Schulte, D. Trout, R.D. Zumwalde, E. Kuempel, C.L. Geraci, V. Castranova, et al., Options for occupational health surveillance of workers potentially exposed to engineered nanoparticles: state of the science. J. Occup. Environ. Med./Am. Coll. Occup. Environ. Med. **50**(5), 517–26 (2008)
2. S.W. Australia, *Australian Work Health and Safety Strategy 2012–2022: Healthy, Safe and Productive Working Lives* (Australian Government - Safe Work Australia, Canberra, 2012)
3. I. Firth, *Simplified Occupational Hygiene Risk Management Strategies*, Tullamarine. (Australian Institute of Occupational Hygienists, Westmeadows, 2006)
4. L. Lewis, D. Fishwick, Health surveillance for occupational respiratory disease. Occup. Med. **63**(5), 322–34 (2013)
5. S. Banerjee, G. Ramachandran, M. Vadali, J. Sahmel, Bayesian hierarchical framework for occupational hygiene decision making. Ann. Occup. Hyg. **58**(9), 1079–93 (2014)
6. J. Lynch, G.D. Smith, A life course approach to chronic disease epidemiology. Annu. Rev. Public Health **26**, 1–35 (2005)
7. C. Duncan, K. Jones, G. Moon, Context, composition and heterogeneity: using multilevel models in health research. Soc. Sci. Med. **46**(1), 97–117 (1998)
8. R.T. McKay, G.K. LeMasters, T.J. Hilbert, L.S. Levin, C.H. Rice, E.K. Borton, et al., A long term study of pulmonary function among US refractory ceramic fibre workers. Occup. Environ. Med. **68**(2), 89–95 (2011)
9. E. Algranti, E.M.C. Mendonça, E. Hnizdo, E.M. De Capitani, J.B.P. Freitas, V. Raile, et al., Longitudinal decline in lung function in former asbestos exposed workers. Occup. Environ. Med. **70**(1), 15–21 (2013)
10. R.D. Szczesniak, G.L. McPhail, L.L. Duan, M. Macaluso, R.S. Amin, J.P. Clancy, A semi-parametric approach to estimate rapid lung function decline in cystic fibrosis. Ann. Epidemiol. **23**(12), 771–7 (2013)
11. E.A.J. Cook, W.A. de Glanville, L.F. Thomas, S. Kariuki, B.M. de Clare Bronsvoort, E.M. Fèvre, Risk factors for leptospirosis seropositivity in slaughterhouse workers in western kenya. Occup. Environ. Med. **74**, 357–365 (2016)
12. F. de Vocht, H. Kromhout, G. Ferro, P. Boffetta, I. Burstyn, Bayesian modelling of lung cancer risk and bitumen fume exposure adjusted for unmeasured confounding by smoking. Occup. Environ. Med. **66**(8), 502–8 (2009)
13. P.S. Bakke, V. Baste, R. Hanoa, A. Gulsvik, Prevalence of obstructive lung disease in a general population: relation to occupational title and exposure to some airborne agents. Thorax **46**(12), 863–70 (1991)

14. G.W. Gibbs, B. Armstrong, M. Sevigny, Mortality and cancer experience of quebec aluminum reduction plant workers. Part 2: mortality of three cohorts hired on or before January 1, 1951. J. Occup. Environ. Med./Am. Coll. Occup. Environ. Med. **49**(10), 1105–23 (2007)

15. J. Hellgren, L. Lillienberg, J. Jarlstedt, G. Karlsson, K. Torén, Population-based study of non-infectious rhinitis in relation to occupational exposure, age, sex, and smoking. Am. J. Ind. Med. **42**(1), 23–8 (2002)

16. K. Radon, U. Gerhardinger, A. Schulze, J.-P. Zock, D. Norback, K. Toren, et al., Occupation and adult onset of rhinitis in the general population. Occup. Environ. Med. **65**(1), 38–43 (2008)

17. A. Olsson, H. Kromhout, M. Agostini, J. Hansen, C.F. Lassen, C. Johansen, et al., A case-control study of lung cancer nested in a cohort of european asphalt workers. Environ. Health Perspect. **118**(10), 1418–24 (2010)

18. K. Shao, B.C. Allen, M.W. Wheeler, Bayesian hierarchical structure for quantifying population variability to inform probabilistic health risk assessments. Risk Anal. **37**, 1865–1878 (2016); An official publication of the Society for Risk Analysis

19. S.M. Bartell, G.B. Hamra, K. Steenland, Bayesian analysis of silica exposure and lung cancer using human and animal studies. Epidemiology **28**(2), 281–7 (2017)

20. W.K. Sieber, C.F. Robinson, J. Birdsey, G.X. Chen, E.M. Hitchcock, J.E. Lincoln, et al., Obesity and other risk factors: the national survey of U.S. Long-haul truck driver health and injury. Am. J. Ind. Med. **57**(6), 615–26 (2014)

21. D. Fekedulegn, C.M. Burchfiel, L.E. Charles, T.A. Hartley, M.E. Andrew, J.M. Violanti, Shift work and sleep quality among urban police officers: the BCOPS study. J. Occup. Environ. Med./Am. Coll. Occup. Environ. Med. **58**(3), e66–71 (2016)

22. D.J. Hendrick, M.J. Connolly, S.C. Stenton, A.G. Bird, I.S. Winterton, E.H. Walters, Occupational asthma due to sodium iso-nonanoyl oxybenzene sulphonate, a newly developed detergent ingredient. Thorax **43**(6), 501–502 (1988)

23. S.C. Stenton, J.H. Dennis, E.H. Walters, D.J. Hendrick, Asthmagenic properties of a newly developed detergent ingredient: sodium iso-nonanoyl oxybenzene sulphonate. Br. J. Ind. Med. **47**(6), 405–10 (1990)

24. S. Steinle, S. Reis, C.E. Sabel, S. Semple, M.M. Twigg, C.F. Braban, et al., Personal exposure monitoring of PM2.5 in indoor and outdoor microenvironments. Sci. Total Environ. **508**, 383–94 (2015)

25. J.P. Hobert, G. Casella, The effect of improper priors on Gibbs sampling in hierarchical linear mixed models. J. Am. Stat. Assoc. **91**(436), 1461–73 (1996)

26. F.J. Rubio, M.F.J. Steel, Flexible linear mixed models with improper priors for longitudinal and survival data. Electron. J. Stat. **12**(1), 572–98 (2018)

27. A. Gelman, Prior distributions for variance parameters in hierarchical models (comment on article by browne and draper). Bayesian Anal. **1**(3), 515–34 (2006).

28. R Core Team, R: A language and environment for statistical computing [Internet] (R Foundation for Statistical Computing, Vienna, 2016). Available from: https://www.R-project.org/

29. H. Wickham, R. Francois, Dplyr: A grammar of data manipulation [Internet] (2017). Available from: https://github.com/hadley/dplyr

30. H. Wickham, Tidyr: Easily tidy data with 'spread()' and 'gather()' functions (2017)

31. H. Wickham, Readxl: Read excel files [Internet] (2016). Available from: https://CRAN.R-project.org/package=readxl

32. H. Wickham, Purrr: Functional programming tools [Internet] (2016). Available from: https://CRAN.R-project.org/package=purrr

33. G. Grolemund, H. Wickham, Dates and times made easy with lubridate. J. Stat. Softw. [Internet] **40**(3), 1–25 (2011). Available from: http://www.jstatsoft.org/v40/i03/

34. M. Dowle, A. Srinivasan, T. Short, R Saporta SL with contributions from, E. Antonyan, Data.table: Extension of data.frame [Internet] (2016). Available from: http://r-datatable.com

35. J. Allaire, J. Cheng, Y. Xie, J. McPherson, W. Chang, J. Allen, et al., R markdown: Dynamic documents for r [Internet] (2017). Available from: http://rmarkdown.rstudio.com

36. Y. Xie, Dynamic documents with R and knitr [Internet], 2nd edn. (Chapman Hall/CRC, Boca Raton/Florida, 2015) Available from: http://yihui.name/knitr/
37. M. Plummer, et al., JAGS: A program for analysis of bayesian graphical models using Gibbs sampling, in *Proceedings of the 3rd International Workshop on Distributed Statistical Computing*, Vienna (2003), p. 125
38. M. Plummer, Rjags: Bayesian graphical models using MCMC [Internet] (2016). Available from: https://CRAN.R-project.org/package=rjags
39. W.A. Link, M.J. Eaton, On thinning of chains in MCMC: thinning of MCMC chains. Methods Ecol. Evol./British Ecol. Soc. **3**(1), 112–115 (2012)
40. Stan Development Team, RStan: The R interface to Stan [Internet] (2018). Available from: http://mc-stan.org/
41. D.J. Lunn, A. Thomas, N. Best, D. Spiegelhalter, WinBUGS - a Bayesian modelling framework: concepts, structure, and extensibility. Stat. Comput. **10**(4), 325–37 (2000)
42. P. de Valpine, D. Turek, C.J. Paciorek, C. Anderson-Bergman, D.T. Lang, et al., Programming with models: writing statistical algorithms for general model structures with nimble. J. Comput. Graph. Stat. **26**(2), 403–13 (2017)
43. N. Golding, Greta: Simple and scalable statistical modelling in R [Internet] (2018). Available from: https://github.com/greta-dev/greta
44. D. Lunn, C. Jackson, N. Best, A. Thomas, D. Spiegelhalter, *The Bugs Book: A Practical Introduction to Bayesian Analysis* (CRC press, Boca Raton, 2012)
45. X. Fernández-i-Marín, "Ggmcmc": analysis of mcmc samples and Bayesian inference. J. Stat. Softw. **70**(9), 1–20 (2016)
46. A. Gelman, J.B. Carlin, H.S. Stern, D.B. Dunson, A. Vehtari, D.B. Rubin, *Bayesian Data Analysis*, vol. 2 (Chapman & Hall/CRC, Boca Raton, 2014)
47. J. Besag, P. Green, D. Higdon, K. Mengersen, Bayesian computation and stochastic systems. Stat. Sci. **10**(1), 3–41 (1995); A Review Journal of the Institute of Mathematical Statistics
48. E.-J. Wagenmakers, T. Lodewyckx, H. Kuriyal, R. Grasman, Bayesian hypothesis testing for psychologists: a tutorial on the Savage–Dickey method. Cogn. Psychol. **60**(3), 158–89 (2010)
49. J.M. Dickey, The weighted likelihood ratio, linear hypotheses on normal location parameters. Ann. Math. Stat. **42**, 204–223 (1971)
50. C.Y. Li, F.C. Sung, A review of the healthy worker effect in occupational epidemiology. Occup. Med. **49**(4), 225–9 (1999)
51. A.J. McMichael, R. Spirtas, L.L. Kupper, An epidemiologic study of mortality within a cohort of rubber workers, 1964–1972. J. Occup. Med. **16**(7), 458–64 (1974); Official Publication of the Industrial Medical Association

Part III
Real World Case Studies in Ecology

Chapter 14
Bayesian Networks for Understanding Human-Wildlife Conflict in Conservation

Jac Davis, Kyle Good, Vanessa Hunter, Sandra Johnson, and Kerrie L. Mengersen

Abstract Human-wildlife conflict is a major threat to survival and viability of many native animal species worldwide. Successful management of this conflict requires evidence-based understanding of the complex system of factors that motivate and facilitate it. However, for many affected species, data on this sensitive subject are too sparse for many statistical techniques. This study considers two iconic wild cats under threat in diverse locations and employs a Bayesian Network approach to integrate expert-elicited information into a probabilistic model of the factors affecting human-wildlife conflict. The two species considered are cheetahs in Botswana and jaguars in the Peruvian Amazon. Results of the individual network models are presented and the relative importance of different conservation management strategies are presented and discussed. The study highlights the strengths of the Bayesian Network approach for quantitatively describing complex, data-poor real world systems.

Keywords Cheetah · Botswana · Jaguar · Peru · Amazon · Conservation · Bayesian network · Human-wildlife conflict

J. Davis
Department of Psychology, University of Cambridge, Cambridge, UK

Institute for Environmental Management, Vrije Universiteit Amsterdam, Amsterdam, Netherlands

K. Good
Cheetah Conservation Botswana, Gaborone, Botswana

V. Hunter
Peruvian Amazon Corridor Trust (PACT), Lima, Peru

S. Johnson · K. L. Mengersen (✉)
Queensland University of Technology, Brisbane, QLD, Australia
e-mail: K.Mengersen@qut.edu.au

K. L. Mengersen et al. (eds.), *Case Studies in Applied Bayesian Data Science*, Lecture Notes in Mathematics 2259, https://doi.org/10.1007/978-3-030-42553-1_14

347

14.1 Introduction

Among the 105,732 species listed in the update to the International Union for the Conservation of Nature (IUCN) update to the Red List of Threatened Species, over a quarter (28,338) are reported to be threatened with extinction [11]. The single most important threat identified in the report is humans. This threat is realised through a range of activities, including but not limited to over-exploitation of the species, habitat loss, spread of disease, environmental mismanagement associated with human activities, and conflict.

In this study, we focus on this last factor, namely conflict, and its impact on a particular set of species, namely wildlife. Although human-wildlife conflict includes both negative impacts of wildlife on humans, and of humans on wildlife (WPC Recommendation [35]), for the purposes of the current paper we confine our attention to direct negative impacts of human behaviour on wildlife.

Many threatened wildlife species have home ranges that extend into modified urban, agricultural and industrial landscapes. The requirements of urban, agricultural and industrial land uses often conflict with the requirements of wildlife biodiversity conservation, contributing to threatening processes which drive wildlife population declines. Proximity of threatened wildlife to urban and rural human populations also leads to diverse and often polarised societal attitudes towards wildlife, thereby threatening agents and conservation efforts to save threatened wildlife. The need to reconcile diverse societal attitudes and conservation imperatives further complicates decision-making processes and conservation efforts.

Substantial resources have been committed to understanding the factors associated with threatened wildlife species, and although resource managers recognize that these factors range across ecological, biological, physical, social and economic perspectives, research and management efforts are typically confined to specific issues. A major reason for this is because it is often difficult to consider the multitude of factors in a coherent, transparent manner.

One approach to modelling the many facets of human-wildlife conflict is through a Bayesian Network (BN). BNs are increasingly being used for ecological, environmental and conservation modelling, among many other applications [8, 14–16, 17]. A key advantage of this method is that it can integrate quantitative information from a variety of sources, including expert knowledge [15]. This is advantageous when there is a lack of observed data, which is the case for many situations involving threatened species. In the case of human-wildlife conflict, a BN based on expert knowledge can help to identify the major factors that are associated with this conflict and their relative impact, as well as quantitatively evaluating the impact of changes to one or more of these factors in light of all the other influences in the system. In this manner, the BN can also be used to prioritise interventions that support the species' continued survival.

In this chapter, we present BN models for two threatened wildlife species, namely cheetahs in Botswana and jaguars in the Peruvian Amazon. Each of these models was developed and quantified using expert information. The intention of these models was to understand the viability of the species from a multi-faceted perspective that not only crosses disciplines but integrates the diversity of stakeholder perspectives. We focus on bringing together the ecological, biological, societal and economic pressures on, and opportunities for the species, in order to facilitate decision-making and conservation initiatives. In addition to illustrating the probabilistic assessments that arise from such models, we also highlight some of the similarities and differences between the factors that were considered to be important for each species.

Cheetah numbers in Botswana are declining, partially as a result of human-cheetah conflict [9, 33]. Human-cheetah conflict can take many forms. Some of the most common are farmers killing cheetahs to protect livestock [25], out of fear for their personal safety, or hunting for skins, meat, and other cheetah products. The scale of this conflict has prompted calls for interventions to prevent local people from killing cheetahs at an unsustainable rate [31]. Implementing these interventions, however, is not a trivial matter, and considering the social context of the intervention is vital [24]. Rural villages are made up of people from either the majority ethnic group (the *Tswana*), which has a strongly hierarchical structure, or minority ethnic groups, who are marginalised and very poor. The livelihoods of rural people, particularly those from these minority groups, are very dependent on hunting and gathering veldproducts, and may be highly impacted by wildlife management interventions [29].

Jaguars are a declared near-threatened species (IUCN) which means that they have the potential to go extinct sometime in the near future. Although it is acknowledged that prime jaguar habitat is the Amazon rainforest, remarkably little is known about jaguar occupancy or abundance in many parts of the jungle. A case in point is the northern part of Peru. Although Peru has the second largest remaining tract of rainforest in the world and an extensive series of national parks and reserves, there have been very few formal studies of jaguars in these areas. Key reasons for this paucity of data include the time required to reach study sites, difficulties in travelling through the jungle, the elusiveness of the target animal and the need to engage with the indigenous residents of the forest.

The chapter proceeds as follows. Section 14.2 provides a description of the BN methodology used to quantitatively evaluate the factors associated with human-wildlife conflict. This is described in general and then for each study in particular. Section 14.3 provides a summary of the results of the BN modelling for each case study, followed by an illustration of the types of inferences that can be made on the basis of these models. These inferences include identification of priority factors and assessment of the sensitivity of the network to hypothetical scenarios of interest. The chapter concludes with a discussion in Sect. 14.4.

14.2   Methods

14.2.1   Bayesian Networks

A Bayesian Network (BN) modelling approach was used to construct a systems model for describing the set of interacting factors that influence the viability of the target wildlife species. The BN model is often represented graphically, with the variables depicted as nodes (circles) and the interactions depicted as directed arrows (arcs). Probabilistic quantification of the model follows, in which the probabilities associated with each factor are conditional on the factors that impact on it (i.e. the parent nodes, connected to the node of interest by directed arrows). These probabilities can be based on a range of available information sources, including observational or experimental data, estimates from published literature or previous studies, expert judgement and so on. Although continuous probability distributions can be employed, it is common practice to discretise the corresponding variable, thereby creating a BN in which each node is quantified by a marginal probability table if it has no parents or a conditional probability table otherwise. The advantages of such a representation include fast computation of marginal probabilities for nodes of interest (including the final outcome node) based on all of the other nodes in the model, and common representation of information as probabilities despite its source.

By evaluating the probabilities in the BN, the model can be used to understand the relative impact of different factors on key nodes in the network, and importantly on the overall outcome node. Sensitivity analyses and scenario assessment can also be undertaken by modifying the underlying marginal and conditional probability tables appropriately. Using the Bayesian formulation, it is also possible to identify conditions for optimum outcomes.

14.2.2   Cheetah Study

Information for the BN for the cheetah case study was gathered via a workshop with twelve experts in cheetah conservation. The experts included local conservationists and ecologists, experts in cheetah biology and ecology, and government agents knowledgeable about relevant policy. The workshop was held over 4 days in Gaborone, the capital city of Botswana.

The BN network structure and the corresponding set of conditional probabilities were elicited using a structured approach that had been validated in other wildlife conservation BN studies [20, 27, 32]. At the workshop, the experts were asked to identify target nodes for the network (the primary outcomes of the model), and then to list all relevant factors that may influence these nodes. The final set of factors and the directed relationships between them was then agreed between the group members via a Delphi selection approach. Finally, the states of the nodes were identified and the underlying conditional probability tables populated, in an iterative process similar to that described in Johnson et al. [15].

14.2.3   Jaguar Study

The initial BN for this study was developed and quantified based on a structured interview with three members of the project team: the leader of a local conservation foundation, an international environmental journalist and a local indigenous representative. The first two members of this group were chosen because they had knowledge of international activities regarding the environment and jaguar protection, close links with the indigenous residents and established links with relevant local and state government agents responsible for the area. The local representative was chosen because he had spent many years living in the deep jungle as well as in the village, and he was highly knowledgeable about the area, the forest and jaguars in the region. As for the cheetah study, the network structure and the corresponding set of conditional probabilities were elicited using a well-tested structured approach.

The draft BN was then refined using results of a survey administered to local indigenous residents in a number of villages in the region. The aim of the survey was to obtain information about jaguar encounters and conflict, perceived trends in jaguar numbers in the past and future, usage of the forest and attitudes with respect to health, culture, environment, food and other benefits, and small and large scale forest clearing and industrial activities such as mining. The survey design and instrument broadly followed that developed by Meijaard et al. [28] for a study of orangutans and attitudes to the forest in Kalimantan, Indonesia, and was conducted in the form of questionnaires administered to local people through personal interviews. The questionnaire was initially drafted in English and subsequently translated into the local language. By necessity, respondents were not chosen randomly. People with a range of duties in the village were interviewed, with a preference for those who had knowledge about local wildlife, in particular jaguars. Steps were also taken to reduce desirability bias and recall bias [1, 28, 30]. The reliability of a respondent's responses about jaguars was determined by asking respondents to identify nine mammal species from a set of photographs, including a number of locally occurring large cat species. Only those respondents who were deemed to be sufficiently reliable were included in the present study.

The refined BN was then presented to and ratified by government representatives in Lima.

14.3   Results

14.3.1   Case Study BNs

14.3.1.1   Cheetahs

The following factors were agreed by the group of experts as important factors in human-cheetah conflict.

Government factors: Governments can affect human-cheetah conflict through enacting key policies, such as commercial hunting laws and conservancies. Governments may also be influenced by international pressure, NGOs, and pressure from citizens who favour cheetah conservation.

Economic factors: Botswana covers a vast area, and is sparsely populated [4], prohibiting timely responses to calls for assistance with cheetahs. Therefore, farmers may take it upon themselves to kill problem cheetahs, rather than waiting for them to be captured and relocated. Since the viability of wild cheetah populations depends on the survival of adult members, it is important that farmers are encouraged not to kill cheetahs to protect their livestock. Other sources of livestock protection, and sustainable management programs, may thus provide economic benefits to decreasing human-cheetah conflict. Farmers, and rural people, are directly affected by the presence of wild cheetahs and likely to be the target groups of interventions aiming to reduce conflict. Therefore, conservation strategies must consult with and engage the local community, take care to comply with village etiquette and politics, and protect disadvantaged members of the community.

Education factors: Cheetah education and rehabilitation programs, and improved access to these programs and facilities, would serve to reinforce other management programs. Rehabilitation of orphaned cheetahs can play an important role in wildlife education in general, and in knowledge of cheetahs in particular. Short- and long-term conflict may be influenced by education strategies aimed at the public such as media and information stalls, farmer education through workshops and site visits, or youth education through training teachers, distributing materials on cheetah conservation or school talks.

The expert group convened to develop the Cheetah BN agreed that interventions may have different impacts in the short and long term. Therefore, the network was designed to predict two main outcomes: a short term decrease in human-cheetah conflict, and a long-term decrease in human-cheetah conflict. The relationships between the factors and these two outcomes are presented in Fig. 14.1. Corresponding subnetworks underpinning some of the major nodes are shown in Fig. 14.2.

14.3.1.2   Jaguars

The three key factors affecting the viability of the jaguar in the wild were determined to be related to human impacts, prey insecurity and habitat loss. These are highly interdependent, as indicated in the BN model described and depicted below.

Four key human impact factors were identified, namely hunting jaguars, monitoring, human settlement and illegal logging. Hunting jaguars was in turn affected by levels of official corruption, effective policing and effective monitoring, with the latter influenced also by effective international monitoring. Human settlement was perceived to be both a benefit and a threat depending on the nature of the settlement, with indigenous villagers potentially protecting or killing the species. Growth in human settlement was perceived as a threat and was in turn affected by squatting, which is a major problem in the Amazon forest since these people are less likely to

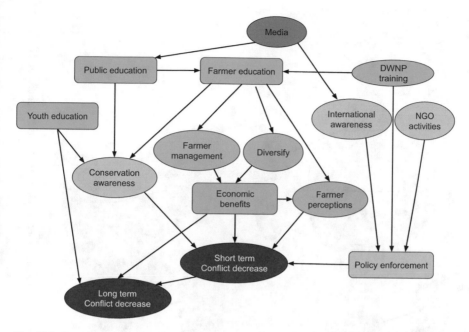

Fig. 14.1 Bayesian network for predicting and managing human-cheetah conflict in Botswana

have a history of co-existence with forest animals such as the jaguar. Illegal logging, which is acknowledged to be one of the most serious factors affecting the forest, was also influenced by effective policing.

Four key factors affecting prey insecurity were also identified. Although one of these, namely weather variability, was not human-induced, the other three were due to human activity. These included illegal logging, hunting for bush meat and harvesting a major forest fruit, aguaje, on which many wildlife species rely. The amount of illegal logging was perceived to be strongly influenced by the degree of effective policing; hunting for bush meat was influenced by human settlement, and aguaje harvest was influenced by economic development.

Drivers of habitat loss were reported to include illegal logging, aguaje harvest (since often the entire tree is cut down to access the fruit), weather variability, agriculture and pollution. The latter was believed to be a major factor and associated with petroleum production and exploration, which were driven by economic development.

The quantified human-jaguar BN is displayed in Fig. 14.3. The figure shows the marginal probabilities derived from the set of conditional probabilities for each node.

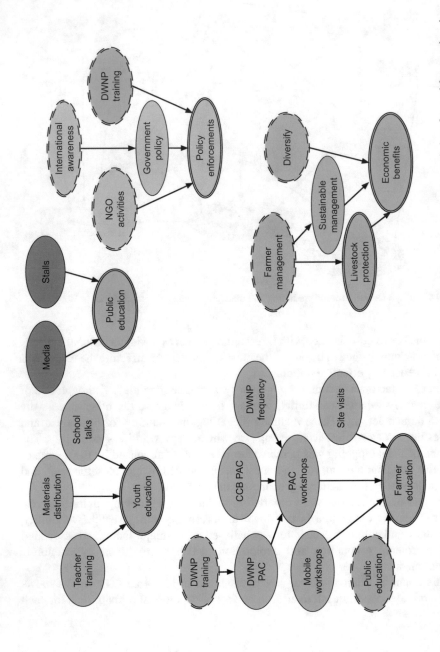

Fig. 14.2 Clockwise from top left, sub-network for youth education, public education, policy enforcements, economic benefits, and farmer education

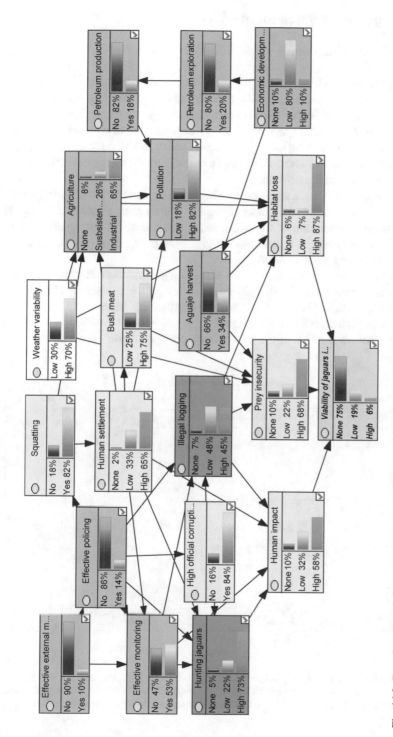

Fig. 14.3 Bayesian network for probabilistically describing the set of interacting factors that influence the viability of jaguars in the Peruvian Amazon. The truncated names of nodes are Effective external monitoring (top left), High official corruption (mid left), and Viability of jaguars in the wild (terminal node)

14.3.2   Inferences Based on the BN

14.3.2.1   Cheetah Case Study

We employ the cheetah case study to highlight inferences that can be made based on the probabilities determined in the BN, as well as the sensitivity of the outcome—the viability of cheetahs in the wild—to specified changes in the system.

The cheetah BN structure revealed four direct predictors of decrease in human-cheetah conflict over the long term: government policy, youth education, economic benefits, and decrease in short-term conflict. Each of these factors was in turn influenced by others in the network.

The BN model provided a set of probabilities—the probability that over the long term, human-cheetah conflict would decrease at a high rate, a low rate, or not at all—conditional on the state of the whole network system. An example of one of the conditional probabilities is presented in Fig. 14.4; plots of the other conditional probability tables are given in the Appendix.

When all factors are optimised—youth education is high, there is a high decrease in short-term human-cheetah conflict, government policy protecting cheetahs is present, and economic benefits to decreasing conflict are high—then the overall probability of a high decrease in long-term human-cheetah conflict is high, as would be expected. An observation of the other conditional probability plots allows us to

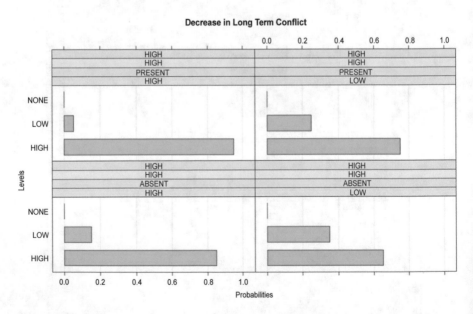

Fig. 14.4 Exemplar representation of the conditional probability table for decrease in long term conflict with cheetahs in the wild in Botswana, based on the levels of the four parent nodes, respectively Youth Education, Short Term Conflict, Government Policy and Economic Benefits

examine more complex scenarios. For example, when short-term conflict decrease is low, and economic benefits are low, but youth education is high and government policy is present, then the probabilities of a low or high decrease in human-cheetah conflict are roughly equal, and when government policy is absent, a low decrease in conflict becomes the most likely outcome.

Examining the network reveals that, surprisingly, if youth education is low and government policy is absent, but short-term conflict decrease is high, and there are high economic benefits to decreasing conflict, then it is very likely that the decrease in human-cheetah conflict will be high in the long term. Furthermore, if government policy is present but all other factors are low, it is very likely that there will be no decrease in human-cheetah conflict in the long term. Together, these scenarios suggest that government policy is less impactful in the long term than other strategies for decreasing human-cheetah conflict.

14.3.2.2   Jaguar Case Study

We employ the jaguar case study to illustrate the ability of the BNs to provide a quantitative assessment of the sensitivity of the network outcomes to hypothetical scenarios.

Based on a sensitivity analysis of all nodes in the system, the strongest links in the jaguar BN were determined to be between the following pairs of factors: Economic development and Aguaje harvest; Effective monitoring and Illegal logging; Human settlement and bush meat; Human settlement and agriculture; Petroleum exploration and petroleum production; Effective policing and squatting; and Effective external monitoring and effective policing.

Six scenarios were evaluated. The first four involved modifying in turn the three nodes that were parents of the target node (Viability of jaguars in the wild), i.e. Human impact = No, Prey insecurity = No, Habitat loss = No, and all three factors = No, respectively. All other factors remained unchanged. The results of these evaluations are shown in Fig. 14.5. The last two scenarios comprised two positive management decisions: 100% effective monitoring and no aguaje harvest; and no official corruption or illegal logging. The results of these evaluations are shown in Table 14.1.

14.4   Discussion

A Bayesian network model was used to synthesise citizen knowledge in a wide variety of domains concerning human-wildlife conflict with two iconic threatened species in two very different locations, namely the plains of Botswana and the jungle of Peru. In each study, the BNs were developed and quantified using the combined expertise from government agents, ecologists and conservationists. In both cases, the BNs reflected the most important factors perceived by the group, the directed

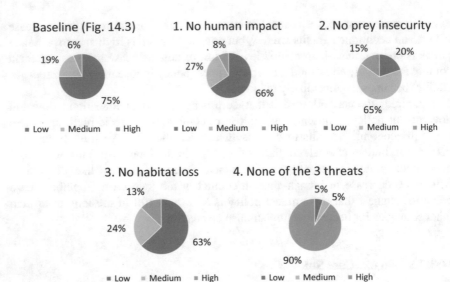

Fig. 14.5 Results of first four hypothetical scenarios for the jaguar case study, varying the three parent nodes of the target node. Last scenario is best case (no human impact, prey insecurity or habitat loss)

connections between these factors, and the quantitative evidence relating to the behaviour of each factor in light of the other impacting influences. Importantly, the BNs were able to describe and predict the outcomes of multiple management strategies at once, incorporating the kind of complex and multifaceted solutions needed to effectively address human-wildlife conflict [3, 7, 19, 21, 23].

No data exist against which to validate the predictions of the BNs developed and reported in this study. However, the system of factors identified by the BNs are echoed in conservation papers elsewhere. For example, in the cheetah study, the experts separated the short-term and long-term effects of conflict management strategies, a distinction which is increasingly recognised as important in conservation biology in general, and human-wildlife conflict in particular [6]. In addition, social factors were well-represented in both BNs, and have been identified as essential to successful conflict management for cheetahs, jaguars and other species [2, 6, 11, 18, 21, 22, 36].

For both systems, data on the respective human-wildlife conflict are scarce and difficult to access, but using the BN allows for the knowledge hidden in experts' heads to be extracted, quantified, and synthesised across domains. The BN supports decision making by identifying most influential factors that impact on the outcome of interest, and allowing various scenarios to be simulated before they are implemented. In this way the BN can provide a key planning tool for managing human-wildlife conflict for big cats in particular, and for other wildlife species in general.

Table 14.1 Results of last two hypothetical scenarios representing positive management decisions for jaguar conservation: Scenario 5 (100% effective external monitoring, no aguaje harvest) and Scenario 6 (no official corruption, no illegal logging)

Factor	Human impact	Prey insecurity	Habitat loss	Hunting jaguars	Viability of jaguars in the wild
Baseline (Fig. 14.4)	0.10, 0.32, 0.58	0.10, 0.07, 0.68	0.06, 0.07, 0.87	0.05, 0.22, 0.73	0.75, 0.19, 0.06
Scenario 5	0.17, 0.32, 0.50	0.12, 0.21, 0.67	0.06, 0.09, 0.85	0.07, 0.37, 0.57	0.73, 0.21, 0.06
Scenario 6	0.41, 0.32, 0.37	0.44, 0.28, 0.28	0.15, 0.22, 0.63	0.08, 0.67, 0.25	0.40, 0.45, 0.15

Probabilities pertain to Low, Medium and High states of the corresponding node in the BN network displayed in Fig. 14.3

The BNs presented here are limited in scope; for example, they do not account for larger forces like poverty reduction or economic forces, which are often closely intertwined and are all important to human-cheetah and human-jaguar conflict. However the focus here is on what can be done to manage conflict; for this reason, the networks emphasised the factors that could reasonably be affected by an intervention. Future work could also improve the decision support utility by adding decision nodes and cost information, to more fully support conservation decisions. Finally, it is important to consider that conflict management strategies can only be successful when implemented with a commitment to ongoing evaluation (e.g., [10, 34]). An example of this is the important role of national and international monitoring in the jaguar BN.

Different people have different interests and want different things from conservation policy. Substantial effort is required to bridge these differences [5]. The tools described here can help identify differences in policy objectives. Moreover, given that there are currently insufficient funds available to support the acknowledged, published recovery actions for threatened species, conservation managers and politicians alike are faced with the difficult task of deciding where those limited funds are best used. This process often works first at the policy decision-making level, and then again at the management level, be that within conservation agencies or non-government organizations in receipt of funds. Access to tools such as the BNs and associated products described in this chapter can make it much easier for those charged with making decisions to see where the greatest impact might be gained from particular actions. The tools are also likely to be useful in other areas of natural resource management [26]. Communication, education and participation will be able to be better integrated as a result, something which Jiménez et al. [13] have identified as necessary for improved participation of multiple stakeholders in developing policy and implementing management strategies in biodiversity projects.

We close this discussion with a few concluding comments about the systems approaches that we have proposed in this paper, along with a call for a cautious application of this approach to managing diverse types of data. First, the systems frameworks in general are not suggested as solutions to the whole issue of conservation evaluation and management. Many other statistical and qualitative tools are highly valuable in highlighting particular aspects of these very complex problems. Examples of such tools are population viability analyses, species distribution models, statistical risk models and predictive models based on field data, surveys, focus group meetings and other evaluations.

The second note aims to highlight the simplicity of the integration of information using the proposed approach. The BN framework can accommodate full (conditional) probability distributions where these are available, or alternatively all it requires is discrete (e.g. high-med-low) descriptors. These probability tables can be quantified using a wide range of information, from observational and experimental data, to literature-based estimates, to expert judgement. The exploitation of expert information in these complex problems, based on careful elicitation and probabilistic representation [27], has strong appeal. This use of a simple common currency is similar to economists' use of monetary measures to compare otherwise incommensurable variables.

Thirdly, we note that any quantitative analysis of social or ecological systems (and especially a socio-ecological system) is necessarily a gross simplification of something that is very complex. However, if it was not simplified it could not be done at all. Moreover, despite the simplicity, it is still not trivial to characterise these systems. It is our experience that attempting to do this in a rigorous, transparent manner results in a deeper, if still incomplete, understanding of the system, and is far better than the alternative which is no representation of the system or integration of information at all.

Acknowledgements The authors acknowledge financial and organisation support from the Cambridge University, Queensland University, Australian Research Council Centre of Excellence in Mathematical and Statistical Frontiers, Cheetah Conservation Botswana and the Lupunaluz Foundation in Peru.

This work is the outcome of two substantive studies undertaken with a range of experts and community members. We thank them all. In particular, for the cheetah study we thank the Mokolodi Nature Reserve for hosting the workshop and Wabotlhe and Brian for reviewing the network, and for the jaguar study we thank the research teams at QUT ACEMS and Vanessa Hunter and Lupunaluz for organising the research trip.

Appendix: Conditional Probability Tables for Long and Short Term Outcomes for Cheetah Case Study

Decrease in Long Term Conflict

Conditional probability tables for decrease in long term conflict with cheetahs in the wild in Botswana, based on the levels of the four parent nodes, respectively Youth Education (violet), Short Term Conflict (light blue), Government Policy (green) and Economic Benefits (orange).

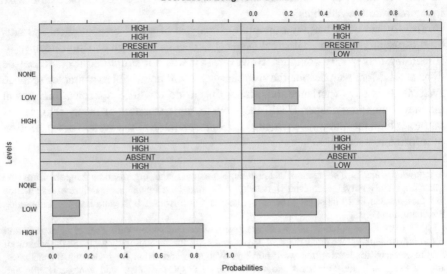

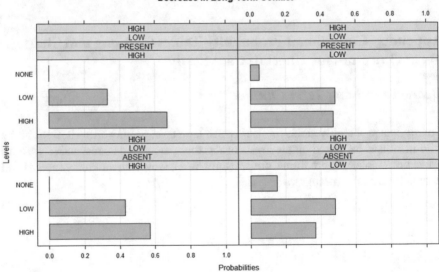

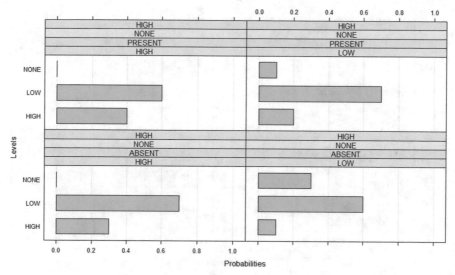

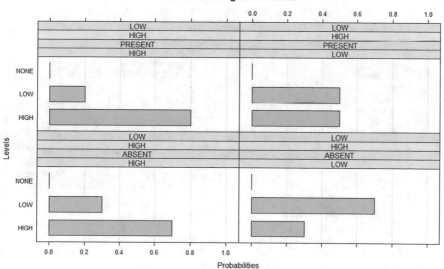

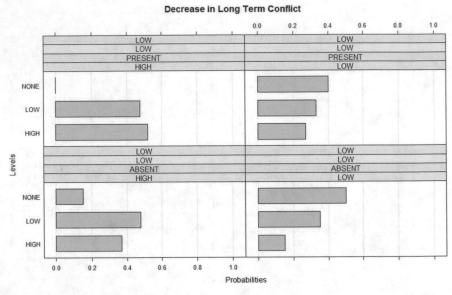

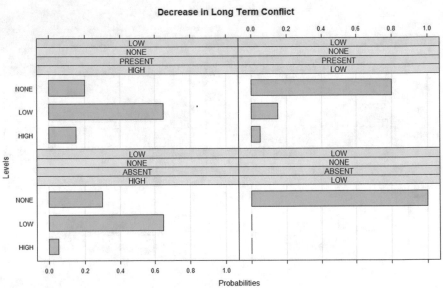

Decrease in Short-Term Conflict

Conditional probability tables for decrease in long term conflict with cheetahs in the wild in Botswana, based on the levels of the four parent nodes, respectively Policy Enforcements (pink), Livestock Protection (violet), Farmer Perceptions (light blue), Economic Benefits (green), Conservation Awareness (orange).

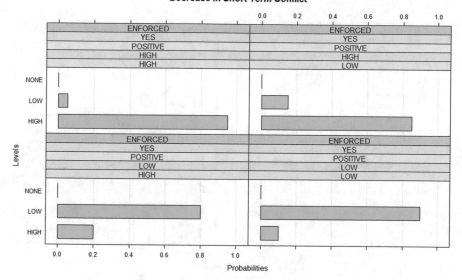

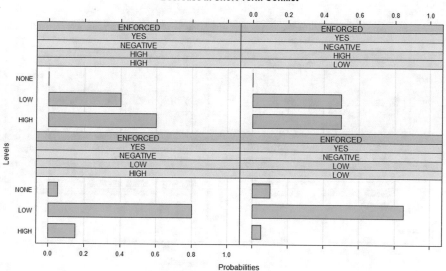

Decrease in Short Term Conflict

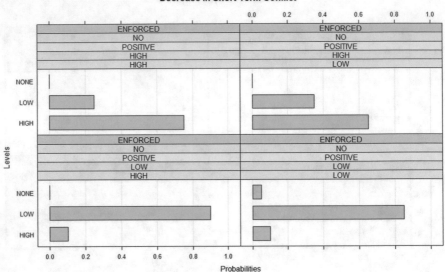

Decrease in Short Term Conflict

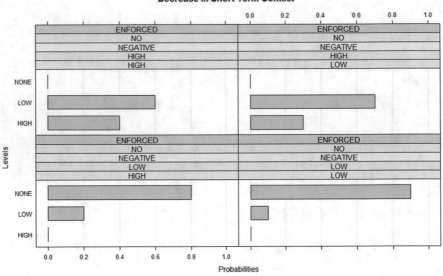

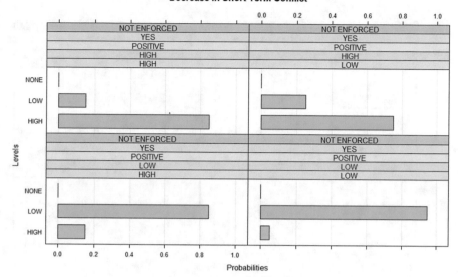

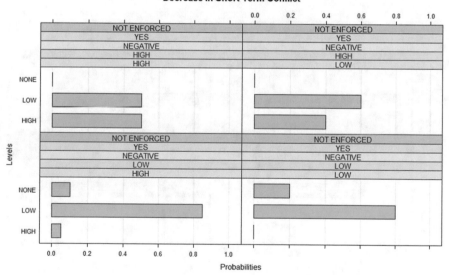

Decrease in Short Term Conflict

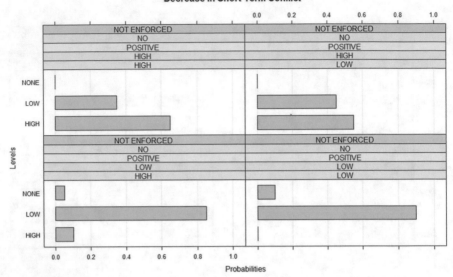

Decrease in Short Term Conflict

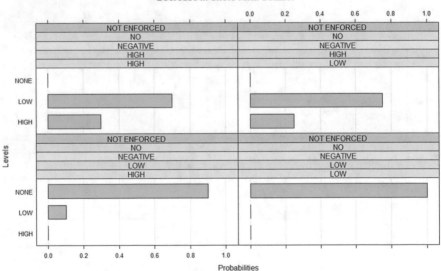

References

1. N. Abram, E. Meijaard, J. Wells, M. Ancrenaz, A.-S. Pellier, R. Runting, D. Gaveau, S. Wich, N. Nardiyono, A. Tjiu, A. Nurcahyo, K. Mengersen, Mapping perceptions of species' threats and population trends to inform conservation efforts: the Bornean orangutan case study. Divers. Distrib. **21**, 487–499 (2015)
2. M. Barua, S.A. Bhagwat, S. Jadhav, The hidden dimensions of human-wildlife conflict: health impacts, opportunity and transaction costs. Biol. Conserv. **157**, 309–316 (2013)

3. Y.K. Bredin, D.C. Linnell, L. Silveira, M. Torres, A.A. Jacomo, J.E. Swenson, Institutional stakeholders' views on jaguar conservation issues in central Brazil. Glob. Ecol. Conserv. **3**, 814–823 (2015)
4. Central Intelligence Agency, Botswana. In The World Factbook (2006). Retrieved from https://www.cia.gov/library/publications/the-world-factbook/geos/br.html
5. T. Darbas, T.F. Smith, E. Jakku, Seeing engagement practitioners as deliberative hinges to improve landholder engagement, in Contested Country: Local and Regional Natural Resources Management in Australia, ed. by M. Lane, C. Robinson, B. Taylor (CSIRO Publishing, Collingwood, VIC, 2009)
6. A.J. Dickman, Complexities of conflict: the importance of considering social factors for effectively resolving human-wildlife conflict. Anim. Conserv. **13**(5), 458–466 (2010)
7. A. Dickman, N.A. Rust, L.K. Boast, M. Wykstra, L. Richmond-Coggan, R. Klein, M. Selebatso, M. Msuha, L. Marker, Chapter 13: The costs and causes of human-cheetah conflict on livestock and game farms, in *Cheetahs: Ecology and Conservation*, ed. by P. J. Nyhus, L. Marker, L. K. Boast, A. Schmidt-Küntzel, (Academic, Cambridge, 2018), pp. 173–198
8. M. Donald, A. Cook, K. Mengersen, Bayesian network for risk of diarrhea associated with use of recycled water. Risk Anal. **29**, 1672–1685 (2009)
9. GCCAP, Global cheetah conservation action plan workshop report. Global Cheetah Conservation Action Plan Workshop, Shumba Valley Lodge, South Africa (2002)
10. C. Inskip, A. Zimmermann, Human-felid conflict: a review of patterns and priorities worldwide. Oryx **43**(1), 18–34 (2009)
11. IUCN Press Release, Unsustainable fishing and hunting bushmeat driving iconic species extinction (2019, July 18). https://www.iucn.org/news/species/201907/unsustainable-fishing-and-hunting-bushmeat-driving-iconic-species-extinction-iucn-red-list. Accessed 22 July 2019
12. R.M. Jackson, R. Wangchuk, A community-based approach to mitigating livestock depredation by snow leopards. Hum. Dimens. Wildl. **9**(4), 1–16 (2010)
13. A. Jiménez, I. Iniesta-Arandia, M. Muñoz-Santos, B. Martín-López, S.K. Jacobson, J. Benayas, Typology of public outreach for biodiversity conservation projects in Spain. Conserv. Biol. **28**, 829–840 (2014). https://doi.org/10.1111/cobi.12220
14. S. Johnson, K. Mengersen, Integrated Bayesian network framework for modeling complex ecological issues. Integr. Environ. Assess. Manag. **8**, 480–490 (2012)
15. S. Johnson, K. Mengersen, A. de Waal, K. Marnewick, D. Cilliers, A.M. Houser, L. Boast, Modelling cheetah relocation success in southern Africa using an Iterative Bayesian Network Development Cycle. Ecol. Model. **221**(4), 641–651 (2010)
16. S. Johnson, F. Fielding, G. Hamilton, K. Mengersen, An Integrated Bayesian Network approach to Lyngbya majuscula bloom initiation. Mar. Environ. Res. **69**, 27–37 (2010)
17. S. Johnson, L. Marker, K. Mengersen, C.H. Gordon, J. Melzheimer, A. Schmidt-Küntzel, M. Nghikembua, E. Fabiano, J. Henghali, B. Wachter, Modeling the viability of the free-ranging cheetah population in Namibia: an object-oriented Bayesian network approach. Ecosphere **4**(7), 1–19 (2013)
18. H.K. Krafte, L.R. Larson, R.B. Powell, Characterizing conflict between humans and big cats *Panthera* spp.: a systematic review of research trends and management opportunities. PLoS One **13**(9), e0203877 (2018)
19. F. Lamarque, J. Anderson, R. Fergusson, M. Lagrange, Y. Osei-Owusu, L. Bakker, Human-wildlife conflict in Africa: causes, consequences and management strategies. Food and Agriculture Organisation of the United Nations, Rome. FAO Forestry Paper 157 (2009)
20. S. Low Choy, R. O'Leary, K. Mengersen, Elicitation by design in ecology: using expert opinion to inform priors for Bayesian statistical models. Ecology **90**(1), 265–277 (2009)
21. F. Madden, Creating coexistence between humans and wildlife: global perspectives on local efforts to address human-wildlife conflict. Hum. Dimens. Wildl. **9**(4), 247–257 (2010)
22. S. Marchini, G. Crawshaw Jr., Human-wildlife conflicts in Brazil: a fast-growing issue. Hum. Dimens. Wildl. **20**, 323–328 (2015)
23. L.L. Marker, L.K. Boast, Human-wildlife conflict 10 years later: lessons learned and their application to cheetah conservation. Hum. Dimens. Wildl. **20**(4), 1–8 (2015)

24. L. Marker, A. Dickman, Human aspects of cheetah conservation: lessons learned from the Namibian farmlands. Hum. Dimens. Wildl. **9**(4), 297–305 (2010)
25. L.L. Marker, A.J. Dickman, M.G.L. Mills, D.W. Macdonald, Aspects of the management of cheetahs, *Acinonyx jubatus jubatus*, trapped on Namibian farmlands. Biol. Conserv. **114**(3), 401–412 (2003)
26. K. Marshall, K.L. Blackstock, J. Dunglinson, A contextual framework for understanding good practice in integrated catchment management. J. Environ. Plan. Manag. **53**(1), 63–89 (2010).https://doi.org/10.1080/09640560903399780
27. T.G. Martin, M.A. Burgman, F. Fidler, P.M. Kuhnert, S. Low-Choy, M. McBride, K. Mengersen, Eliciting expert knowledge in conservation science. Conserv. Biol. **26**(1), 29–38 (2012)
28. E. Meijaard, K. Mengersen, D. Buchori, A. Nurcahyo, M. Ancrenaz, et al., Why don't we ask? A complementary method for assessing the status of great apes. PLoS One **6**, e180 (2011)
29. T. Mompati, G. Prinsen, Ethnicity and participatory development methods in Botswana: some participants are to be seen and not heard. Develop. Pract. **10**(5), 625–637 (2000). https://doi.org/10.1080/09614520020008805
30. E. Meijaard, K. Mengersen, N. Abram, A.-S. Pellier, J. Wells, et al., People's perceptions on the importance of forests for people's livelihoods and health in Borneo. PLoS One **8**(9), e73008 (2013)
31. K. Nowell, *Namibian Cheetah Conservation Strategy*. Report to the Ministry of Environment and Tourism, Government of Namibia (1996)
32. J. Pitchforth, K. Mengersen, A proposed validation framework for expert elicited Bayesian Networks. Expert Syst. Appl. **40**, 162–167 (2013)
33. A. Treves, K.U. Karanth, Human-carnivore conflict and perspectives on carnivore management worldwide. Conserv. Biol. **17**, 1491–1499 (2003)
34. A.D. Webber, C.M. Hill, V. Reynolds, Assessing the failure of a community-based human-wildlife conflict mitigation project in Budongo Forest Reserve, Uganda. Oryx **41**(2), 177–184 (2007)
35. WPC, Preventing and mitigating human-wildlife conflicts: world parks congress recommendation. Hum. Dimens. Wildl. **9**(4), 259–260 (2004).https://doi.org/10.1080/10871200490505684
36. A. Zimmerman, Jaguars and people: a range-wide review of human-wildlife conflict. PhD Thesis, University of Oxford, 2014

Chapter 15
Bayesian Learning of Biodiversity Models Using Repeated Observations

Ana M. M. Sequeira, M. Julian Caley, Camille Mellin, and Kerrie L. Mengersen

Abstract Predictive biodiversity distribution models (BDM) are useful for understanding the structure and functioning of ecological communities and managing them in the face of anthropogenic disturbances. In cases where their predictive performance is good, such models can help fill knowledge gaps that could only otherwise be addressed using direct observation, an often logistically and financially onerous prospect. The cornerstones of such models are environmental and spatial predictors. Typically, however, these predictors vary on different spatial and temporal scales than the biodiversity they are used to predict and are interpolated over space and time. We explore the consequences of these scale mismatches between predictors and predictions by comparing the results of BDMs built to predict fish species richness on Australia's Great Barrier Reef. Specifically, we compared a series of annual models with uninformed priors with models built using the same predictors and observations, but which accumulated information through

Ana M. M. Sequeira and M. Julian Caley are co-first authors.

Electronic Supplementary Material The online version of this chapter (https://doi.org/10.1007/978-3-030-42553-1_15) contains supplementary material, which is available to authorized users.

A. M. Sequeira (✉)
IOMRC, UWA Oceans Institute and School of Biological Sciences, M096, The University of Western Australia, Crawley, WA, Australia

M. Caley · K. L. Mengersen
School of Mathematical Sciences, Queensland University of Technology, Brisbane, QLD, Australia

Australian Research Council Centre of Excellence for Mathematical and Statistical Frontiers, Melbourne, VIC, Australia

C. Mellin
Australian Institute of Marine Science, Townsville, QLD, Australia

The Environment Institute and School of Biological Sciences, University of Adelaide, Adelaide, SA, Australia

K. L. Mengersen et al. (eds.), *Case Studies in Applied Bayesian Data Science*, Lecture Notes in Mathematics 2259, https://doi.org/10.1007/978-3-030-42553-1_15

371

time via the inclusion of informed priors calculated from previous observation years. Advantages of using informed priors in these models included (1) down-weighting the importance of a large disturbance, (2) more certain species richness predictions, (3) more consistent predictions of species richness and (4) increased certainty in parameter coefficients. Despite such advantages, further research will be required to find additional ways to improve model performance.

15.1   Introduction

Estimating biodiversity metrics is a central pursuit in ecological research and management. These metrics inform our understanding of the states and trends of ecosystems [13, 19], their responses to biotic and abiotic factors [11, 22, 29], and the best options for their management, conservation, and the on-going provision of ecosystem services [4, 13, 16, 18, 31].

Estimating biodiversity metrics, however, is often challenging because of the high costs of surveying and monitoring coupled with limited available resources, and because ecological systems are often highly diverse and respond in complex ways to myriad biotic and abiotic interacting factors [13]. Consequently, the data available to estimate biodiversity metrics are often insufficient to address current needs [7]. In some cases though, long-running monitoring programs provide extensive repeated measures of biological communities and can contribute to robust estimates of these metrics. Such data, however, are typically most useful for estimating the status and trends of observed biological communities, whereas estimates and their associated uncertainties are often required for entire communities across a hierarchy of spatial scales [30]. For example, the Australian Institute of Marine Science's (AIMS) Long Term Monitoring Program (LTMP) of the Great Barrier Reef (GBR) has monitored individual reefs annually for more than three decades using a spatial design that samples representative cross-shelf habitats, latitudinal sectors, and management regimes [28]. Although it is one of the most spatially extensive long-term monitoring programs on Earth, it only monitors a small fraction of all the reefs present (<2% of all GBR reefs). Consequently, biodiversity estimation beyond this relatively small set of reefs must rely on predictions for unmonitored reefs (e.g. [15]) and requires good predictability into unsampled space (a component of spatial statistical modeling: [25]).

Where it is desirable to predict biodiversity into a larger domain than a series of observed communities, biodiversity distribution models (BDM), a general case of species distribution models (SDM) where the response variable may be a composite metric such as species richness or total abundance across species, can be constructed using combinations of environmental and spatial variables. These models can then be useful to predict biodiversity metrics across domains where values are not observed [15, 24, 26, 32, 33]. While such models have proven effective to varying degrees, there is commonly a mismatch between the states and dynamics of the ecological communities being predicted and the environmental and spatial observations used to predict them. For example, ecological communities are typically sampled at regular intervals (e.g., yearly). Spatial predictors, such as a community's location relative to geological features, vary over geological time and can be assumed to be invariant with respect to ecological prediction, whereas environmental predictors vary on a variety of temporal and spatial scales. For the purposes of predictive modelling, these environmental predictors are often available only as long-term annual averages and spatially interpolated to common scales (e.g. marinehub.org). Consequently, repeated measures of ecological metrics often rely on a set of predictors that are inherently less variable than the metric they are being used to predict, either because of the characteristics of the predictor (e.g. spatial predictors) or the way it was collected and processed (e.g. environmental predictors). These characteristics of predictors may in turn compromise the performance of predictive models given that a diversity metric might vary through time and space at rates unrelated to the variables used to predict it.

The consequences of these scale mismatches between predictors and predictions on the performance of such models are likely to vary from year to year, as the response variable changes but the predictor variables do not. These changes have the potential to affect the predictive performance of a model in a number of ways including the ability to predict true values and their associated uncertainties, the coefficients of the predictors estimated for the model, and the structure of the best performing model. Despite the potential importance of such mismatches, understanding of their influence on the construction and application of BDMs is poor. To begin addressing this knowledge gap, we explore ways in which analytical approaches to building and applying BDMs affect their predictive performance, the estimation of the coefficients of model parameters, and the selection of the best model structure in cases where recorded values of the response variables vary in space or time but observations of predictor variables vary less over time.

When repeated observations from a monitoring program are available, common approaches to their analysis include considering each repeated set of observations separately and then making post-hoc comparisons between them to infer community states through time, or averaging all data across replicates and then estimating the best model. A disadvantage of such approaches is that they fail to use all the information available from repeated observations to help understand how a predictive model might improve as monitoring continues through time. Comparisons between models can also be difficult as the importance of predictors change between years. Moreover, understanding how such information accumulates through time can facilitate more efficient and effective allocation of limited and valuable monitoring resources through the implementation of adaptive sampling designs [12]. By adopting a Bayesian learning approach to this problem, it should be possible to better understand how the performance of such models changes with the addition of information through time. In such an approach, the results obtained from a previous survey or surveys can be used as prior information in the analysis of the data for the latest survey. Adopting this approach results in an iterative updating of information as it becomes available which has theoretical and computational advantages. Theoretically, compared with the independent analysis of surveys described above, the obtained estimates should move closer to true values more quickly and smoothly, any trend in the replicate estimates should be smoother, the estimates obtained in each replicate should be more precise (i.e., have narrower credible intervals), and post-hoc analysis across the time series should no longer be needed. Similarly, the estimates of the coefficients of the model and the model structure should converge to the true values and be associated with progressively decreasing uncertainties. Computationally, compared with analysing all available data each time new observations become available, the Bayesian learning approach does not require reanalysis of the entire dataset during updating but instead requires a simpler and less computationally costly analysis of the current data and the prior which encapsulates information from past surveys.

Bayesian modelling of data with repeated measures is now commonplace and offers advantages over other approaches in terms of estimation, model flexibility, and inference [3]. Bayesian learning, also known as recursive Bayesian estimation or Bayesian filtering, is commonly employed for a wide variety of problems that require iterative updating of information from quality monitoring and control [1] to analyses of streaming data [23]. To explore the comparative benefits of using a Bayesian learning approach for estimating biodiversity using typical monitoring data, we analysed species richness patterns of fishes on Australia's Great Barrier Reef (GBR). We analyse the annual LTMP data using uninformed priors for each year analogous to the frequentist analyses of individual repeated observations, and compare these results to those obtained using informed priors derived from previous observations. Based on the results of these analyses we make recommendations for improved learning where a set of temporally less variable predictors are used to make predictions from observations that vary to a greater extent through time.

15.2 Methods

15.2.1 Fish Species Counts and Environmental and Spatial Predictors

We used counts of fish species on the GBR collected by the Australian Institute of Marine Science's (AIMS) Long-Term Monitoring Program (LTMP) [28] for the years 2003–2013. For this period, annual survey data were available for each year from 2003 to 2005 and every second year after 2005 due to a change of sampling design. A total of 46 reefs were monitored across six latitudinal sectors (Cooktown-Lizard Island, Cairns, Townsville, Whitsunday, Swain and Capricorn-Bunker) spanning 150,000 km^2 of the GBR. In each sector, with the exception of the Swain and Capricorn-Bunker sectors, at least two reefs were sampled in each of three shelf positions (i.e., inner, mid- and outer). At each reef, 5 transects in each of 3 sites were sampled and we analysed observations from the same 133 locations in each of these 7 years. Observations were made using transect-based underwater visual survey. Transects were randomly selected, permanently marked, and ran roughly parallel to the reef crest, each separated by at least 10 m along the 6–9 m depth contour. Counts of 251 fish species from across 10 taxonomic families were recorded. This set of species excluded cryptic and nocturnal species. Larger mobile species were counted first along a 5 m wide transect, and smaller, less mobile species (e.g. damselfishes: Pomacentridae) were counted in a 1-m wide strip along the same transect during the return swim (for detailed methods and species counted, see [10]). To prevent potential systematic bias in the fish counts associated with different observers, calibration of all divers occurred annually [10]. To predict these fish species counts (i.e. species richness), we used both environmental and spatial predictors. We used a set of environmental predictors available for Australia at a national scale and at a 0.01° resolution (marinehub.org) including sea surface temperature (SST), chlorophyll-a (Chl a), salinity, nutrients (NO3, PO4, and SI), light (K490av), depth (as a proxy for habitat), oxygen, and sediment characteristics including percentages of carbonates, gravel, sand, and mud [14]. To account for geographical effects on the distributional patterns of reef fishes, we also included two spatial predictors: the shortest distances to coast (*coast*) and to the outer limit of the reefs (*barrier*), which have been used to successfully predict fish species richness and abundances on the GBR [15, 24]. We calculated these distances for each sampled site and node on the 0.01° national grid using the Near tool in ArcGIS10.1 (ESRI, Redlands, CA, USA) and an equidistant cylindrical coordinate system. We then assigned each sampling site to the closest node on the 0.01° national grid and used the environmental and spatial predictors corresponding to these locations.

15.2.2 Bayesian Models

Using *reef* as a random effect to account for the hierarchical nature of the dataset with sites nested within reefs, we developed Bayesian generalized linear mixed-effects models (GLMM) of fish species richness assuming a Poisson distributed response S_{ij} for the ith location in the jth reef, with a log-link and linear and quadratic regression terms for the covariates X_{ij}. Seven separate models were developed using each yearly dataset of fish species richness observations from the GBR as a response variable (Table 15.1). Allowing for extra-Poisson variation through a residual $\varepsilon_i{\sim}N(0, \sigma^2)$, the likelihood is thus given by

$$S_{ij} \sim \text{Poisson}\left(\mu_{ij}\right)$$

$$\log\left(\mu_{ij}\right) = \alpha_j + X\beta + \varepsilon_i.$$

An uninformative Gaussian prior (i.e., zero mean and relatively large variance) was specified for the random effect for reef, α_j. Twelve combinations of covariates were considered for each set of yearly models (Table 15.1). Univariate priors for each of the regression coefficients in the vector β were specified. Two sets of such priors were considered. First, we used independent uninformative Gaussian priors

Table 15.1 Description of fitted models

No.	Model description	Covariates included in model
1	Comprehensive model	Reef + Coast + Coast^2 + Barrier + Barrier^2 + Depth + Depth^2 + Slope + O$_2$ + SST + SST^2 + Light
2	Distance to domain boundaries	Reef + Coast + Coast^2 + Barrier + Barrier^2
3	Physical predictors	Reef + Depth + Depth^2 + Slope + Aspect
4	Particular sediment type	Reef + Gravel
5	Particular sediment type	Reef + Sand
6	Particular sediment type	Reef + Mud
7	Nutrients	Reef + NO$_3$ + PO$_4$ + Silica
8	Oxygen and salinity	Reef + O$_2$ + Salinity
9	Productivity	Reef + Chl a
10	Temperature	Reef + SST + SST^2
11	Light availability	Reef + Light
12	Intercept only	Reef

All models included a random effect for reef. Coast: distance to coast; Barrier: distance to the reef's outer limit; Gravel, Sand and Mud represent percentage of gravel, sand, and mud, respectively; average concentrations of NO$_3$: nitrate, PO$_4$: phosphate, SI: silicate, O$_2$: dissolved oxygen, Chl a: chlorophyll a; Sal: salinity; SST: average annual sea surface temperature; Light: coefficient of light attenuation at 490 nm. All predictors were mean centred, and superscript 2 indicates predictors included as quadratic terms. Sediment variables (gravel, mud and sand) were included in separate models (4–6) due to collinearity

as above, assuming no prior knowledge of the relationships between the response variable and the set of predictors being included in each model. This approach provided a baseline against which we compared a second set of models using informed priors and which therefore could exploit potential benefits of Bayesian methods. In this second approach, we modelled the first year's observations using uninformative Gaussian priors as described above. Consequently, the results of both approaches will be the same for the first year. In each subsequent year, we used the posterior mean and variance from the previous year(s) to construct an informed Gaussian prior, which was then used to model the responses of the current year of observation (Table S2).

We used a Markov Chain Monte Carlo (MCMC) algorithm with 100,000 iterations, a burn in of 10,000, a thinning rate of 2 (i.e., discarding every second simulated value to reduce autocorrelation and Monte Carlo error), and ran three chains to check convergence. To ensure the behaviour of the chains would not differ for larger MCMC runs, we also compared results from 600,000 iterations after burn in. Due to limits to computational power, we ran this larger MCMC in steps of 20,000 iterations by updating the chains with the last value obtained in each of the previous iterations. The modelling results shown here are derived from the last 10,000 iterations in each MCMC, having ensured convergence had been reached based on the Gelman-Brooks-Rubin diagnostic (i.e., rhat < 1.1). The retained MCMC samples were used to obtain posterior means and 95% credible intervals (CrI). CrIs were estimated for parameters of interest and a posterior predictive check of their individual contributions made using the sum of squared Pearson residuals, the raw residual divided by the square root of the variance.

To understand the effects of the different modelling methods on our results, we used wBIC and wAICc for comparison, as the use of DIC and wDIC can be inconsistent for GLMM [17]. Moreover, the wAICc diagnostic provided a more straightforward comparison with previous published results obtained using a frequentistic approach (e.g. [15]). We also included a posterior predictive check and report Bayesian p-values to assess the resulting predictions from our Bayesian models. To predict species richness across the entire GBR, we used a model-averaging procedure using wAICc to average the set of model formulations included in each model run.

15.3   Results

Observed species richness of fishes varied among years with the greatest species richness densities and variation among reefs recorded in 2011 (range: 10–80 species). The peak density of this year was also shifted left compared to the other years of observations (Figs. 15.1 and 15.2a), which displayed less variability (range: 25–70 species) and lower peak densities (Figs. 15.1 and 15.2a). With independent priors, the Bayesian models identified Chl a, SST, or light as influential in models 9, 10 and 11, respectively, for all yearly datasets, with emphasis differing between linear and quadratic terms for SST in different years. PO_4 (model 7) in years 2005, 2007 and 2013, and salinity (model 8) in all years also showed substantively non-zero effects in that the 95% CrIs excluded zero. The analogous CrIs for the coefficients of all other predictors overlapped zero (Table S1). Results for models and datasets using informative priors were similar but with some additional effects observed for predictors included in models 4–7 in early years only (Table S2).

Chl a and Light in models 9 and 11, respectively, were the only two predictors for which the 95% CrIs for the coefficient estimates excluded zero for all

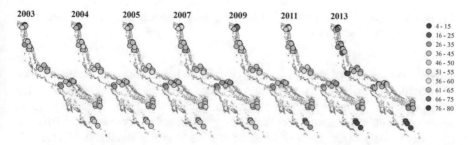

Fig. 15.1 Map of sampled species richness in the GBR across six latitudinal sections of the Great Barrier Reef (GBR) and three shelf positions (outer, mid and inner). Legend indicates number of species recorded per site after pooling counts made using 5×50-m long transects per site

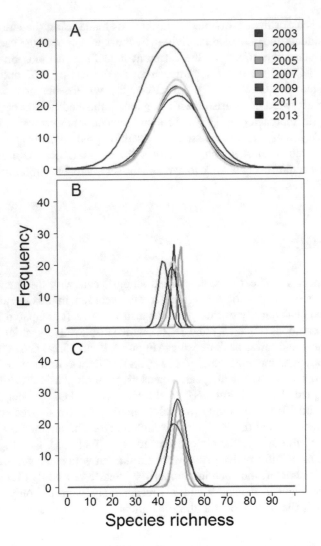

Fig. 15.2 Density plot of observed (**a**) and predicted species richness across datasets when using independent (**b**) or informative priors (**c**)

datasets irrespective of the application of informative or non-informative priors. The coefficient estimates for these two predictors averaged across models within years tended to be more negative toward the end of this time series (Fig. 15.3). Models with uninformative priors varied in goodness of fit according to wBIC, but generally models 2 (reef and reef position relative to spatial domains), 10 (reef and temperature), and 11 (reef and light) were among the best-fitting models (Table 15.2). As expected, when using informative priors, wBIC values were generally

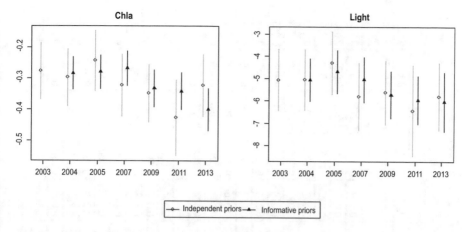

Fig. 15.3 Estimates of coefficient for Chl *a* and Light. Results are shown for each model run across datasets and when using independent (open circle with grey standard deviation lines) and informative priors (filled triangle with black standard deviation line)

lower. Goodness of fit according to wAICc showed similar patterns to those obtained with wBIC (Table 15.2).

15.3.1 Model Predictions

Density plots of species richness predictions demonstrate that each model set differed across datasets, but were more consistent when using informative priors (cf. Fig. 15.2b, c). Posterior predictive checks resulted in Bayesian *p*-values close to 0.5, indicating good predictive performance only for years 2011 and 2009, ranging respectively from 0.349–0.511 and 0.605–0.682 for uninformed priors and from 0.354–0.503 and 0.625–0.689 for informed priors. For all other datasets across all model runs, the Bayesian *p-value* was always close to one (0.916–0.989 for uninformative, and 0.916–0.989 for informative priors), indicating poor fits between observed and predicted species richness distribution, and hence, poor predictive performance. In no case, however, was the Bayes *p*-value >0.99, which would indicate major failure of model fit [8]. Irrespective of the use of independent or informative priors, higher fish species richness was predicted mostly in the northern-central offshore reefs, with the difference between inner and outer reefs being more marked in some years (e.g., 2003, 2005 and 2011 with independent priors) (Fig. 15.4).

Table 15.2 wBIC and wAICc calculated for each model and for all datasets considered when using independent or informative priors

Model		wBIC							wAICc						
		2003	2004	2005	2007	2009	2011	2013	2003	2004	2005	2007	2009	2011	2013
Independent priors	1	0.006	0.001	0.003	–	0.018	0.073	–	0.011	0.003	0.005	0.001	0.024	**0.104**	0.001
	2	**0.289**	0.037	**0.676**	**0.132**	**0.150**	0.012	**0.132**	**0.263**	0.052	**0.579**	**0.146**	**0.168**	0.017	**0.146**
	3	–	–	–	–	–	–	–	–	–	–	–	–	–	–
	4	0.002	0.004	0.004	0.009	0.006	0.025	0.009	0.006	0.009	0.009	0.015	0.013	0.033	0.015
	5	0.002	0.008	0.007	0.010	0.006	0.014	0.010	0.005	0.017	0.013	0.016	0.012	0.020	0.016
	6	0.003	0.012	0.009	0.018	0.007	0.022	0.018	0.008	0.025	0.016	0.027	0.014	0.029	0.027
	7	0.001	–	0.003	0.002	–	–	0.002	0.002	0.001	0.008	0.003	0.001	–	0.003
	8	0.009	0.023	0.020	0.059	0.016	0.015	0.059	0.016	0.030	0.033	0.084	0.023	0.016	0.084
	9	0.022	**0.105**	0.013	0.058	**0.176**	**0.101**	0.058	0.030	0.108	0.018	0.055	**0.159**	**0.108**	0.055
	10	**0.466**	**0.279**	**0.196**	**0.193**	**0.147**	**0.543**	**0.193**	**0.446**	**0.300**	**0.234**	**0.217**	**0.171**	**0.457**	**0.217**
	11	**0.196**	**0.523**	0.060	**0.502**	**0.464**	**0.148**	**0.502**	**0.205**	**0.440**	0.069	**0.408**	**0.397**	**0.160**	**0.408**
	12	0.004	0.008	0.009	0.018	0.010	0.047	0.018	0.010	0.016	0.016	0.027	0.019	0.058	0.027
Informative priors	1	–	**0.393**	**0.347**	**0.274**	**0.529**	**0.975**	0.038	–	**0.328**	**0.264**	**0.228**	**0.450**	**0.975**	0.034
	2	–	0.073	**0.307**	**0.225**	**0.110**	0.001	0.080	–	0.084	**0.320**	**0.203**	**0.134**	0.000	0.089
	3	–	–	–	–	–	–	–	–	–	–	–	–	–	–
	4	–	0.001	0.002	0.004	0.002	–	0.021	–	0.004	0.004	0.008	0.006	0.001	0.029
	5	–	0.003	0.003	0.009	0.003	–	0.029	–	0.008	0.007	0.014	0.008	0.001	0.037
	6	–	0.003	0.004	0.009	0.005	–	0.021	–	0.008	0.008	0.016	0.011	0.001	0.028
	7	–	–	0.006	0.004	–	–	0.003	–	0.001	0.015	0.008	0.001	–	0.006
	8	–	0.009	0.027	**0.076**	0.010	0.001	**0.224**	–	0.014	0.042	**0.106**	0.016	0.002	**0.238**
	9	–	0.029	0.015	0.027	0.063	0.009	0.048	–	0.040	0.018	0.032	0.071	0.014	0.042
	10	–	**0.313**	**0.225**	**0.128**	0.062	0.007	**0.193**	–	**0.323**	**0.250**	**0.147**	0.087	0.011	**0.188**
	11	–	**0.172**	0.061	**0.240**	**0.212**	0.005	**0.318**	–	**0.183**	0.064	**0.228**	**0.207**	0.009	**0.276**
	12	–	0.002	0.003	0.005	0.003	0.001	0.026	–	0.006	0.007	0.009	0.008	0.001	0.034

Shown only for wBIC and wAICc > 0.001, with bold indicating values >0.01. Model descriptions are given in Table 15.1

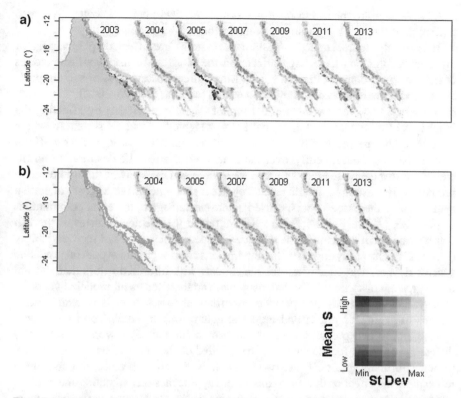

Fig. 15.4 Model predictions of species richness across the entire GBR for each model set run from yearly datasets when using independent (**a**) and informed (**b**) priors. Figure shows model-averaged results when using wAICc to average the contribution of each model in the model set

15.4 Discussion

Despite the difficulty and expense of observing complex natural ecosystems [13], the need to estimate ecosystem states and trajectories through time as they are influenced by increasingly frequent and severe disturbances is becoming more urgent. It is important, therefore, to understand how best to use information currently available to estimate these states and trajectories, understand their causes, and how to optimize the design of survey and monitoring programs to improve our understanding of these dynamics. In light of these information needs, we have compared here a series of independent annual analyses using uninformative priors with a recursive approach using informative priors based on previous data. The models were evaluated in the context of constructing and testing BDMs with specific reference to the prediction of fish species richness on Australia's Great Barrier Reef. The performances of the models, constructed using either of these approaches, indicate further room for improvement in how such models are constructed and some advantages to the use of informed priors. We detail below knowledge gained that would not have been possible using uninformed priors alone.

It is widely appreciated that natural ecosystems are affected by a variety of disturbances that can have large effects on the states and subsequent trajectories of the biological communities they host. Many direct observations of such impacts and recovery are now recorded in the literature (e.g., [5, 6, 9, 16, 20, 21, 27]). Such disturbances, however, can also affect model selection of BDMs by affecting the inclusions of particular predictors that are upweighted in specific situations where extreme values are reached because of either the immediate or longer-term effects of disturbances. Consequently, over the longer term, it may be desirable to down-weight or average these effects to achieve a more general view of the role of these predictors. For example, in early 2011 cyclone Yasi, one of the largest Australian cyclones over the past 20 years, caused extensive damage to the coral communities of the GBR [2]. Yasi also seems to have affected the species richness of the fish communities both in terms of the densities of species observed and their variability (Fig. 15.2a). Our study suggests that the impacts from such rare events on predicted species richness was much less for the models with informed priors that showed much greater consistency in both density and variation across all modeled years.

Bayesian approaches also provide opportunities to assess estimates and uncertainties in parameter values and model structure. In our study, Chl a and Light, were the only two predictors for which coefficient estimates were substantively different from zero for all datasets, irrespective of the application of informative or non-informative priors. These two parameters, therefore, provide an opportunity to examine the effect of these two modelling approaches on estimating their values and uncertainties across years as the Bayesian priors contained progressively more information. In all cases, estimates based on informed priors were more certain, however, modal values were not consistently greater or smaller, nor were the credible intervals progressively smaller as information accumulated in the informed priors and these intervals overlapped extensively. The two modelling frameworks also nominated different model structures as best. The recursive analysis, using informed priors based on previous survey data, indicated statistical contributions from more predictors than did the independent analysis of each time period. Accordingly, more information was harnessed by using informed priors indicating that greater investment in observations of these predictors may have additional utility.

While much was learned here by comparing Bayesian models with informed and uniformed priors, neither model performed very well with respect to prediction, indicating much is still to be learned regarding how best to increase their performance. The options for improvement here are many and will depend on the interests and opportunities of individual researchers and their groups. In contrast to this study, previous studies of predictive models based on these and similar data were better able to predict species richness and abundance by averaging these responses across the time-series of observations [15, 24]. Therefore, the challenge of better predictive performance identified here appears to be in generating finer-scale temporal predictions. Where predictions at this scale are desirable, our results suggest Bayesian models with informed priors can be useful for better selection of model structures, estimation of their parameters, and the down weighting of rare but significant events. Nonetheless, even though the time series used here to build these predictive models was spatially extensive and long compared to many ecological data series, the complexity of the processes, and potentially their non-stationarity, that can configure a metric such as species richness are likely to remain challenging. This challenge is likely to be exacerbated when predicting other biodiversity metrics such as abundances of individual species or abundances summed across species.

It is also clear that much longer time series may be required before prior probabilities can become sufficiently informed to facilitate more substantial reductions in parameter uncertainty. In the meantime, however, even a modest number of repeated observations appears to inform priors sufficiently to obtain quite consistent 95% CrIs and modal predicted values.

References

1. H. Assareh, I. Smith, K.L. Mengersen, Bayesian change point detection in monitoring clinical outcomes, in *Case Studies in Bayesian Statistical Modelling and Analysis*, ed. by C. L. Alston, K. L. Mengersen, A. N. Pettitt, (Wiley, Chichester, 2012)
2. R. Beeden, J. Maynard, M. Puotinen, P. Marshall, J. Dryden, J. Goldberg, G. Williams, Impacts and recovery from severe tropical cyclone Yasi on the Great Barrier Reef. PLoS One **10**, e0121272 (2015)
3. L. Broemeling, *Bayesian Methods for Repeated Measures* (Chapman and Hall/CRC, New York, 2016)
4. B.J. Cardinale, J.E. Duffy, A. Gonzalez, D.U. Hooper, C. Perrings, P. Venail, A. Narwani, G.M. Mace, D. Tilman, D.A. Wardle, A.P. Kinzig, G.C. Daily, M. Loreau, J.B. Grace, A. Larigauderie, D.S. Srivastava, S. Naeem, Biodiversity loss and its impact on humanity. Nature **486**, 59–67 (2012)
5. J.H. Connell, Diversity in tropical rain forests and coral reefs. Science **199**, 1302–1310 (1978)
6. P.K. Dayton, Competition, disturbance, and community organization: the provision and subsequent utilization of space in a rocky intertidal community. Ecol. Monogr. **41**, 351–389 (1971)
7. R. Fisher, B.T. Radford, N. Knowlton, R.E. Brainard, F.B. Michaelis, M.J. Caley, Global mismatch between research effort and conservation needs of tropical coral reefs. Conserv. Lett. **4**, 64–72 (2011)
8. A. Gelman, J. Carlin, H. Stern, D. Dunson, A. Vehtari, D. Rubin, *Bayesian Data Analysis* (Chapman and Hall, New York, 2015)
9. A.R. Halford, M.J. Caley, Towards an understanding of resilience in isolated coral reefs. Glob. Chang. Biol. **15**, 3031–3045 (2009)
10. A.R. Halford, A.A. Thompson, *Visual Census Surveys of Reef Fish. Long Term Monitoring of the Great Barrier Reef Standard Operational Procedure Number 3* (Australian Institute of Marine Science, Townsville, 1996)
11. A.R. Ives, S.R. Carpenter, Stability and diversity of ecosystems. Science **317**, 58–62 (2007)
12. S.Y. Kang, J.M. McGree, C.C. Drovandi, M.J. Caley, K.L. Mengersen, Bayesian adaptive design: improving the effectiveness of monitoring of the Great Barrier Reef. Ecol. Appl. **26**, 2635–2646 (2016)
13. G.M. Lovett, D.A. Burns, C.T. Driscoll, J.C. Jenkins, M.J. Mitchell, L. Rustad, J.B. Shanley, G.E. Likens, R. Haeuber, Who needs environmental monitoring? Front. Ecol. Environ. **5**, 253–260 (2007)

14. S.A. Matthews, C. Mellin, A. MacNeil, S.F. Heron, W. Skirving, M. Puotinen, M.J. Devlin, M. Pratchett, High-resolution characterization of the abiotic environment and disturbance regimes on the Great Barrier Reef, 1985–2017. Ecology **100**, e02574 (2019)

15. C. Mellin, C.J.A. Bradshaw, M.G. Meekan, M.J. Caley, Environmental and spatial predictors of species richness and abundance in coral reef fishes. Glob. Ecol. Biogeogr. **19**, 212–222 (2010)

16. C. Mellin, M. Aaron MacNeil, A.J. Cheal, M.J. Emslie, M. Julian Caley, Marine protected areas increase resilience among coral reef communities. Ecol. Lett. **19**, 629–637 (2016)

17. R.B. Millar, Comparison of hierarchical Bayesian models for overdispersed count data using DIC and Bayes' factors. Biometrics **65**, 962–969 (2009)

18. N. Myers, R.A. Mittermeier, C.G. Mittermeier, G.A.B. da Fonseca, J. Kent, Biodiversity hotspots for conservation priorities. Nature **403**, 853–858 (2000)

19. J.D. Nichols, B.K. Williams, Monitoring for conservation. Trends Ecol. Evol. **21**, 668–673 (2006)

20. R.T. Paine, S.A. Levin, Intertidal landscapes: disturbance and the dynamics of pattern. Ecol. Monogr. **51**, 145–178 (1981)

21. S.T.A. Pickett, P.S.E. White, *The Ecology of Natural Disturbance and Patch Dynamics* (Academic, London, 1985)

22. S.L. Pimm, The complexity and stability of ecosystems. Nature **307**, 321–326 (1984)

23. S. Sarkka, *Bayesian Filtering and Smoothing* (Cambridge University Press, Cambridge, 2013)

24. A.M.M. Sequeira, C. Mellin, H.M. Lozano-Montes, M.A. Vanderklift, R.C. Babcock, M.D.E. Haywood, J.J. Meeuwig, M.J. Caley, Transferability of predictive models of coral reef fish species richness. J. Appl. Ecol. **53**, 64–72 (2016)

25. A.M.M. Sequeira, P.J. Bouchet, K.L. Yates, K. Mengersen, M.J. Caley, Transferring biodiversity models for conservation: opportunities and challenges. Methods Ecol. Evol. **9**, 1250–1264 (2018)

26. A.M.M. Sequeira, C. Mellin, H.M. Lozano-Montes, J.J. Meeuwig, M.A. Vanderklift, M.D.E. Haywood, R.C. Babcock, M.J. Caley, Challenges of transferring models of fish abundance between coral reefs. PeerJ **6**, e4566 (2018)

27. W.P. Sousa, The role of disturbance in natural communities. Annu. Rev. Ecol. Syst. **15**, 353–391 (1984)

28. H. Sweatman, A. Cheal, G. Coleman, M. Emslie, K. Johns, M. Jonker, I. Miller, K. Osborne, *Long-Term Monitoring of the Great Barrier Reef. Status Report No 8* (Australian Institute of Marine Science, Townsville, 2008)

29. D. Tilman, J.A. Downing, Biodiversity and stability in grasslands. Nature **367**, 363 (1994)

30. J. Vercelloni, K. Mengersen, F. Ruggeri, M.J. Caley, Improved coral population estimation reveals trends at multiple scales on Australia's Great Barrier Reef. Ecosystems **20**, 1337–1350 (2017)

31. B. Worm, E.B. Barbier, N. Beaumont, J.E. Duffy, C. Folke, B.S. Halpern, J.B.C. Jackson, H.K. Lotze, F. Micheli, S.R. Palumbi, E. Sala, K.A. Selkoe, J.J. Stachowicz, R. Watson, Impacts of biodiversity loss on ocean ecosystem services. Science **314**, 787–790 (2006)

32. K.L. Yates, C. Mellin, M.J. Caley, B.T. Radford, J.J. Meeuwig, Models of marine fish biodiversity: assessing predictors from three habitat classification schemes. PLoS One **11**, e0155634 (2016)

33. K.L. Yates, P.J. Bouchet, M.J. Caley, K. Mengersen, C.F. Randin, S. Parnell, A.H. Fielding, A.J. Bamford, S. Ban, A.M. Barbosa, C.F. Dormann, J. Elith, C.B. Embling, G.N. Ervin, R. Fisher, S. Gould, R.F. Graf, E.J. Gregr, P.N. Halpin, R.K. Heikkinen, S. Heinänen, A.R. Jones, P.K. Krishnakumar, V. Lauria, H. Lozano-Montes, L. Mannocci, C. Mellin, M.B. Mesgaran, E. Moreno-Amat, S. Mormede, E. Novaczek, S. Oppel, G. Ortuño Crespo, A.T. Peterson, G. Rapacciuolo, J.J. Roberts, R.E. Ross, K.L. Scales, D. Schoeman, P. Snelgrove, G. Sundblad, W. Thuiller, L.G. Torres, H. Verbruggen, L. Wang, S. Wenger, M.J. Whittingham, Y. Zharikov, D. Zurell, A.M.M. Sequeira, Outstanding challenges in the transferability of ecological models. Trends Ecol. Evol. **33**, 790–802 (2018)

Chapter 16
Thresholds of Coral Cover That Support Coral Reef Biodiversity

Julie Vercelloni, M. Julian Caley, and Kerrie L. Mengersen

Abstract Global environmental change, such as ocean warming and increased cyclone activity, is driving widespread and rapid declines in the abundance of key ecosystem engineers, reef-building corals, on the Great Barrier Reef. Our ability to understand how coral associated species, such as reef fishes, respond to coral loss can be impeded by uncertainty surrounding natural spatio-temporal variability of coral populations. To address this issue, we developed a semi-parametric hierarchical Bayesian model to estimate long-term trajectories of habitat-forming coral cover as a function of three spatial scales (sub-region, habitat and site) and environmental disturbances. The relationships between coral cover trajectories and fish community structure were examined using posterior predictive distributions of estimated coral cover from the statistical model. In the absence of direct observations of fish community structure, we used the probability of coral cover being above some ecological threshold values as a proxy for potential disruptions of fish community structure. Threshold values were derived from published field studies that estimated changes in the structure of coral-reef fish communities and coral cover after major disturbances. In these studies, fish community structure did not change where post-disturbance coral cover was >20%. Disruptions in the structure of these communities were observed when coral cover dropped to between 10–20% and declines in fish diversity were typical where coral cover ranged from between 5 and 10%. Based on these thresholds values, posterior probabilities of coral cover being above 20% and between 10 and 20% and between 5 and 10% were calculated across spatial scales on the Great Barrier Reef (GBR) from 1995 to 2011. At the GBR scale, probabilities of coral cover being above these thresholds remained relatively stable through time. Across years, probabilities of coral cover being at least >20% remained null for the sub-regions of Cairns, Townsville,

J. Vercelloni (✉) · M. J. Caley · K. L. Mengersen
School of Mathematical Sciences and ARC Centre of Mathematical and Statistical Frontiers,
Science and Engineering Faculty, Queensland University of Technology, Brisbane, QLD,
Australia
e-mail: j.vercelloni@qut.edu.au

385

K. L. Mengersen et al. (eds.), *Case Studies in Applied Bayesian Data Science*,
Lecture Notes in Mathematics 2259, https://doi.org/10.1007/978-3-030-42553-1_16

Whitsundays and Swain but highly variable between reef sites within these sub-regions, with the exception of Townsville. In the Townsville area, probabilities of coral cover being between 10–20% and 5–10% declined from 0.75 to 0 during the study period. This finding highlights potential sub-regional fish community structure disruptions which have not yet been observed at this spatial scale. As frequency and intensity of disturbance events continue to rise, and consequently, as coral cover declines further, the probabilistic Bayesian approach presented in this chapter could be used to help provide early warnings of major ecological shifts at management relevant scales in the absence of direct observations.

16.1 Introduction

Healthy functioning of tropical coral reefs depends on corals, sponges and other sessile species to create three-dimensional structure that provides shelter, and a space to live [2, 5, 13, 25]. Indeed, the biogenic structure of coral reefs host in the order of one million multicellular species [12]. These valuable marine ecosystems are degrading rapidly, typified by the substantial loss of habitat-forming corals [4, 10, 34]. Similar trends around the world portend the degradation of coral reefs toward non coral-dominated ecosystems [17] and associated losses of biodiversity [6].

The percentage of hard coral cover is closely related to fish community structure with specific types of corals being strong predictors of fish species richness [21, 22]. A single colony of branching coral can provide habitat and refuge from predators for several fish species that vary in abundance as a function of the available space between the branches of the coral. These species-specific and community patterns, however, can be disrupted when percent cover of habitat-forming corals (e.g. branching corals) drop below 20% [13] and the loss of fish species has been observed when coral cover drops below 10% [21]. On some coral reefs, such as those of the Great Barrier Reef (GBR), changes in the species composition of fish communities in past decades [7] and recent loss of reef fish richness following disturbance events [22] suggest an influence of declining habitat-forming corals on fish communities.

By generating finer-scale understanding through monitoring changes in coral cover, our ability to understand the broader impacts of declining habitat-forming corals on reef biodiversity should also improve. With recent advances in the analyses of big-data, learning from long-term coral-reef monitoring data can help address these critical conservation challenges for coral reefs. For example, the Long-Term Monitoring Program (LTMP) of the GBR, one of the longest and most extensive coral reef survey in the world, monitors ~1.5% of the individual reefs of the GBR [30]. The resulting datasets, while not large by big data standards, are highly complex with myriad interacting biotic and abiotic ecological processes affecting trajectories in coral cover across space and time. Indeed, non-linear trajectories in coral cover can differ substantially on coral reefs situated only a few hundreds meters apart [32]. These differences can be attributed to fine-scale responses of corals to disturbances that vary as a function of their nature and intensity but also reef topography, differential susceptibility of coral taxa to disturbances, and legacy of past disturbance events. Consequently, inferences derived from these data, without sufficient regard to the variability of coral cover trajectories across scales of space and time, has fueled scientific debate regarding the origins of coral decline and recovery [16, 29, 31]. Moreover, uncertainty regarding states and trajectories of coral cover has the potential to compromise effective reef management strategies [3, 11] and impede learning about how changes in habitat-forming coverage might affect the biodiversity hosted by corals.

To address these knowledge gaps, a semi-parametric Bayesian hierarchical model [9] was developed for the LTMP data to identify and locate sources of uncertainty when modelling this complex dataset [32], estimate long-term trajectories in *Acropora* spp. cover, the primary provider of habitat-forming on the GBR, across spatial scales [34], and quantify the cumulative effect of disturbances on these trajectories [33]. Extending from these previous analyses, we investigated potential relationships between changes in coral cover trajectories and fish community structure by using posterior predictive distributions of estimated coral cover from the statistical model. The benefit of using these predictive distributions is the preservation of the observed spatial and temporal structure of the LTMP data during the estimation of coral cover. In the absence of direct observations of fish community structure, we used the probability of *Acropora* spp. cover being above three ecological threshold values (20% and between 10–20% and 5–10%) as a proxy for the effects of changes in coral cover on associated fish communities. These probabilities were calculated from the posterior predictive distributions of coral cover across spatial scales on the Great Barrier Reef (GBR) from 1995 to 2011. This approach to investigating the relationships between coral cover and proxies of fish community structure provides a novel framework for future assessments of early warnings of major ecological shifts resulting from the decline of habitat-forming corals across a wide range of spatial scales.

16.2   Methods

Coral Cover Data

Observations of coral cover from the LTMP from 1995 to 2011 were used to model coral cover trajectories [30]. The LTMP sampled benthic cover annually from 1995 to 2005, and then every second year, for 141 reef sites between six and nine meters depth. The program was designed to track changes in benthic coral reef communities over time across 47 coral reefs within six management sub-regions (aka sectors) of the GBR. Within these sub-regions, the LTMP monitors three reef habitats defined by the position of reefs on the continental shelf (aka shelf positions), except for the Swain and Capricorn-Bunker sub-regions in which only two (mid-shelf and outer-shelf) and one (outer-shelf) habitats, respectively, are represented. Inner-reefs, being closest to the coast, are most exposed to terrestrial influences. The mid-shelf habitat extends over a large part of the GBR lagoon, with reefs situated at various distances between the inner and outer habitats of the GBR. Outer-reef habitat extends into oceanic conditions. The survey is spatially replicated on two to four reefs per habitat and sub-region, each reef being itself sampled at three distinct sites using five permanent transects. Hard coral cover was estimated at the genus level and expressed as a percentage of transect area based on observations taken at 200 random points along each transect. The five transects per site were pooled within sites for the purposes of this study. The percentage of coral cover of *Acropora* spp. (hereafter referred to as coral cover) was used to model the trajectories as these taxa are the most abundant on the GBR and responsible for most of its annual and decal variability in coral cover changes [23].

Disturbance Data

The LTMP also recorded disturbances by matching observed variations in coral cover on sites with observations of particular phenomena [30]. Since 1995, coral bleaching, crowns-of-thorns starfish outbreaks (CoTS), storms/cyclones, coral diseases, and multiple and unknown disturbances were systematically monitored. Distinct signatures of these disturbances allowed disturbance types to be identified in the field [23]. Also, the availability of meteorological data allowed estimation of storm and cyclones tracks at a relative fine-scale. Following the LTMP methodology, disturbances were recorded as having had an impact if total coral cover (*Acropora* spp. plus all other coral genera) at the reef scale decreased by more than 5% on a scale of 0–100% between two survey periods. To calibrate disturbance data at the site scale, we assumed the presence of a disturbance (coded as 1) with a >5% decline of coral cover between two consecutive years pooled for the three sites within a reef and absence (coded as 0) otherwise. The cumulative effect of disturbances was estimated by summing the presence of disturbances of all types over the previous years for each reef site. We assumed that a disturbance influenced the coral cover trajectories uniformly during the entire surveyed years, and that all possible combinations of disturbances, irrespective of their nature and intensity, acted in isolation and are added in the same way [33].

The Semi-parametric Bayesian Hierarchical Model

The semi-parametric Bayesian hierarchical model (SPa-BaHM) was developed to examine different ecological aspects of the long-term trajectories of the habitat-forming corals across spatial scales (Fig. 16.1). These previous investigations focused on the different stages of the model such as the parameters stage to identify sources of uncertainty when estimating coral cover trajectories within a sub-region [32] and across the GBR [34] or the process stage in order to estimate the cumulative effect of disturbances on coral cover [33]. For model selection, values of Deviance Information Criterion (DIC) were computed for each model formulation per sub-region. Smaller values of the DIC indicated more preferable models with respect to goodness-of-fit to the observed data and model parsimony. DIC associated with the three model versions showed a better model fit with the presence of disturbances and

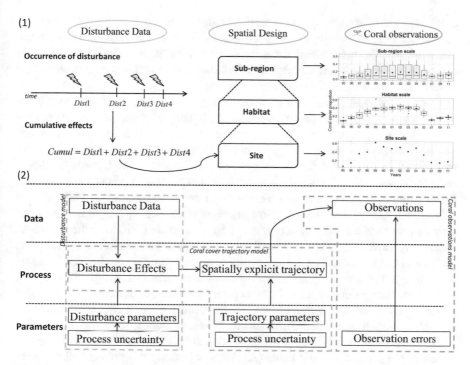

Fig. 16.1 Conceptual representation of the three-tiered semi-parametric Bayesian hierarchical model [33]

Table 16.1 Values of DIC for the three published versions of the semi-parametric Bayesian hierarchical models

Sub-region	[32]	[34]	[33]
Cooktown-Lizard Island	−1178	−1176	−1417
Cairns	−1194	−1394	−1761
Townsville	−934	−1202	−1528
Whitsundays	−1085	−1198	−1561
Swain	−726	−799	−1120
Capricorn-Bunker	−360	−380	−459

their cumulative effect into the model formulation (Table 16.1). This result guided the development of model presented here.

In this chapter, we examine relationships between coral cover trajectories and proxies of fish community structure using posterior predictive distributions of coral cover from the third stage of the model, the data stage.

Within each sub-region, the cover of *Acropora* spp., y_{ij}, at time t indexed by site i ($i=1$–141) and time period j ($j=1$–17) was modelled using a three-tiered hierarchical model (Fig. 16.1). For the first tier of the model (the data stage), values of coral cover y_{ij} were arcsine square root transformed to be normally distributed with an expected value μ_{ij} and a sampling variance σ^2. For the second tier (the process stage), the expected value μ_{ij} was described by the contribution of coral trajectories at three spatial scales: $\mathbf{f}_s(.)$ describes the overall mean trajectory at the scale of the sub-region, and deviations from this overall curve represent hierarchically, the habitat and site-specific dynamics, indexed by $h(s)$ and $i(sh)$ respectively. For the last tier (the parameters stage), the unknown parameters of the model and associated uncertainty were described.

Coral cover trajectories were modelled using a semi-parametric approach composed of a linear and non-linear component [9]. This approach benefits from estimating ecological trends from the data while allowing for increases and decreases in coral cover among years resulting from coral recovery and disturbances, respectively [32]. In order to estimate changes in coral cover through time, the linear component was estimated using slope and intercept parameters at each spatial scale. Fixed parameters were used to estimate coral cover trajectories at the scales of habitat and sub-region. These parameters were described by vague normal prior distributions with a mean equal to 0 and a large variance of 10^3. At the site scale, coral cover trajectories were estimated using random slope and intercept parameters which allowed us to explore the variability in coral cover changes between the reef sites through time. We also modelled the cumulative effect of disturbances (δ_{2i}) of coral cover changes at this scale. Gamma priors were used to describe the associated variance parameters for each reef site. The variability in coral cover was modelled by adding smooth non-linear functions to the linear components for each spatial scale. These penalized splines [9] were described by the matrices z_t and z_Occ that indicated the positions of the knots ($K_1, K_2, K_3=4$) along the year and disturbance (at the site scale only) variables. These matrices were associated with additional model parameters that constrained the smoothing effect. Therefore, the number of knots used in the model is not critical unlike other non-linear regression approaches such as Generalized Additive Models which are commonly implemented to model coral cover trajectories.

Statistical diagnostics including deviance information criteria (DIC), posterior predictive checks, and residual analysis were used to identify the preferred model. Convergence from the MCMC runs was assessed by performing tests available in the CODA R packages [26]. Gelman and Rubins and Geweke's convergence diagnostics, trace and density-plots of parameters and autocorrelation plots between MCMC draws confirmed the convergence of MCMC chains. Three MCMC chains were simultaneously run to confirm convergence to stationarity. Model convergence was reached after 800,000 iterations, of which 500,000 values were discarded as burn in and using a thinning rate of 50 iterations. Inferences presented here, therefore, are based on 6000 values from three different MCMC chains. The model was fitted using the Bayesian software analysis WinBUGS [27] from the R package R2WinBUGS [28].

$$\text{arcsin} \sqrt{y_{ij}} \sim \mathcal{N}(\mu_{ij}, \sigma^{-2})$$

$$\mu_{ij} = \mathbf{f}_s + \mathbf{f}_{h(s)} + \mathbf{f}_{i(hs)}$$

$$\mathbf{f}_s = \beta_0 + \beta_1 \times t_{ij} + \sum_{k=1}^{K1} c_k z^s_{t_{ijk}}$$

$$\mathbf{f}_{h(s)} = \gamma_0 + \gamma_{1_h} \times t_{ij} + \sum_{k=1}^{K2} d_{hk} z_{t_{ijk}}^h \qquad (16.1)$$

$$\mathbf{f}_{i(hs)} = \delta_{0_i} + \delta_{1_i} \times t_{ij} + \delta_{2_i} \times Cumul_{ij} + \sum_{k=1}^{K3} g_{ik} z_{t_{ijk}}^i + \sum_{k=1}^{K3} h_{ik} z_{Occ_{ijk}}^i$$

$$\beta_0, \beta_1, \gamma_0, \gamma_{1_h} \sim \mathcal{N}(0, \ 10^3)$$

$$c, d, g, h, \delta_{0_i}, \delta_{1_i}, \delta_{2_i} \sim \mathcal{N}(0, \ \sigma_{(.)}^{-2})$$

$$\sigma_{(.)}^{-2} \sim \mathcal{G}(10^{-3}, \ 10^{-3})$$

The variability of coral cover trajectories was well captured by the model fit with an average of 85.6% of coral cover observations included in the 95% credible intervals of the coral cover posterior distributions. Coral cover predictions (number of times that coral observations were included in the 95% credible intervals) were the most accurate for the southern sub-region of Capricorn-Bunker (88.4%) and the less accurate in Cooktown-Lizard Island (83.6%).

Posterior Probabilities and Thresholds
Posterior predictive distributions of coral cover $\hat{y}_{ij}$ were previously examined in order to estimate long-term coral cover trajectories and associated uncertainty at the scale of individual reef site (Fig. 16.2, [33]). In this chapter, these distributions were used to calculate proxies of fish community structure defined as the probabilities of coral cover being above 20% and between 10–20% and 5–10%. These thresholds were chosen to broadly represent degrees of stability of associated reef fish communities. Posterior probabilities were calculated by estimating the number of times, $\hat{y}_{ij}$, was greater than the proxies using the 6000 values from the MCMC simulations. Probabilities were estimated for each surveyed year at the scales of coral reef, sub-region, and GBR. Proxies for fish community structure were arcsine squared-root transformed to align with coral cover prediction outputs.

16.3   Results

Disturbance Regimes and Coral Cover Trajectories
For the period 1995–2011, 294 disturbances were recorded by the LTMP. During this period, Unknown and Storm/Cyclone were the two most frequent and spatially widespread disturbances. Coral Bleaching was recorded in the sub-regions of Townsville and Whitsundays and Coral Disease in the two most northern and southern sub-regions. Coral decline associated with CoTS outbreaks were the most frequent in the sub-regions of Cairns, Townsville, and Swain sub-regions. Storm/Cyclone disturbance was the most prevalent during the last 2 years of this period [33]. The two category 5 tropical cyclones [8] were associated with a 68%

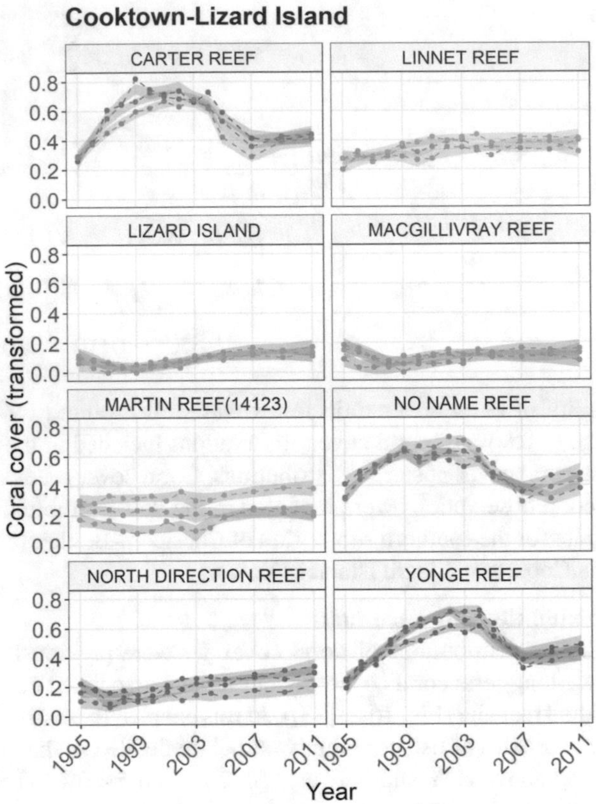

Cooktown-Lizard Island

Fig. 16.2 Example of coral cover trajectories for the sub-region Cooktown-Lizard Island. Coloured dots and dotted lines indicate the measurements of coral cover at site and shaded areas the 95% credible intervals from the posterior predictive distributions of coral cover estimated by the model

decline in coral cover across >1000 km of the central-southern part of the GBR [8]. Coral decline associated with Coral Bleaching was recorded in 1998–1999 and 2003 (the survey year following bleaching in late 2002) and match with records of heat stress [18].

The accumulation of disturbances of all types increased from 4 records in 1995 to 100 in 2000, 181 in 2005 and 286 in 2011 (Fig. 16.3). The presence of disturbances was recorded in all of the six sub-regions from 1999. Disturbances were the most recorded in the sub-region of Townsville with a total of 77 accumulated events in 16 years data followed by Swain (46), Cairns and Whitsundays (45), Cooktown-Lizard Island (41) and Capricorn-Bunker (38).

Posterior Probabilities and Thresholds

The proxies used to examine potential disruptions to fish community structure as a function of coral cover changes through time did not reveal any trends at the GBR

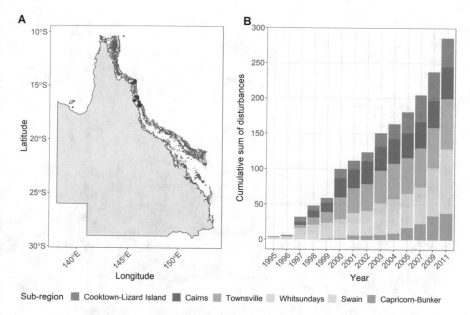

Fig. 16.3 (**a**) Spatial locations of surveyed reefs within region along the Great Barrier Reef. (**b**) Cumulative sum of disturbances across all types per sub-region

scale. The probabilities of coral cover being >20% and between the ranges of 10–20% and 5–10% remained stable through time with the exception of a decline from 0.5 to 0.25 for the probabilities of coral cover being between 5–10% from 2009 to 2011 (Fig. 16.4).

Probability trajectories were variable between and within sub-regions ranging from 0 to 1 through the studied period. At the sub-region scale, the probabilities of a coral cover >20% was close to 0 during the entire period for the sub-regions of Cairns, Townsville and Whitsundays. In the Cairns sub-region, few reef sites displayed varying probability trajectories with some probabilities close to 1, whereas, the decline in posterior probabilities was homogeneous across all the reef sites within the Townsville sub-region. In the Cooktown-Lizard Island sub-region, probabilities of coral cover between 10–20% increased from 0.10 in 1995 to 0.50 in 1999 and remained stable throughout the year. This latest trend was also detected for the probabilities coral cover between 5–10% but with variations in probabilities ranging between 0 and 1 at the reef site scale within this sub-region. The southern most sub-region of Capricorn-Bunker displayed similar probabilities across its reef sites and thresholds through time. The central sub-regions of Cairns, Townsville, Whitsundays, and Swain also showed similar differences in regional probability trajectories between the thresholds but at different magnitudes. Probabilities were typically higher for the 5–10% coral cover threshold especially for the sub-regions

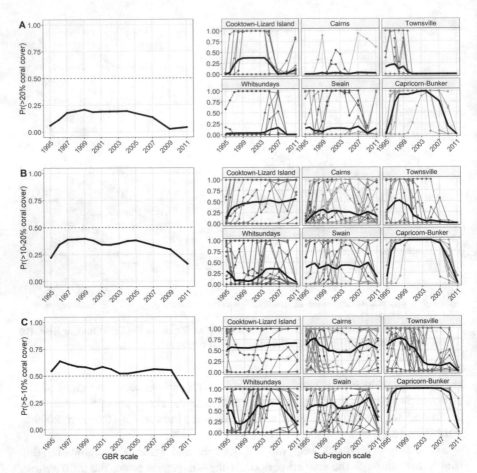

Fig. 16.4 Posterior probabilities of percent of coral cover strictly above a threshold of >20% (**a**), between >10 and 20% (**b**) and between >5 and 10% (**c**), used as proxies for potential disruptions in fish community structure. Broad-spatial scales probabilities are indicated by black tick lines at the scale of the Great Barrier Reef (left panels) and sub-regions (right panels). Coloured dots and lines show posterior probabilities at the reef site scale within each sub-region

of Cairns and Whitsundays with a different magnitudes ranging from 0.25 to 0.50 throughout the years. The sub-region of Townsville displayed long-term declines in probabilities of coral cover being above any of the three thresholds at different rates starting from 1999. Probabilities of coral cover being between 10 and 20% decreased from 0.50 to 0 between 1997–2003, whereas the decline in probabilities of coral cover between 5 and 10% ranged from 0.75 in 1999 to 0.25 in 2003. From 2003 onward, these probabilities remained stable.

16.4   Discussion

Modern Bayesian approaches to the study of biological communities are able to help address gaps in monitoring data required for conservation and management across communities and spatial scales [20]. Our application of the Bayesian statistical framework to data for habitat-forming corals allowed us to understand different aspects of the trajectories of these communities while controlling for several sources of uncertainties inherent in monitoring data, ecological processes, and model parameter estimation [32]. By retaining information at different hierarchical spatial scales in a non-linear context while considering the effects of disturbances, we were able to estimate 16 years of variation in coral cover trajectories with associated estimates of uncertainty across different spatial scales [33, 34]. Using posterior predictive distributions of estimated coral cover, the investigation of probabilities of coral cover beyond certain critical thresholds demonstrates that information contained in coral cover data may be used to indirectly infer the structure of other organisms that depend on corals to provide habitat space. In the field of coral-reef ecology, this approach could help understand flow-on effects of changes in coral cover on the dynamics of biodiversity dependant on it without the need for direct observations. Considering that only 15% of the Great Barrier Reef is regularly monitored (>2 surveyed years) by several organizations with different purposes, it is essential that the future coral reef research acknowledges these data gaps and look for innovative approaches to monitor changes in coral reef biodiversity across the entire GBR [1, 19, 24].

The probabilistic approach adopted here to assume changes in fish community structure as a function the spatio-temporal variability in coral cover reveals that the conditions favourable to the maintenance in fish community structure (i.e. >20% coral cover) remained stable throughout the study period and never exceed $p = 0.25$ at the scale of the GBR. The probabilities of fish community structure being disrupted (i.e. when coral cover is between >10 and 20%) or decline in fish diversity (i.e. when coral cover is between >5 and 10%) varied the most at the regional and reef levels. This variability in probabilities through time and space within and between sub-regions suggests that the processes responsible for changes in coral cover mostly acted at fine spatial scales with the exception of Capricorn-Bunker and Townsville sub-regions.

The LTMP surveys recorded numerous disturbances along the GBR including early CoTS outbreaks in the northern parts [30] and storms in the south [14]. These disturbance events were reflected in the low probabilities of coral cover being between 10 and 20% in the Capricorn-Bunker and Cooktown-Lizard Island regions in 1995, respectively. Large and widespread effects of storms in 2008 and tropical cyclone Hamish in 2009 dramatically reduced the probabilities coral cover being between 5 and 10% for all the coral reefs in the Capricorn-Bunker sub-region. From 1999, the decline in probabilities of coral cover being above the critical thresholds suggests potential disruptions of fish community structures in the Townsville sub-region. The numerous disturbances including tropical cyclone

Justin in 1997, coral bleaching in 1998 and other types of disturbances recorded almost every surveyed year resulted in the accumulation of 77 disturbances in 16 years and likely prevented recovery at the sub-regional scale. Nonetheless, no sub-regional trends in fish community disruptions were reported [30]. Therefore, the choice of the thresholds based on the literature may be conservative and warrant further investigation to more carefully define how, and at what levels, they operate.

In this chapter, we demonstrate the benefits of carefully modelling of habitat forming corals in order to extract information on their trajectories at different spatial scales and their potential implications for coral reef associated biodiversity. This information has the potential to inform management of coral reefs in the face of increasing anthropogenic disturbances. For example, the sub-regions of Townsville and Whitsundays may require specific intervention as a result of very low probabilities coral cover being between 10 and 20%. Also, the decline in probabilities of coral cover being between 5 and 10% used as a proxy for fish diversity loss for all the sub-regions, with the exception of the Cooktown-Lizard Island, should be carefully monitored. In the absence of signs of recovery (i.e. increase in probabilities of coral cover being between 5 and 10%) options should be considered regarding how best to arrest any further declines and support recovery to reduce the risks of major ecological shifts. Operational methods for managing the Great Barrier Reef using information extracted from statistical models is not yet well developed or deployed. As the Anthropocene unfolds, associated with unprecedented rapid decline of habitat-forming corals [15, 18], a more robust quantitative framework, at management relevant scales such as the entire GBR, is urgently required.

Acknowledgments We thank members of the Australian Institute of Marine Science Long Term Monitoring Program that have collected the data used in these analyses; and Emma Kennedy for providing helpful comments.

References

1. D.R. Bellwood, M.S. Pratchett, T.H. Morrison, G.G. Gurney, T.P. Hughes, J.G. Álvarez-Romero, J.C. Day, R. Grantham, A. Grech, A.S. Hoey, et al., Coral reef conservation in the anthropocene: confronting spatial mismatches and prioritizing functions. Biol. Conserv. **236**, 604–615 (2019)
2. S.J. Brandl, L. Tornabene, C.H. Goatley, J.M. Casey, R.A. Morais, I.M. Côté, C.C. Baldwin, V. Parravicini, N.M. Schiettekatte, D.R. Bellwood, Demographic dynamics of the smallest marine vertebrates fuel coral-reef ecosystem functioning. Science **364**, aav3384, 1189–1192 (2019)
3. J. Brodie, J. Waterhouse, A critical review of environmental management of the 'not so great' barrier reef. Estuar. Coast. Shelf Sci. **104**, 1–22 (2012)
4. J.F. Bruno, E.R. Selig, Regional decline of coral cover in the indo-pacific: timing, extent, and subregional comparisons. PLoS One **2**(8), e711 (2007)
5. M.J. Caley, J. St John, Refuge availability structures assemblages of tropical reef fishes. J. Anim. Ecol. **65**, 414–428 (1996)

6. K.E. Carpenter, M. Abrar, G. Aeby, R.B. Aronson, S. Banks, A. Bruckner, A. Chiriboga, J. Cortés, J.C. Delbeek, L. DeVantier, et al., One-third of reef-building corals face elevated extinction risk from climate change and local impacts. Science **321**(5888), 560–563 (2008)
7. A.J. Cheal, S.K. Wilson, M.J. Emslie, A.M. Dolman, H. Sweatman, Responses of reef fish communities to coral declines on the great barrier reef. Mar. Ecol. Prog. Ser. **372**, 211–223 (2008)
8. A.J. Cheal, M.A. MacNeil, M.J. Emslie, H. Sweatman, The threat to coral reefs from more intense cyclones under climate change. Glob. Chang. Biol. **23**(4), 1511–1524 (2017)
9. C. Crainiceanu, D. Ruppert, M.P. Wand, Bayesian analysis for penalized spline regression using winbugs. J. Stat. Softw. **14**(14), 1–24 (2005)
10. G. De'ath, K.E. Fabricius, H. Sweatman, M. Puotinen, The 27–year decline of coral cover on the great barrier reef and its causes. Proc. Natl. Acad. Sci. **109**(44), 17995–17999 (2012)
11. R. Fisher, B.T. Radford, N. Knowlton, R.E. Brainard, F.B. Michaelis, M.J. Caley, Global mismatch between research effort and conservation needs of tropical coral reefs. Conserv. Lett. **4**(1), 64–72 (2011)
12. R. Fisher, R.A. O'Leary, S. Low-Choy, K. Mengersen, N. Knowlton, R.E. Brainard, M.J. Caley, Species richness on coral reefs and the pursuit of convergent global estimates. Curr. Biol. **25**(4), 500–505 (2015)
13. A.R. Halford, M.J. Caley, Towards an understanding of resilience in isolated coral reefs. Glob. Chang. Biol. **15**(12), 3031–3045 (2009)
14. A. Halford, A. Cheal, D. Ryan, D.M. Williams, Resilience to large-scale disturbance in coral and fish assemblages on the great barrier reef. Ecology **85**(7), 1892–1905 (2004)
15. O. Hoegh-Guldberg, P.J. Mumby, A.J. Hooten, R.S. Steneck, P. Greenfield, E. Gomez, C.D. Harvell, P.F. Sale, A.J. Edwards, K. Caldeira, et al., Coral reefs under rapid climate change and ocean acidification. Science **318**(5857), 1737–1742 (2007)
16. T. Hughes, D. Bellwood, A. Baird, J. Brodie, J. Bruno, J. Pandolfi, Shifting base-lines, declining coral cover, and the erosion of reef resilience: comment on sweatman et al. (2011). Coral Reefs **30**(3), 653–660 (2011)
17. T.P. Hughes, M.L. Barnes, D.R. Bellwood, J.E. Cinner, G.S. Cumming, J.B. Jackson, J. Kleypas, I.A. Van De Leemput, J.M. Lough, T.H. Morrison, et al., Coral reefs in the anthropocene. Nature **546**(7656), 82 (2017)
18. T.P. Hughes, J.T. Kerry, M. Álvarez-Noriega, J.G. Álvarez-Romero, K.D. Anderson, A.H. Baird, R.C. Babcock, M. Beger, D.R. Bellwood, R. Berkelmans, et al., Global warming and recurrent mass bleaching of corals. Nature **543**(7645), 373 (2017)
19. S.Y. Kang, J.M. McGree, C.C. Drovandi, M.J. Caley, K.L. Mengersen. Bayesian adaptive design: improving the effectiveness of monitoring of the great barrier reef. Ecol. Appl. **26**(8), 2637–2648 (2016)
20. H.K. Kindsvater, N.K. Dulvy, C. Horswill, M.J. Juan-Jordá, M. Mangel, J. Matthiopoulos, Overcoming the data crisis in biodiversity conservation. Trends Ecol. Evol. **33**(9), 676–688 (2018)
21. V. Komyakova, P.L. Munday, G.P. Jones. Relative importance of coral cover, habitat complexity and diversity in determining the structure of reef fish communities. PLoS One **8**(12), e83178 (2013)
22. E.C. McClure, L.E. Richardson, A. Graba-Landry, Z. Loffler, G.R. Russ, A.S. Hoey, Cross-shelf differences in the response of herbivorous fish assemblages to severe environmental disturbances. Diversity **11**(2), 23 (2019)
23. K. Osborne, A.M. Dolman, S.C. Burgess, K.A. Johns, Disturbance and the dynamics of coral cover on the great barrier reef (1995–2009). PloS One **6**(3), e17516 (2011)
24. E.E. Peterson, E. Santos-Fernández, C. Chen, S. Clifford, J. Vercelloni, A. Pearse, R. Brown, B. Christensen, A. James, K. Anthony, et al., Monitoring through many eyes: integrating scientific and crowd-sourced datasets to improve monitoring of the great barrier reef (2018). Preprint arXiv:180805298
25. L. Plaisance, M.J. Caley, R.E. Brainard, N. Knowlton, The diversity of coral reefs: what are we missing? PLoS One **6**(10), e25026 (2011)

26. M. Plummer, N. Best, K. Cowles, K. Vines, CODA: convergence diagnosis and output analysis for MCMC. R News **6**(1), 7–11 (2006)
27. D. Spiegelhalter, A. Thomas, N. Best, D. Lunn, Winbugs user manual: Version 1.4. (MRC Biostatistics Unit, Cambridge, 2003)
28. S. Sturtz, U. Ligges, A.E. Gelman, R2WinBUGS: a package for running winBUGS from R. J. Stat. Softw. **12**(3), 16 (2005)
29. H. Sweatman, C. Syms, Assessing loss of coral cover on the great barrier reef: a response to hughes et al. (2011). Coral Reefs **30**(3), 661 (2011)
30. H.H. Sweatman, A.A. Cheal, G.G. Coleman, M.M. Emslie, K.K. Johns, M.M. Jonker, I.I. Miller, K.K. Osborne, et al., Long-term monitoring of the great barrier reef, Townsville: Australian Institute of Marine Science, Status Report no 5, pp. 106 (2008)
31. H. Sweatman, S. Delean, C. Syms, Assessing loss of coral cover on australia's great barrier reef over two decades, with implications for longer-term trends. Coral Reefs **30**(2), 521–531 (2011)
32. J. Vercelloni, M.J. Caley, M. Kayal, S. Low-Choy, K. Mengersen, Understanding uncertainties in non-linear population trajectories: a Bayesian semi-parametric hierarchical approach to large-scale surveys of coral cover. PloS One **9**(11), e110968 (2014)
33. J. Vercelloni, M.J. Caley, K. Mengersen, Crown-of-thorns starfish undermine the resilience of coral populations on the great barrier reef. Glob. Ecol. Biogeogr. **26**(7), 846–853 (2017)
34. J. Vercelloni, K. Mengersen, F. Ruggeri, M.J. Caley, Improved coral population estimation reveals trends at multiple scales on australia's great barrier reef. Ecosystems **20**(7), 1337–1350 (2017)

Chapter 17
Application of Bayesian Mixture Models to Satellite Images and Estimating the Risk of Fire-Ant Incursion in the Identified Geographical Cluster

Insha Ullah and Kerrie L. Mengersen

Abstract Bayesian non-parametric mixture models have found great success in the statistical practice of identifying latent clusters in data. However, fitting such models can be computationally intensive and of less practical use when it comes to tall datasets, such as Landsat imagery. To overcome this issue, we propose to obtain multiple samples from data using stratified random sampling to enforce adequate representation in each sample from sub-populations that may exist in data. The non-parametric model is then fitted to each sample dataset independently to obtain posterior estimates. Label correspondence across multiple estimates is achieved using multivariate component densities of a chosen reference partition followed by pooling multiple posterior estimates to form a consensus posterior inference. The labels for pixels in the entire image are inferred using the conditional posterior distribution given pooled estimates, thereby substantially reducing the computational time and memory requirement.

The method is tested on Landsat images from the Brisbane region in Australia, which were compiled as a part of the national program for the eradication of the imported red fire-ant that was launched in September 2001 and which continues to the present date. The aim is to estimate the risk of fire-ant incursion in each of the identified geographical cluster so that the eradication program focuses on high risk areas.

Keywords Dirichlet process mixture models · Stratified random sampling · Satellite imagery data · Fire-ants habitat · Consensus posterior

I. Ullah (✉) · K. L. Mengersen
School of Mathematical Sciences, ARC Centre of Mathematical and Statistical Frontiers, Science and Engineering Faculty, Queensland University of Technology, Brisbane, QLD, Australia
e-mail: insha.ullah@qut.edu.au

399

K. L. Mengersen et al. (eds.), *Case Studies in Applied Bayesian Data Science*, Lecture Notes in Mathematics 2259, https://doi.org/10.1007/978-3-030-42553-1_17

17.1 Introduction

Imported red fire-ant have been a cause for concern in Brisbane, Australia. They are an invasive species and their spread could have serious social, environmental and economic impacts throughout Australia. They were first discovered in February 2001 in surrounding areas of the Port of Brisbane but are believed to have been imported a couple of decades prior to 2001. Despite the eradication program, which was launched in September 2001, spread from the initial Brisbane infestation has led to infestations around the greater Brisbane area. Isolated incursions have been found even beyond the greater Brisbane area.

In order to prioritize the use of the surveillance budget and to promote better decision making, modelling is performed to estimate the risk of fire-ant incursion in each area so that the eradication program focuses on high risk areas. As part of the surveillance program the colony locations were recorded prior to their eradication. The analysis of imagery data in combination with the location observations helps identify the preferred habitats of fire-ants [29]. However, the field data are presence-only data [9]: information on observed absences is not available and it is not reasonable to assume that areas where the pest has not been observed are absences since they are known to have very wide potential habitat. Hence supervised learning models such as logistic regression to predict occurrence probability are too arbitrary for the presence-only data and are not justifiable in this situation [11].

In light of the above, unsupervised clustering methods are more appealing for these presence-only data. These methods involves dividing the whole region into smaller clusters based on available covariate data and determining the possibility of presence in each cluster. In the context of our case study, the covariates are obtained from the satellite imagery and the presence if interest is fire-ants. However, this requires model selection that is the pre-specification of the number of clusters, K.

Dirichlet process Gaussian mixture models (DPGMMs) have been widely adopted as a data-driven cluster analysis technique. The main attraction of these models lies in sidestepping model selection by assuming that data are generated from a distribution that has a potentially infinite number of components. However, for a limited amount of data, only a finite number of components is detected and an appropriate value for the number of components has to be determined directly from data in a Bayesian manner (hence the term, 'data-driven'). These infinite, non-parametric representations allow the models to grow in size to accommodate the complexity of the data dynamically. However, they are computationally demanding and do not scale well to the satellite imagery data, each image of which is usually made up of millions of pixels. This is because they need to iterate through the full dataset at each iteration of the MCMC algorithm [see, e.g., 1]. The computational time per iteration increases with the increasing sizes of the datasets.

How to scale Bayesian mixture models up to massive data comprises a significant proportion of contemporary statistical research. One way to speed up computations is to use graphics processing units [see, e.g., 18, 30] and parallel programming approaches [see, e.g., 4, 8, 31]. Relatively less computationally demanding methods for fitting the mixture models include approximate Bayesian inference techniques such as variational inference [3, 13, 23, 25] and approximate Bayesian computation [22, 24]. Other strategies to speed up computations are the sampling based approaches. This is adopted by Huang and Gelman [14] who partition the data at random and perform MCMC independently on each subset to draw samples from the posterior given the data subset. They suggested methods based on normal approximation and importance re-sampling to make consensus posteriors. A similar idea has been proposed in [27] with a different rule for combining posterior draws. Manolopoulou et al. [21] improve inference about the parameters of the component of interest in the mixture model. An initial sub-sample is analysed to guide selection from targeted components in a sequential manner using Sequential Monte Carlo sampling. This approach depends critically on an adequate representation of the component of interest in the initial random sample. However, in a massive dataset, a low probability component of interest is likely to escape the initial random sample, which will lead to unreliable inference.

In satellite imagery, most of the data are replications. For example, all water pixels should appear similar while pixels from the land covered with the same crop should produce similar observations. Thus, inference based on a stratified random sample of the data should be representative of the whole image. This is possible in the case of supervised learning where the training data is labelled a priori. In the case of unsupervised learning one could use a computationally faster method such as k-means clustering to first label the data. These labels could then be used to obtain a stratified random sample (hence enforcing representation from each sub-population). A much more reliable inference based on a stratified random sample can be obtained using more flexible and sophisticated mixture models which allow incorporation of additional available information and also take into account the correlation between variables rather than imposing a simple model, such as k-means clustering, just because of computational problems.

In this article, we fit a Bayesian mixture model to stratified samples that have been selected from pre-clustered images. Importantly, we make use of the strengths of two clustering methods: the computationally less demanding method of k-means clustering and the more sophisticated DPGMMs, which not only account for correlations between variables, but also learn K in a data-driven fashion. Our method is explained in Sects. 17.2 and 17.3 and applied to a case study in Sects. 17.4 and 17.5. Conclusions are presented in Sect. 17.6.

17.2 Dirichlet Process Gaussian Mixture Models

Assume that we are interested in clustering real-valued observations contained in $X = (x_1, \ldots, x_n)$, where x_i is a p-dimensional sample realization made independently over n objects. Denoting the p-dimensional Gaussian density by

$\mathcal{M}(\cdot)$, a mixture of K Gaussian components takes the form

$$f(x|\theta_1, \ldots, \theta_K) = \sum_{k=1}^{K} \pi_k \mathcal{M}(x|\theta_k), \tag{17.1}$$

where $\theta_k = \{\mu_k, \Sigma_k\}$ contains the unknown mean vector μ_k and the covariance matrix Σ_k is associated with component k. The parameters $\pi = \{\pi_1, \ldots, \pi_K\}$ are the unknown mixing proportion, which satisfies $0 \leq \pi_k \leq 1$ and $\sum_{k=1}^{K} \pi_k = 1$.

In Dirichlet process Gaussian mixture models [26], the number of components K is an unknown parameter without any upper bound and inference algorithms are used to facilitate learning K from the observed data. Therefore, with every new data observation, there is a chance for the emergence of an additional component.

Define a latent indicator z_i, $i = 1, \ldots, n$, such that the prior probability of assigning a particular observation x_i to a cluster k is $p(z_i = k|\pi) = \pi_k$. Given the cluster assignment indicator z_i and the prior distribution G on the component parameters, the model in (17.1) can be expressed as:

$$x|z_i = k, \theta_k \sim \mathcal{M}(x|\theta_k),$$

$$\theta_k|G \sim G,$$

$$G|\alpha, G_0 \sim DP(\alpha, G_0),$$

where G_0 is the base distribution for the Dirichlet process prior such that $E(G) = G_0$ and α is the concentration parameter. Integrating out the infinite dimensional G from the posterior allows the application of Gibbs sampling to DPGMM [6, 7, 19]. By integrating out G, the predictive distribution for a component parameter follows a Pólya urn scheme [2]

$$\theta_k|\theta_1, \ldots, \theta_{k-1} \sim \frac{\alpha}{k-1+\alpha} G_0 + \frac{1}{k-1+\alpha} \sum_{i=1}^{k-1} \delta_{\theta_i}(\cdot).$$

Specifying a Gamma prior over the Dirichlet concentration parameter α, $\alpha \sim Ga(\eta_1, \eta_2)$, allows the drawing of posterior inference about the number of components, K.

Simpler and more efficient methods have been developed to fit the DPGMM. Consider two independent random variables $V_k \sim Beta(1, \alpha)$ and $\theta_k \sim G_0$, for $k = \{1, 2, \ldots\}$. The stick-breaking process formulation of G is such that

$$\pi_k = \begin{cases} V_k & (k = 1) \\ V_k \prod_{i=1}^{k-1}(1 - V_i) & (k > 1) \end{cases},$$

and

$$G = \sum_{k=1}^{\infty} \pi_k \delta_{\theta_k}(\cdot),$$

where $\delta_{\theta_i}(\cdot)$ is a discrete measure concentrated at θ_k [28]. In practice, however, the Dirichlet process is truncated by fixing K to a large number such that the number of active clusters remains far less than K [15]. A truncated Dirichlet process is achieved by letting $V_K = 1$, which also ensures that $\sum_{k=1}^{K} \pi_k = 1$. The base distribution G_0 is specified as a bivariate normal-inverse Wishart

$$G_0(\mu_k, \Sigma_k) = \mathcal{N}(\mu_k | \mu_0, a_0 \Sigma_k) IW(\Sigma_k | s_0, S_0),$$

where μ_0 is the prior mean, a_0 is a scaling constant to control variability of μ around μ_0, s_0 denotes the degrees of freedom and S_0 represent our prior belief about the covariances among variables. The data generating process can be described as follows:

1. For $k = 1, \ldots, K$: draw $V_k | \alpha \sim Beta(1, \alpha)$ and $\theta_k | G_0 \sim G_0$.

2. For the nth data point: draw $z_i | V_1, \ldots, V_k \sim Mult(\pi)$ and draw $x_i | z_i = k, \theta_k \sim \mathcal{N}(x | \theta_k)$

17.2.1 Blocked Gibbs Sampling Scheme to Fit DPGMM

A blocked Gibbs sampler [15] avoids marginalization over the prior G, thus allowing G to be directly involved in the Gibbs sampling scheme. The algorithm is described as follows:

1. Update z by multinomial sampling with probabilities

$$p(z_i = k | x, \pi, \theta) \propto \pi_k \mathcal{N}(x_i | \mu_k, \Sigma_k)$$

2. Update the stick breaking variable V by independently sampling from a beta distribution

$$p(V | x) \sim Beta\left(1 + n_k, \alpha + \sum_{i=k+1}^{K} n_i\right),$$

where $V_k = 1$ and n_k is the number of observations in component k. Obtain π by setting $\pi_1 = V_1$ and $\pi_k = V_k \prod_{i=1}^{k-1}(1 - V_i)$ for $k > 1$.

3. Update α by sampling independently from

$$p(\alpha|V) \sim Ga\left(\eta_1 + K - 1, \eta_2 - \sum_{i=1}^{K-1} \log(1 - V_i)\right),$$

4. Update Σ_k by sampling from

$$p(\Sigma_k|x, z) \sim IW(\Sigma_k|s_k, S_k),$$

where

$$s_k = s_0 + n_k,$$

$$S_k = S_0 + \sum_{z_i=k} (x_i - \bar{x}_k)(x_i - \bar{x}_k)^t + \frac{n_k}{1 + n_k a_0}(\bar{x}_k - \mu_0)(\bar{x}_k - \mu_0)^t$$

and

$$\bar{x}_k = \frac{1}{n_k} \sum_{z_i=k} x_i.$$

5. Update μ_k by sampling from

$$p(\mu_k|x, z, \Sigma_k) \sim \mathcal{N}(\mu_k|m_k, a_k \Sigma_k),$$

where

$$m_k = \frac{a_0 \mu_0 + n_k \bar{x}_k}{a_0 + n_k}$$

and

$$a_k = \frac{a_0}{1 + a_0 n_k}.$$

17.3   The Method

As noted earlier, in satellite imagery, many of the pixels are exact replicates or at least provide similar information (different up to a level of noise). To reduce computational time and memory storage requirements, it is sufficient to obtain an adequate representation from each group of similar observations rather than analysing data that include tens of thousands of duplicate copies of observations. The full dataset can be mapped onto the cluster obtained based on sample data.

Many authors have resorted to this option; for example, [17] used a 0.1% sample of 500 million documents and extended results to cluster the rest of the documents. However, if a sample is selected at random, it is likely that some smaller clusters of interest are not sampled. This eventually will produce results that are biased towards a small number of larger clusters, which may in turn lead to lower quality clusters [5].

We use a similar sampling based strategy but select a sample of size n in a way that potentially ensures representation from very small clusters that may exist in the data. This is made possible by first arbitrarily clustering the N pixels of the whole image into a large number, say C, of smaller clusters (C is much larger than the actual number of clusters one can expect in the whole image) using computationally faster k-means clustering [20], which is a popular clustering algorithm because of its scalability and efficiency in large data sets [16]. The pre-clustered image is then sampled using stratified sampling with proportional allocation; that is, a sample of size $n_i = n(N_i/N)$ is chosen from the ith cluster, where $i = 1, \ldots, C$ and N_i denotes the number of pixels in ith cluster. Note that the total sample size, $n = n_1 + \ldots + n_C$, should be large enough to contain a reasonable number of observations from the smallest cluster obtained via k-means clustering. Another way to ensure adequate representation from the smallest cluster is, for example, by increasing each n_i by the size of smallest cluster, say n_s, that is $n_i = n(N_i/N) + n_s$ or by a fraction of n_s if n_s is large. The sample of size n thus obtained is clustered using DPGMM.

To control for sampling variation we obtain M samples each of size n using the above process and apply DPGMM independently to each sample. The label correspondence across mixture components from the multiple samples is created using multivariate component densities of a chosen reference partition and the mean vectors from the rest of M partitions as data. This is followed by pooling posterior estimates based on multiple samples to form consensus posterior estimates. Denote the mth stratified random sample obtained from X by $X_{(m)}$, $m = 1, \ldots, M$, the respective sample-data posterior by $p(\theta|X_{(m)})$ and the sample-posterior estimate of the kth component parameter by $\hat{\theta}_{k(m)} = \{\hat{\mu}_{k(m)}, \widehat{\Sigma}_{k(m)}\}$. Then the pooled estimates of the parameters of the kth component are obtained using the following identities [14]:

$$\hat{\mu}_k = \widehat{\Sigma}_k \left(\sum_{m=1}^{M} \widehat{\Sigma}_{k(m)}^{-1} \hat{\mu}_{k(m)} \right)$$

and

$$\widehat{\Sigma}_k = \left(\sum_{m=1}^{M} \widehat{\Sigma}_{k(m)}^{-1} \right)^{-1}.$$

The labels for the N pixels in the entire image are inferred using the conditional posterior distribution given the pooled estimates.

17.4 The Data

Since the launch of the fire-ant eradication program in September 2001, data have been collected on the location of each colony that has been found. The dataset used in this case study comprises 17,717 locations where nests of fire-ants were identified during the years 2001–2013. These locations are indicated on a Google image snapshot provided in Fig. 17.1. The proportion of colonies identified for each year are provided in Fig. 17.2. A sudden rise in the number of identified nests during 2009–2010 and then a drop back to normal in the following years is surprising. There may be a number of factors responsible for this phenomenon, such as flooding events, changes in surveillance processes or major developmental projects, but definitive reasons for it still require further investigation.

A Landsat image is also available for each year of the study. These were acquired on days of low cloud coverage, generally in the period between May and September, most commonly in July. These images were chosen as being typical winter images, and sufficiently near to the date required to be included in the winter planning period for summer surveillance. The images were converted into workable data files using the 'raster' package [12] in R. Note that we use 6 Landsat spectral bands (variables): visible blue, visible green, visible red, near infrared, middle infrared, and thermal

Fig. 17.1 Google image snapshot of the study area and the observed location of fire-ant colonies (indicated by red dots) over the study period 2001–2013

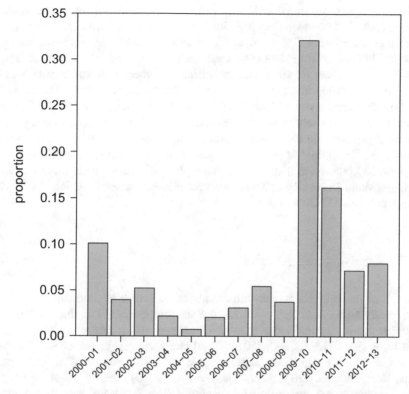

Fig. 17.2 Proportions of fire-ant colonies detected each year from 2000–2013

infrared. The Landsat variables were centred at mean zero and scaled to a unit variance.

We also used R for the substantive statistical analysis. To solve the k-means problem, we used the algorithm in [10], which is a default option in the R function *kmeans()*, available from the 'stats' package. Since it is recommended to make repeated runs with different random starting points and choose the run that gives the minimum within-class variance, we used 8 random starting points in our analysis. Note that the function *kmeans()* also allows to specify multiple random starting points. A larger number of starting points, however, increases computation cost, particularly when the number of clusters is larger, which is due to multiple runs of the algorithm. We avoided this by using the parallel processing facility in R provided by *foreach* loop from the 'foreach' package. Since the k-means clustering is intended to include small strata in order to acquire a representative sample (rather than final clustering), we did not find noticeable differences in terms of visual interpretation when used a single random starting point.

To fit a DPGMM, we translated Matlab code, available at http://ftp.stat.duke. edu/WorkingPapers/09-26.html, into R code [for details about Matlab codes, see, 21]. Due to having no formal convergence guarantee, we did experiments with different images (considered in this study) to decide on the total number of iterations of the blocked Gibbs sampling algorithm including the burn-in iterations. In our experiments we found that the algorithm provided visually interpretable solutions after 1000 iterations (our final results can be visualized and checked with Google maps) and we did not see any noticeable difference when used with a larger number of iterations (30,000 iterations excluding 5000 burn-in iterations). Therefore, we used 10,000 iterations of a blocked Gibbs sampler excluding the first 2000 burn-in iterations in all the analyses whose results are shown here. The overall computation time averaged over the 13 images considered in this study was 8 h and 11 min when we set $n = 100,000$ and $M = 10$. This computation time increased to 14 h and 5 min for $n = 200,000$. Note that we used the high performance computing facility at the Queensland University of Technology for our computations which has 2.6 GHz processors with 251 Gb memory.

17.5   Analysis and Results

The aim of the analyses was to find out about the potential characteristics of fire-ants' preferred habitats by classifying the satellite images. The images were first clustered arbitrarily into large number of clusters using k-means clustering. We tried $C = 50, 100, 150, 200, 250, 300$ and show the results for $C = 100$, since we did not notice significant improvement for larger values of C in terms of visual interpretation. Ten stratified samples ($M = 10$) each of size $n = 100,000$ were selected using proportional allocations. The DPGMM was then fitted to each sample independently in parallel and the pooled estimates were obtained by combining the posterior estimates across the multiple samples. The results based on different samples were very consistent apart from the labels correspondence issue. For example, the component-1 represented water in the partitioning based on the sample-1 but it represented forest areas in the partitioning based on the sample-2. We dealt with this problem by using density of a chosen reference partition (the one that gave the maximum number of components) and considering the mean vectors from the rest of the partitions as data. In this way, the water component in all partitions had a high probability to correspond with the water component in the reference partition; therefore, we re-labelled them the same across different partitions. In our analysis we used $M = 10$ because of the availability of high performance computing facility. However, we did not notice any visually interpretable change when we used a smaller value of $M = 5$ on a personal 8-core Intel platform with processor speed 3.4 GHz and 16 GB of memory. The whole process for this experiment took 5 h with 5000 iterations of a blocked Gibbs sampler excluding 1000 burn-in iterations. The reference partition was chosen, among M partitions, as the one with the largest number of components. The labels for the whole image are inferred using obtained posterior distribution given the pooled estimates. This process was performed independently for each image from the year 2001 to 2013. We tested a range of values of n (between 10,000 and 300,000, inclusive) and found that the number of components and their structure did not change (in terms of visual interpretation) as we increased the value of n beyond 100,000. Therefore, we set $n = 100,000$ for all the results shown here.

The classification based on the images from years 2003 and 2010 are shown, respectively, in Figs. 17.3 and 17.4. The proportion of observed fire-ants identified in each cluster are presented in Tables 17.1 and 17.2. Note that each of these tables is based on a single year image; however, the proportions of the observed fire-ants for the rest of the study period that falls in a particular class are also provided for prediction purpose. The figures for other years and their respective tables are diverted to the supplementary material due to the compatibility of the results across different years.

The final number of components per image varied across different years but stayed below 36. The variation in the number of components was mainly due to a number of very small clusters that each contained less than 1% of the total pixels. However, the number of components that consisted of more than 1% of the pixels were quite consistent across different years and remained around 20. Some of the variation in the number of clusters across different years could possibly be attributed to the time of the day the image was acquired. For example, the mountainous area was broken into a various number of components in images from different years possibly because of shadows (see components 10 and 12 in Fig. 17.3 and components 10, 13 and 18 in Fig. 17.4). In the image from 2001, clouds over the mountains were well separated (image not shown here). Other variations are because of the changes in the landscape over time. For example, Wyaralong Dam cannot be seen in Fig. 17.3 but can be seen in Fig. 17.4 since it was built in 2009–2010.

The large components were materially similar across different years and were visually interpretable into different land cover classes, namely, hills, forest, water, residential areas, warehouses, roads, parks and play grounds, plain areas with natural non-forest vegetation (scrub-land) and some impervious surfaces, and new development sites or land with recent deforestation. Other smaller clusters (each consisting of less than 1% of the pixels and visually not interpretable) are found to be of less interest and are therefore merged together in the figures.

The water component in the image was always well separated from the rest of the components and was often partitioned into shallow and deep water (see components 6, 18 and 20 in Fig. 17.3 and components 12 and 14 in Fig. 17.4). Although this component is not of interest to us, it helps in identifying and interpreting other components. The parks and playgrounds were found to be consistently at risk of infestation over time (see components 17 and 9, respectively in Tables 17.1 and 17.2). The components that represent the scrub land with thinner forest and the land with natural vegetation are generally the largest by area and are found to be consistently at risk of fire-ant incursion (see components 1 and 2 in Table 17.1; components 1, 3, and 4 in Table 17.2); in particular, the incursion in component 3 of Table 17.2 has increased over time.

The residential area (see component 3 in Table 17.1 and components 7 and 19 in Table 17.2) including the areas with commercial buildings (see component 21 in Table 17.1 and component 15 in Table 17.2) were found to be at high risk in the

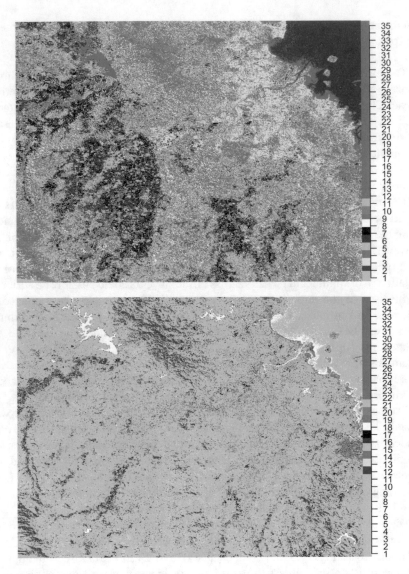

Fig. 17.3 Cluster analysis of satellite image of the Brisbane area taken in 2004. For clarity, some of the clusters are merged together, in dark-orange (top) and gray colours (bottom), and the results are presented in two plots: (top panel) 1: scrub-land with thinner forest, 2: scrub-land with natural vegetation, 3: residential area 4: Dense forest, 5: mountainous areas with scarce forest, 6: water, 7: mix of impervious surfaces and scrub-land, 8: mountainous areas, 9: mountainous areas, 10: hills and forest, 11: mountainous areas with scars forest; (bottom panel) 12: hilly areas, 13: hard for visual interpretation, 14: residential area, 15: hard for visual interpretation, 16: fields, 17: mix of parks, playgrounds and grassland, 18: shallow water, 19: fields with crops, 20: seashore and shallow water, and 21: commercial buildings. Cluster 22 to cluster 35 are too small to be visually interpreted

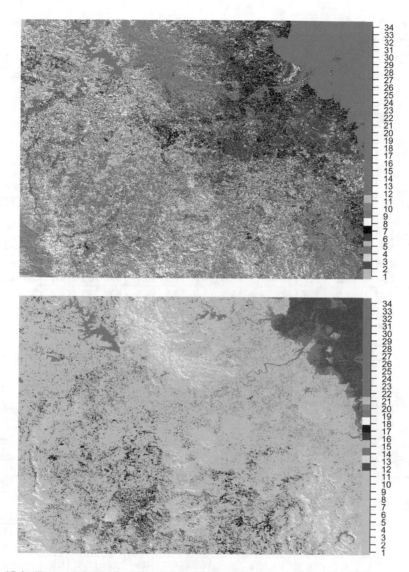

Fig. 17.4 Cluster analysis of satellite image of the Brisbane area taken in 2011. For clarity, some of the clusters are merged together, in dark-orange (top) and gray colours (bottom), and the results are presented in two plots: (top panel) 1: scrub-land with thinner forest, 2: dense forest, 3: impervious surfaces and scrub-land, 4: scrub-land with natural vegetation, 5: mountainous areas with thinner forest, 6: forest, 7: residential area, 8: mountainous areas scarce forest, 9: parks and playgrounds, 10: hills with dense forest, 11: mountainous areas with scarce forest; (bottom panel) 12: deep water, 13: hilly areas with dense forest, 14: shallow water, 15: mix of commercial buildings and fields without crops, 16: hard for visual interpretation, 17: hard for visual interpretation, 18: hills, 19: residential areas, 20: hard for visual interpretation, and 21: fields with crops. Cluster 22 to cluster 34 are too small to be visually interpreted

Table 17.1 The percentages of fire-ant colonies identified in each of the spatial components (shown in Fig. 17.3) over the period of 13 years conditional on the image acquired in 2004 (highlighted in italics)

C.No	C.Size	2001	2002	2003	*2004*	2005	2006	2007	2008	2009	2010	2011	2012	2013
1	17.48	13.9	11.2	16.2	*16.7*	20.6	16.8	5.5	9.5	10.2	11.8	10.8	18.5	19.9
2	9.77	1.1	1.9	3.3	*10.7*	3.1	0.5	4.7	9.8	11.5	36.1	4.3	10.7	8.7
3	8.75	53.1	47.7	50.8	*23.4*	53.3	56.4	35.9	33.8	26.3	9.6	30.2	18.6	14.4
4	8.06	1.0	1.9	1.1	*1.0*	0.8	1.6	0.7	1.2	3.9	2.2	6.2	12.9	13.0
5	6.02	1.7	1.3	1.4	*5.7*	1.5	0.8	0.7	2.0	4.0	5.4	2.6	2.8	3.9
6	5.58	0.0	0.0	0.0	*0.0*	0.0	0.0	0.0	0.0	0.0	0.0	0.0	0.0	0.0
7	5.49	1.3	2.1	2.9	*27.1*	0.0	0.3	14.4	18.6	19.6	21.4	12.8	11.2	7.4
8	5.37	0.0	0.3	0.5	*0.0*	0.0	0.0	0.0	0.0	0.2	0.2	0.3	1.4	1.5
9	4.65	0.1	0.0	0.3	*0.3*	0.8	0.0	0.0	0.0	0.2	0.0	0.1	0.5	1.3
10	4.16	0.0	0.0	0.0	*0.8*	0.0	0.0	0.0	0.0	0.0	0.0	0.0	0.1	0.4
11	4.04	0.3	0.6	1.0	*2.7*	0.0	0.0	0.0	0.2	0.8	0.3	0.2	0.6	1.1
12	3.78	0.0	0.0	0.0	*0.0*	0.0	0.0	0.0	0.0	0.0	0.0	0.0	0.2	0.0
13	2.94	0.2	1.0	1.1	*0.3*	0.0	0.0	0.0	0.0	0.2	0.5	1.0	1.4	1.4
14	2.37	9.9	10.7	10.1	*4.4*	5.4	5.9	3.1	3.4	6.5	4.1	4.2	4.3	6.9
15	2.01	0.0	0.0	0.1	*0.5*	0.8	0.0	0.0	0.0	0.2	0.6	0.3	0.9	0.9
16	1.99	1.1	3.6	1.5	*1.0*	0.8	7.6	25.9	15.0	6.6	5.0	22.3	8.7	12.0
17	1.86	0.7	2.7	2.0	*1.6*	3.1	2.2	0.5	0.9	1.7	0.5	1.0	2.2	2.0
18	1.78	0.0	0.0	0.0	*0.0*	0.0	0.0	0.0	0.0	0.0	0.0	0.0	0.0	0.0
19	1.09	0.7	0.3	0.6	*0.6*	2.3	2.2	0.5	1.0	4.5	0.2	0.7	1.5	0.9
20	0.64	0.0	0.0	0.1	*0.0*	0.0	0.0	0.0	0.3	0.0	0.0	0.0	0.0	0.2
21	0.61	9.5	10.9	3.7	*2.1*	4.6	1.6	1.9	1.0	1.1	1.5	1.7	1.3	1.8
22	0.48	4.6	3.1	2.3	*0.5*	0.0	2.2	3.8	2.5	0.6	0.1	0.1	0.9	0.6
23	0.36	0.1	0.3	0.3	*0.0*	3.1	1.5	1.8	0.3	1.6	0.1	0.7	1.2	1.0
24	0.34	0.2	0.1	0.3	*0.0*	0.0	0.0	0.2	0.1	0.0	0.1	0.1	0.0	0.1
25	0.23	0.0	0.0	0.0	*0.0*	0.0	0.0	0.0	0.0	0.5	0.0	0.0	0.2	0.2
26	0.06	0.0	0.0	0.0	*0.5*	0.0	0.3	0.2	0.0	0.0	0.0	0.0	0.0	0.0
27	0.03	0.3	0.3	0.0	*0.0*	0.0	0.0	0.2	0.0	0.0	0.1	0.0	0.0	0.1
28	0.03	0.2	0.1	0.2	*0.0*	0.0	0.0	0.0	0.0	0.0	0.0	0.0	0.0	0.0
29	0.02	0.0	0.0	0.0	*0.0*	0.0	0.0	0.0	0.0	0.0	0.0	0.0	0.0	0.0
30	0.01	0.0	0.0	0.0	*0.0*	0.0	0.0	0.0	0.0	0.0	0.2	0.0	0.0	0.0
31	0.01	0.0	0.0	0.0	*0.3*	0.0	0.0	0.0	0.3	0.0	0.0	0.0	0.0	0.1
32	0.00	0.1	0.0	0.2	*0.0*	0.0	0.0	0.0	0.0	0.0	0.0	0.0	0.0	0.0
33	0.00	0.0	0.0	0.0	*0.0*	0.0	0.0	0.0	0.0	0.0	0.0	0.0	0.0	0.0
34	0.00	0.0	0.0	0.0	*0.0*	0.0	0.0	0.0	0.0	0.0	0.0	0.0	0.0	0.0
35	0.00	0.0	0.0	0.0	*0.0*	0.0	0. 0	0.0	0.0	0.0	0.0	0.0	0.0	0.0
Total incursions		1788	701	928	*387*	130	365	547	965	664	5690	2866	1272	1414

The C.No indicates component numbers corresponding to the component numbers in Fig. 17.3. The C.Size (in %) indicates the size of a cluster relative to image. The clusters are sorted in descending order with respect to their sizes

Table 17.2 The percentages of fire-ant colonies identified in each of the spatial components (shown in Fig. 17.4) over the period of 13 years conditional on the image acquired in 2011 (highlighted in italics)

C.No	C.Size	2001	2002	2003	2004	2005	2006	2007	2008	2009	2010	*2011*	2012	2013
1	12.05	2.5	3.9	6.0	14.8	6.9	8.9	3.6	3.8	2.4	14.7	*3.2*	1.8	4.5
2	9.59	1.4	1.6	2.4	4.5	3.1	1.4	0.0	0.2	0.0	0.1	*0.1*	0.3	0.5
3	7.93	9.0	10.2	9.0	30.1	21.3	7.4	27.9	24.2	40.1	24.2	*49.2*	29.8	30.9
4	7.40	0.2	0.4	0.1	3.8	0.0	0.0	4.7	0.5	4.1	27.7	*11.3*	0.9	3.9
5	6.01	0.1	0.3	0.8	0.8	0.0	0.0	0.0	0.0	0.0	0.0	*0.0*	0.1	0.8
6	5.28	10.3	9.8	11.0	4.8	7.7	7.9	1.9	5.6	7.0	2.3	*2.0*	1.2	2.5
7	5.27	27.6	27.4	28.1	12.7	13.9	23.4	13.6	16.6	14.2	5.0	*10.1*	23.6	20.5
8	4.85	0.2	0.3	0.1	0.8	0.0	2.1	1.6	0.4	0.2	0.5	*0.5*	0.6	0.7
9	4.79	6.7	7.4	9.3	7.2	15.3	23.1	3.2	4.3	13.5	5.3	*5.1*	6.3	6.3
10	4.72	0.0	0.0	0.1	0.0	0.0	0.0	0.0	0.0	0.0	0.0	*0.0*	0.0	0.0
11	4.36	0.1	0.1	0.1	0.3	0.0	0.0	0.2	0.2	0.0	0.0	*0.0*	0.1	0.1
12	4.10	0.0	0.0	0.0	0.0	0.0	0.0	0.0	0.0	0.0	0.0	*0.0*	0.0	0.0
13	3.65	0.0	0.0	0.2	0.0	0.0	0.0	0.0	0.0	0.0	0.0	*0.0*	0.0	0.0
14	3.51	0.0	0.0	0.0	0.0	0.0	0.0	0.0	0.0	0.0	0.0	*0.0*	0.0	0.0
15	3.38	31.6	29.6	21.0	10.5	14.3	18.7	27.2	31.5	6.2	3.1	*6.5*	9.2	9.4
16	3.12	0.2	0.0	0.4	1.3	1.5	0.0	0.7	0.0	0.2	1.6	*0.9*	0.1	0.6
17	2.81	0.1	1.1	0.2	4.5	0.0	0.0	4.5	1.5	3.8	11.1	*2.2*	3.4	1.5
18	1.66	0.0	0.0	0.0	0.0	0.0	0.0	0.0	0.0	0.0	0.0	*0.0*	0.0	0.0
19	1.43	8.7	5.9	9.6	1.8	7.7	5.7	1.9	1.4	3.2	0.9	*2.2*	2.2	1.7
20	1.25	0.1	0.1	0.2	0.0	0.8	0.8	0.4	0.6	1.5	2.3	*0.9*	0.2	0.6
21	0.99	0.3	0.4	0.1	0.8	0.0	0.0	0.0	0.7	0.5	0.1	*0.5*	0.7	0.9
22	0.48	0.2	0.0	0.1	0.0	0.0	0.0	0.0	1.0	0.0	0.1	*0.0*	0.0	0.0
23	0.44	0.0	0.0	0.0	0.0	0.0	0.0	0.0	0.0	0.0	0.0	*0.0*	0.0	0.0
24	0.27	0.2	0.6	0.4	1.0	3.8	0.0	6.6	2.4	2.4	0.4	*3.5*	10.9	9.9
25	0.24	0.0	0.0	0.0	0.0	0.0	0.0	0.0	0.0	0.0	0.0	*0.0*	0.0	0.0
26	0.24	0.2	0.6	0.3	0.0	3.7	0.5	2.2	4.8	0.9	0.2	*1.6*	8.5	4.6
27	0.12	0.1	0.0	0.0	0.0	0.0	0.0	0.0	0.0	0.0	0.0	*0.0*	0.0	0.0
28	0.04	0.5	0.4	0.3	0.3	0.0	0.0	0.0	0.0	0.0	0.0	*0.1*	0.0	0.0
29	0.01	0.0	0.0	0.0	0.0	0.0	0.0	0.0	0.3	0.0	0.2	*0.1*	0.0	0.1
30	0.01	0.0	0.0	0.0	0.0	0.0	0.0	0.0	0.0	0.0	0.0	*0.0*	0.0	0.0
31	0.00	0.0	0.0	0.0	0.0	0.0	0.0	0.0	0.0	0.0	0.0	*0.0*	0.0	0.0
32	0.00	0.0	0.0	0.0	0.0	0.0	0.0	0.0	0.0	0.0	0.0	*0.0*	0.1	0.0
33	0.00	0.0	0.0	0.0	0.0	0.0	0.0	0.0	0.0	0.0	0.0	*0.0*	0.0	0.0
34	0.00	0.0	0.0	0.0	0.0	0.0	0.0	0.0	0.0	0.0	0.0	*0.0*	0.0	0.0
Total incursions	1788	701	928	387	130	365	547	965	664	5690	2866	*1272*	1414	

The C.No indicates component numbers corresponding to the component numbers in Fig. 17.4. The C.Size (in %) indicates the size of a cluster relative to image. The clusters are sorted in descending order with respect to their sizes initial years when the eradication program started. However, the risk of incursion declined soon after the launch of eradication program in this class, which probably shows that the eradication program has been more effective in the residential areas. A potential reason could be swift reporting once the incursion has been observed. The risk of incursion increased in the components that represent agricultural fields and in the components that represents impervious surfaces and scrub-land (see, respectively, component 16 and 7 in Table 17.1).

The components with forest areas were found to be consistently at low risk of fire-ant incursion (see component 4 in Table 17.1 and component 2 Table 17.2). Similarly, the mountainous areas were also found to be at low risk (see components 10 and 12 in Table 17.1 and components 10, 13 and 18 in Table 17.2).

As mentioned above, Tables 17.1 and 17.2 also present the proportions of fire-ant nests observed in the years other than the one in which the analysed image was acquired. In general, the classes with high proportions of fire-ant nests in the image year calibrate well with the proportions in a few years that follow. For example, in Table 17.1 areas in component 1 were at risk of fire-ant incursions in 2003 (contained 16.2% of the observed nests) remained at similar risk in the following year (component 1 contained 16.7% of the observed nests in 2004). The risk of infestation in component 3 of Table 17.1 is consistent for a few years following 2003 (2004 is an exception that contained 23.4% of the observed nests) with a gradual decreasing trend in the later years. Similarly, in Table 17.2, which is based on classification of image from 2011, component 3 was found to be at highest risk in 2011 (contained 49.2% of the observed nests) and remained at high risk in the following 2 years (contained 29.8% in 2012 and 30.9% in 2013). The risk of incursion in component 7 was almost doubled in the following years (contained 10.1% of the observed nests in 2011, 23.6% of the observed nests in 2012 and 20.5% of the observed nests in 2013). The component 15 contained 6.5% in 2011 and 9.2% in 2012. Some of the potential factors for anomalous changes could possibly be attributed climatic events such as floods or drought.

The above results indicate that image classification provides useful information for operational projects. The classification can be produced routinely at a low cost, which when combined with the observed data helps in learning about the high risk areas. These high risk areas could be prioritized in order to satisfy budgetary constraints. For example, as mentioned above the infestation of fire-ants has declined in residential areas over a period of 13 years and probably reflects the success of the eradication program but has increased in other components such as scrub-land and agricultural fields that needs to be prioritized in future.

17.6 Conclusions and Recommendations

DPGMM are computationally prohibitive for tall datasets such as satellite imagery data. We used computationally faster k-means clustering to pre-cluster the data into a large number of clusters and obtain stratified samples of suitable sizes to ensure representation from very small clusters. These samples are partitioned using DPGMM and the posterior estimates are pooled across multiple samples. The labels for all the pixels in the image are predicted using the posterior distribution given the pooled estimates of components parameters. The proposed method enables classification of a dataset with millions of observations in a matter of hours and minutes.

We clustered satellite images to identify the land cover classes that are at high, medium, and low risk of infestation of fire-ants. Residential areas were found to be at a high risk of infestation in the initial years of the eradication program that started in 2001. However, the risk of incursion in the residential area declined within a few years after the start of eradication program. The scrub-land with natural vegetation and the classes that represent agricultural fields have seen high incursions in the later half of the study period. Parks, playgrounds and some impervious surfaces were also found to be at risk of infestation.

The overall analyses show that clustering satellite images could be very useful to make rational decisions about where the eradication program needs to focus next. For example, the eradication program is found to be successful in the residential areas perhaps due to prompt response from the residents and businesses. However, as mentioned earlier, other clusters such as scrub-land with natural vegetation, agricultural fields, parks, playgrounds and roads have seen high incursions in the later years of the study periods. Since having fire-ants at home or at the commercial places are more threatening as compared to having encountered them at the park or on a road, people are more likely to report them when they are posing a threat to their personal comfort. This makes it important to create awareness among the public about these high risk areas, in order to better support the collective effort to detect and manage this pest.

Note that this is was an initial exploratory study and has some limitations pertaining to it. First, the information from both the images and fire-ant incursions data have spatio-temporal structures, we fitted a separate model for each year and calculated the proportions of presence-only data observed in that year in the clusters found by the model. A more principled way would be to embed the presence-only data in the fitted model. This would require a hierarchical model that in one level performs the clustering based on the spectral bands and in the other level uses the clusters as predictors in a model for the presence-only data. One need to account for spatial dependence in such model, which could potentially play an important role in the problem being tackled. Second, we did not account for the temporal effects in our model and calculated the proportions of the observed presence-only data for other years assuming no significant temporal changes in the land-cover over a period of few years. A more sophisticated model that take into account the temporal variation in the land cover would be required. We leave these extensions for future research.

Acknowledgments This research was supported by an ARC Australian Laureate Fellowship for project, Bayesian Learning for Decision Making in the Big Data Era under Grant no. FL150100150. The authors also acknowledge the support of the Australian Research Council Centre of Excellence for Mathematical and Statistical Frontiers (ACEMS) and the support of QUT's high-performance computing and Research Support (HPC) group.

References

1. R. Bardenet, A. Doucet, C. Holmes, On markov chain monte carlo methods for tall data. J. Mach. Learn. Res. **18**(1), 1515–1557 (2017)
2. D. Blackwell, J.B. MacQueen, Ferguson distributions via pólya urn schemes. Ann. Stat. **1**, 353–355 (1973)
3. D.M. Blei, A. Kucukelbir, J.D. McAuliffe, Variational inference: a review for statisticians. J. Am. Stat. Assoc. **112**(518), 859–877 (2017)
4. J. Chang, J.W. Fisher III, Parallel sampling of DP mixture models using sub-cluster splits, in *Advances in Neural Information Processing Systems* (2013), pp. 620–628
5. C.M. De Vries, L. De Vine, S. Geva, R. Nayak, Parallel streaming signature em-tree: a clustering algorithm for web scale applications, in *Proceedings of the 24th International Conference on World Wide Web, International World Wide Web Conferences Steering Committee* (2015), pp. 216–226
6. M.D. Escobar, Estimating normal means with a dirichlet process prior. J. Am. Stat. Assoc. **89**(425), 268–277 (1994)
7. M.D. Escobar, M. West, Bayesian density estimation and inference using mixtures. J. Am. Stat. Assoc. **90**(430), 577–588 (1995)
8. S. Guha, R. Hafen, J. Rounds, J. Xia, J. Li, B. Xi, W.S. Cleveland, Large complex data: divide and recombine (D&R) with RHIPE. Stat **1**(1), 53–67 (2012)
9. G. Guillera-Arroita, J.J. Lahoz-Monfort, J. Elith, A. Gordon, H. Kujala, P.E. Lentini, M.A. McCarthy, R. Tingley, B.A. Wintle, Is my species distribution model fit for purpose? Matching data and models to applications. Glob. Ecol. Biogeogr. **24**(3), 276–292 (2015)
10. J.A. Hartigan, M.A. Wong, Algorithm as 136: a k-means clustering algorithm. J. R. Stat. Soc. C **28**(1), 100–108 (1979)
11. T. Hastie, W. Fithian, Inference from presence-only data; the ongoing controversy. Ecography **36**(8), 864–867 (2013)
12. R.J. Hijmans, J. van Etten, J. Cheng, M. Mattiuzzi, M. Sumner, J.A. Greenberg, O.P. Lamigueiro, A. Bevan, E.B. Racine, A. Shortridge, et al., Package 'raster'. R package (2016). https://cranr-projectorg/web/packages/raster/indexhtml. Accessed October 1, 2016
13. M.D. Hoffman, D.M. Blei, C. Wang, J. Paisley, Stochastic variational inference. J. Mach. Learn. Res. **14**(1), 1303–1347 (2013)
14. Z. Huang, A. Gelman, Sampling for Bayesian computation with large datasets. Technical Report (2005)
15. H. Ishwaran, L.F. James, Approximate dirichlet process computing in finite normal mixtures: smoothing and prior information. J. Comput. Graph. Stat. **11**(3), 508–532 (2002)
16. A.K. Jain, Data clustering: 50 years beyond k-means. Pattern Recognit. Lett. **31**(8), 651–666 (2010)
17. A. Kulkarni, J. Callan, Document allocation policies for selective searching of distributed indexes, in *Proceedings of the 19th ACM International Conference on Information and Knowledge Management* (ACM, New York, 2010), pp. 449–458
18. A. Lee, C. Yau, M.B. Giles, A. Doucet, C.C. Holmes, On the utility of graphics cards to perform massively parallel simulation of advanced monte carlo methods. J. Comput. Graph. Stat. **19**(4), 769–789 (2010)

19. S.N. MacEachern, Estimating normal means with a conjugate style dirichlet process prior. Commun. Stat. Simul. Comput. **23**(3), 727–741 (1994)
20. J. MacQueen, et al., Some methods for classification and analysis of multivariate observations, in *Proceedings of the Fifth Berkeley Symposium on Mathematical Statistics And Probability, Oakland*, vol. 1 (1967), pp. 281–297
21. I. Manolopoulou, C. Chan, M. West, Selection sampling from large data sets for targeted inference in mixture modeling. Bayesian Anal. **5**(3), 1 (2010)
22. J.M. Marin, P. Pudlo, C.P. Robert, R.J. Ryder, Approximate Bayesian computational methods. Stat. Comput. **22**(6), 1167–1180 (2012)
23. C.A. McGrory, D. Titterington, Variational approximations in Bayesian model selection for finite mixture distributions. Comput. Stat. Data Anal. **51**(11), 5352–5367 (2007)
24. M.T. Moores, C.C. Drovandi, K. Mengersen, C.P. Robert, Pre-processing for approximate Bayesian computation in image analysis. Stat. Comput. **25**(1), 23–33 (2015)
25. J.T. Ormerod, M.P. Wand, Explaining variational approximations. Am. Stat. **64**(2), 140–153 (2010)
26. C.E. Rasmussen, The infinite gaussian mixture model, in *Advances in Neural Information Processing Systems* (MIT Press, Cambridge, 2000), pp. 554–560
27. S.L. Scott, A.W. Blocker, F.V. Bonassi, H.A. Chipman, E.I. George, R.E. McCulloch, Bayes and big data: the consensus Monte Carlo algorithm. Int. J. Manage. Sci. Eng. Manage. **11**(2), 78–88 (2016)
28. J. Sethuraman, A constructive definition of dirichlet priors, in *Statistica Sinica* (1994), pp. 639–650
29. D. Spring, O.J. Cacho, Estimating eradication probabilities and trade-offs for decision analysis in invasive species eradication programs. Biol. Invasions **17**(1), 191–204 (2015)
30. M.A. Suchard, Q. Wang, C. Chan, J. Frelinger, A. Cron , M. West, Understanding GPU programming for statistical computation: studies in massively parallel massive mixtures. J. Comput. Graph. Stat. **19**(2), 419–438 (2010)
31. S. Williamson, A. Dubey, E.P. Xing, Parallel Markov chain Monte Carlo for nonparametric mixture models, in *Proceedings of the 30th International Conference on Machine Learning (ICML-13)*, (2013), pp. 98–106

LECTURE NOTES IN MATHEMATICS

Editors in Chief: J.-M. Morel, B. Teissier;

Editorial Policy

1. Lecture Notes aim to report new developments in all areas of mathematics and their applications – quickly, informally and at a high level. Mathematical texts analysing new developments in modelling and numerical simulation are welcome.

   Manuscripts should be reasonably self-contained and rounded off. Thus they may, and often will, present not only results of the author but also related work by other people. They may be based on specialised lecture courses. Furthermore, the manuscripts should provide sufficient motivation, examples and applications. This clearly distinguishes Lecture Notes from journal articles or technical reports which normally are very concise. Articles intended for a journal but too long to be accepted by most journals, usually do not have this "lecture notes" character. For similar reasons it is unusual for doctoral theses to be accepted for the Lecture Notes series, though habilitation theses may be appropriate.

2. Besides monographs, multi-author manuscripts resulting from SUMMER SCHOOLS or similar INTENSIVE COURSES are welcome, provided their objective was held to present an active mathematical topic to an audience at the beginning or intermediate graduate level (a list of participants should be provided).

   The resulting manuscript should not be just a collection of course notes, but should require advance planning and coordination among the main lecturers. The subject matter should dictate the structure of the book. This structure should be motivated and explained in a scientific introduction, and the notation, references, index and formulation of results should be, if possible, unified by the editors. Each contribution should have an abstract and an introduction referring to the other contributions. In other words, more preparatory work must go into a multi-authored volume than simply assembling a disparate collection of papers, communicated at the event.

3. Manuscripts should be submitted either online at www.editorialmanager.com/lnm to Springer's mathematics editorial in Heidelberg, or electronically to one of the series editors. Authors should be aware that incomplete or insufficiently close-to-final manuscripts almost always result in longer refereeing times and nevertheless unclear referees' recommendations, making further refereeing of a final draft necessary. The strict minimum amount of material that will be considered should include a detailed outline describing the planned contents of each chapter, a bibliography and several sample chapters. Parallel submission of a manuscript to another publisher while under consideration for LNM is not acceptable and can lead to rejection.

4. In general, **monographs** will be sent out to at least 2 external referees for evaluation.

   A final decision to publish can be made only on the basis of the complete manuscript, however a refereeing process leading to a preliminary decision can be based on a pre-final or incomplete manuscript.

   Volume Editors of **multi-author works** are expected to arrange for the refereeing, to the usual scientific standards, of the individual contributions. If the resulting reports can be forwarded to the LNM Editorial Board, this is very helpful. If no reports are forwarded or if other questions remain unclear in respect of homogeneity etc, the series editors may wish to consult external referees for an overall evaluation of the volume.

5. Manuscripts should in general be submitted in English. Final manuscripts should contain at least 100 pages of mathematical text and should always include

   – a table of contents;
   – an informative introduction, with adequate motivation and perhaps some historical remarks: it should be accessible to a reader not intimately familiar with the topic treated;
   – a subject index: as a rule this is genuinely helpful for the reader.
   – For evaluation purposes, manuscripts should be submitted as pdf files.

6. Careful preparation of the manuscripts will help keep production time short besides ensuring satisfactory appearance of the finished book in print and online. After acceptance of the manuscript authors will be asked to prepare the final LaTeX source files (see LaTeX templates online: https://www.springer.com/gb/authors-editors/book-authors-editors/manuscriptpreparation/5636) plus the corresponding pdf- or zipped ps-file. The LaTeX source files are essential for producing the full-text online version of the book, see http://link.springer.com/bookseries/304 for the existing online volumes of LNM). The technical production of a Lecture Notes volume takes approximately 12 weeks. Additional instructions, if necessary, are available on request from lnm@springer.com.

7. Authors receive a total of 30 free copies of their volume and free access to their book on SpringerLink, but no royalties. They are entitled to a discount of 33.3 % on the price of Springer books purchased for their personal use, if ordering directly from Springer.

8. Commitment to publish is made by a *Publishing Agreement*; contributing authors of multiauthor books are requested to sign a *Consent to Publish form*. Springer-Verlag registers the copyright for each volume. Authors are free to reuse material contained in their LNM volumes in later publications: a brief written (or e-mail) request for formal permission is sufficient.

Addresses:
Professor Jean-Michel Morel, CMLA, École Normale Supérieure de Cachan, France
E-mail: moreljeanmichel@gmail.com

Professor Bernard Teissier, Equipe Géométrie et Dynamique,
Institut de Mathématiques de Jussieu – Paris Rive Gauche, Paris, France
E-mail: bernard.teissier@imj-prg.fr

Springer: Ute McCrory, Mathematics, Heidelberg, Germany,
E-mail: lnm@springer.com

Printed in the United States
By Bookmasters